Contractor Health and Safety Compliance for Small to Medium-Sized Construction Companies

Contractor Health and Safety Compliance for Small to Medium-Sized Construction Companies

Zakari Mustapha
Clinton Aigbavboa
Wellington Thwala

CRC Press
Taylor & Francis Group
Boca Raton London New York

CRC Press is an imprint of the
Taylor & Francis Group, an **informa** business

CRC Press
Taylor & Francis Group
6000 Broken Sound Parkway NW, Suite 300
Boca Raton, FL 33487-2742

Printed on acid-free paper

International Standard Book Number-13: 978-1-138-08155-0 (Hardback)

Library of Congress Cataloging-in-Publication Data

Names: Mustapha, Zakari, author. | Aigbavboa, Clinton, author. | Thwala, Wellington, author.
Title: Contractor health and safety compliance for small to medium-sized construction companies / Zakari Mustapha, Clinton Aigbavboa, Wellington Thwala.
Description: Boca Raton : Taylor & Francis, CRC Press, 2017. | Includes bibliographical references.
Identifiers: LCCN 2017010579 | ISBN 9781138081550 (hardback : alk. paper) | ISBN 9781315112817 (ebook)
Subjects: LCSH: Building--Developing countries--Safety measures.
Classification: LCC TH443 .M87 2017 | DDC 690/.22--dc23
LC record available at https://lccn.loc.gov/2017010579

Visit the Taylor & Francis Web site at
http://www.taylorandfrancis.com

and the CRC Press Web site at
http://www.crcpress.com

This book on contractors' health and safety is dedicated to all construction workers who have lost their lives and to those who have suffered harm due to negligence on construction sites; and to all other construction workers who are devoted to construction health and safety.

Contents

SECTION I Background Information

SECTION II Theoretical and Conceptual Perspectives of Health and Safety Research

SECTION III Review of the Health and Safety Compliance Literature

SECTION IV Construction Industry: International Literature

SECTION V Occupational Health and Safety Issues in Developing Countries

SECTION VI Research Methodology

SECTION IX Results from the Questionnaire Survey: Findings Part II

SECTION X Discussion of Findings

SECTION XI Conclusions and Recommendations

List of Tables

List of Figures

Preface

This book explores the formation of small and medium-sized construction companies' compliance with health and safety (H&S) issues in developing countries; the Ghanaian construction industry is a case study. There have been a lot of publications on health and safety policy and implementation in the developed nations and as proof in the body of knowledge. However, little has been written about the formation of small and medium-sized enterprise (SME) contractors' H&S compliance for developing countries, especially, in the sub-Saharan regions where construction/infrastructure development activities have significantly increased in order to serve the development mandate of the countries. This book provides readers with three major practical insights. The first focuses on the theory underpinning SME contractors' H&S compliance by developing a conceptual framework. The second is focused on SME contractors and the current trends of H&S in the construction industry. The third is focused on the development and validation of a conceptual model on SME compliance with H&S in the construction industry. A comparative overview of SME contractors is provided on two West African countries (Ghana and Nigeria). Further emphasis is provided on the philosophical basis for SME contractors' development in these countries. The book investigates and models H&S compliance using the following variables of safe environment features, safe acts of workers, safe working conditions, the reaction of workers to safe conditions, government support, and the contractor's organisational culture. A conceptual SME contractors integrated H&S compliance model was based on the theory developed from literature review findings and the Delphi study.

Empirical data were collected through a Delphi and a field questionnaire survey. Analysis of results from the Delphi study was done to inform on consensus reached by the group of selected experts for the study. Thereafter, structural equation modelling (SEM) using the software EQS, version 6.2, was used in the analysis of the field questionnaire. Before the use of SEM in modelling the construction, exploratory factor analysis (EFA) using Statistical Package for the Social Sciences (SPSS), version 20, using maximum likelihood with promax rotation was used to determine the validity and reliability of the six H&S constructs of the priori (conceptual model). At the end of the EFA on the six factor constructs, fourteen factors were realised and sixty-four statements were retained as valid and reliable measures of H&S for SME contractors at project level. A further validity and reliability test was conducted using confirmatory factor analysis (CFA) with EQS, version 6.2. Findings from the first set of results pertain to the literature on H&S studies. The findings revealed the theory that H&S practices and the latent variables lead to H&S compliance. Findings from the second set of results pertaining to the Delphi study revealed that several factors (safe environmental features, safe acts of workers features, safe working conditions features, government support features, and contractor's organisational culture features) were considered to be the most important determinants of H&S compliance among SME contractors in the Ghanaian construction industry. Further findings from the literature and the Delphi study showed that H&S compliance could be considered as a six-factor model defined by the influence of H&S practices among SME contractors. The third set of findings relates to the field questionnaire survey. Generally, findings were that the hypothesis on H&S compliance had an influence on H&S practices and could not be rejected. Hence, it was found that the SEM results on the model's goodness of fit and statistical significance of parameter estimates met the cut-off criteria for the hypothesised model's fit to the sample data.

This book's contribution to the body of knowledge is significant because it addresses the lack of theoretical information (historical literature data) about which factors are most significant in predicting H&S compliance among SME contractors. Hence, the book develops a new compliance model for the prediction of SME contractors' H&S compliance. The current integrated model advances that H&S compliance is a six-factor construct. Previous studies have tried to model compliance using other variables without the inclusion of important exogenous variables as advanced in this book. The book thus shows that there is more than one factor that influences H&S compliance. Another noteworthy contribution to the body of knowledge is in the methodology adopted. The literature review revealed a lack of evidence, suggesting that a mixed method of using the Delphi study and SEM had been used before in H&S studies in the Ghanaian construction industry. This book offers a base for other researchers to use as a follow-up for future studies. Therefore, the book recommends that government institutions and policy makers should consider the empirically tested constructs as they plan for and implement H&S compliance programmes to enhance the quality of H&S practices among SME contractors. The book should constitute a reference of guidance in Ghanaian contractors' H&S policies. Hence, stakeholders and institutions that are involved in the planning process should consider the contemporary factors that reveal SME contractors' preferences about H&S compliance as part of the planning input. The book makes a significant contribution towards understanding H&S practices and should be seen as a critical area for improvement in H&S compliance. The central aim of this book is to provide readers with ideas on SME contractors' health and safety compliance, and policy implementation trends and formation. The book is of interest to researchers in the construction industry, building science researchers, urban and regional planning, and estate management researchers. Furthermore, the authors confirm that the text utilised in this work reflects original work, and, where necessary, materials have benefited from relevant context-setting/referencing.

About the Authors

Zakari Mustapha is a lecturer at the Department of Building Technology, School of Engineering, Cape Coast Technical University, Cape Coast, Ghana. His research interests are in the fields of construction health and safety, environmental management, and construction industry development (education and training).

Clinton Aigbavboa, PhD, is an associate professor at the University of Johannesburg, Johannesburg, South Africa. His research interests are situated in the fields of sustainable human development, with a focus on sustainable housing regeneration (urban renewal and informal housing), life cycle assessment in the construction industry, remanufacturing, leadership in low-income housing, biomimicry, postoccupancy evaluation, construction project management, construction health and safety, and green job creation.

Wellington Thwala, PhD, is one of the researchers rated by the National Research Foundation (NRF). He is a professor of construction project management and the masters programme coordinator at the Department of Construction Management and Quantity Surveying, University of Johannesburg. Thwala has varied research interests, including project management, construction health and safety, engineering design management, economic and social infrastructure delivery, operations and production management, and business competitive intelligence.

List of Abbreviations

ADF	African Development Fund
ANOVA	analysis of variance
ARCTM	accident root causes tracing model
BLS	Bureau of Labour Statistics
BRTF	Better Regulation Task Force
BSc	bachelor of science
BTech	bachelor of technology
CBI	Confederation of British Industries
CCOHS	Canadian Centre for Occupational Health and Safety
CCRT	Center for Construction Research and Training
CDM	Construction Design and Management Regulations
CEDI	Ghanaian cedi
CFA	confirmatory factor analysis
CFI	comparative factor index
CI	confidence interval
CIDB	Construction Industry Development Board
CIOB	Chartered Institute of Building
COC	contractor's organisational culture
DF	degree of freedom
DFR	Department of Feeder Roads
DME	Department of Minerals and Energy
DoL	Department of Labour
DOSH	Department of Safety and Health
DPhil/PhD	doctor of philosophy
DSc	doctor of science
DTI	Department of Trade and Industry
DUR	Department of Urban Roads
EASHW	European Agency for Safety and Health at Work
EFA	exploratory factor analysis
EPA	Environmental Protection Agency
EU	European Union
FV	full variable model
GDP	gross domestic product
GFI	goodness-of-fit index
GHA	Ghana Highways Authority
GH¢	Ghanaian cedi
GLM	general linear modelling
GS	government support
H&S	health and safety
HND	Higher National Diploma
HRIRC	Human Right Impact Resource Centre

HSA	Health and Safety Authority
HSE	Health and Safety Executive
IGM	interacting group method
ILO	International Labour Organization
IOHA	International Occupational Hygiene Association
IOSH	Institution of Occupational Safety and Health
IQD	interquartile deviation
LFS	Labour Force Survey
LM	Lagrange multiplier
MAD	average (mean) absolute deviation
MANOVA	multivariate analysis of variance
MBAWC	Master Builder Association of the Western Cape
MBSAT	Master Builder South Africa Audit Tool
MCAR	missing completely at random
MES	Ministry of Environment and Science
MMDE	Ministry of Manpower Development and Employment
MRT	Ministry of Roads and Transport
MSc	master of science
MTech	master of technology
MWRWH	Ministry of Water Resources, Works and Housing
n.d.	no date
NFI	normed fit index
NGT	nominal group technique
NOHSP	Nemours Human Research Protection Plan
OECD	Organisation for Economic Co-operation and Development
OHS	occupational health and safety
OHS Act	Occupational Health and Safety Act
OHSA	Occupational Health and Safety Act
OSHA	Occupational Safety and Health Administration
PPE	personal protective equipment
PWD	Public Works Department
RML	robust maximum likelihood
RMSEA	root mean square error of approximation
RWSC	reaction of workers to safe condition
S-Bχ^2	Satorra-Bentler scaled chi-square
SAW	safe act of workers
SE	safe environment
SEM	structural equation modelling
SME	small and medium-sized enterprise
SPSS	Statistical Package for the Social Sciences
SRMR	standardised root mean square residual
SSNIT	Social Security and National Insurance Trust
STATKON	Statistical Consultation Service
SWC	safe working condition
TLI	Tucker-Lewis index
UJ	University of Johannesburg

UK	United Kingdom
UKGBL	United Kingdom Government Business Links
UMIST	University of Manchester Institute of Science and Technology
US$	U.S. dollar
USA	United States of America
WHO	World Health Organization

Section I

Background Information

1 General Introduction

1.1 INTRODUCTION

This chapter discusses the processes followed in writing the book. It began with the milieu of the problem, aim, and motivation for the book. The book will substantiate whether compliance with health and safety (H&S) in the construction industry will reduce the rate of accidents on construction sites and enhance their performances.

The construction industry contributes to the national socioeconomic development and the physical infrastructure of every nation's economic backbone as well as constituting a large part of the economy in every country (Ofori, 2012). The construction industry's contribution to the gross domestic product (GDP) is between 5 and 10 percent in all countries. It also employs up to 10 percent of the working population and handles about half of the gross fixed capital formation. At the same time, a period of low construction output can adversely affect the growth of the economy (Ofori, 2012). For instance, the GDP growth rate of Ghana from the 2014 figures of the Bank of Ghana was pegged at 4.2 percent (Economic Sector Report, 2015; Monetary Policy Committee Press Release, 2015; Senzu, 2015). Apart from the construction industry playing a vital role in boosting the economy of developing countries, the construction industry also provides the infrastructure required for other sectors of the economy to flourish. Construction industries, in general, have been noted all over the world to have a poor H&S record, and the construction industries of developing countries are no exception. A lack of stringent measures in safety and construction laws has contributed to poor performance of construction H&S in developing countries (Shibani, Saidani & Alhajeri, 2013). According to Murie (2007:5), 'the construction industry employs about 180 million people or constitutes about seven percent of global employment and yet contributes to the highest rate of accidents'. Many large construction organisations have it as their top priority to improve the performance of H&S in the construction industry.

Health is the protection of the bodies and minds of people from illness resulting from materials, processes or procedures used at the workplace (Hughes & Ferrett, 2008). Safety has been defined by Hughes and Ferrett (2008) as the protection of people from physical injury. The adopted definitions of health and safety in this book include: Health and safety are used together and can thus be defined as protecting people from illnesses and injuries (i.e. harm) triggered by work-related conditions or activities (Health and Safety Executive [HSE], 2005). Accidents are unplanned events that result in injury or ill health or harm to people. Mansingh and Haupt (2008) assert that accidents do not only happen as a result of operatives' unsafe actions and unsafe site conditions. Accidents can also happen due to lack of management control and organisational failures, as viewed from the perspective of the domino theory by Heinrich (to be discussed later).

Health and safety improvement has received considerable attention in recent years. This is partly due to the introduction and pressure from the legislative environment, coupled with increased personal responsibility of senior managers and organisations for H&S (Fitzgerald, 2005). This new development has been partly as a result of the poor H&S performance of the construction industry. The need to develop a better image of the construction industry has also been another reason (Misnan, Mohammed, Mahmood, Mahmud & Abdullah, 2008). The complex nature of the construction industry has been a challenging issue for the improvement of H&S performance. The improvement of the H&S performance remains a vital issue as indicated by several studies (Hoonakker, Loushine, Carayon, Kallman, Kappa & Smith, 2005; Lee, Halpin & Chang, 2006). A report by the South African Construction Industry Development Board (CIDB, 2008) has shown that 60,000 fatal accidents occur in the construction sector in industrialized countries. These accidents account for 6 percent to 10 percent of its total employment, but contribute more than 25 percent to 40 percent of work-related deaths. A report from the United Kingdom Government Business Link (UKGBL) (in Arewa & Farrell, 2012) asserted that all organizations are compelled, regardless of their nature or size, to comply with H&S rules. Compliance with health and safety entails carrying out thorough health and safety risk assessments and drawing up health and safety policies for businesses with more than five employees. Workplaces should meet minimum standards of conformity and cleanliness. There must be a record of serious injuries, diseases or dangerous accidents. Windapo and Oladapo (2012) stated that the profit maximisation motive was due to the competitive nature of the construction industry. They further indicated negligence or attitude of the contractor as being the other reasons for non-compliance in the construction industry. Gibb and Bust (in Kheni, Dainty & Gibb, 2007) asserted that poor infrastructure, extreme climatic conditions and inappropriate work practices have a negative impact on H&S management. Construction workers face various H&S issues at their workplaces, which include many hazardous tasks and conditions such as working at heights, excavations, noise, dust, power tools and equipment, confined spaces and electricity. They argued that these activities lead to the occurrence of accidents on construction sites and are very common around the world despite various occupational health and safety (OHS) laws, rules and regulations that are in place. They emphasised that the fatal occupational injury rate among private construction workers was nearly three times that of all workers in the United States in the year 2009.

The number of fatal injuries in construction was said to have declined from 975 in 2008 to 816 in 2009 according to the Center for Construction Research and Training (CCRT, 2013). The distribution of these fatalities across construction occupations has also changed slightly. The proportion of fatalities among labourers was also noted to have increased from 22 percent in 2008 to fully one-quarter in 2009. First-line supervisors, carpenters, and electricians also saw their share of fatal injuries increase from 9 to 12 percent, from 7 to 9 percent, and from 5 to 7 percent, respectively. Construction managers, equipment operators, painters and truck drivers all saw small decreases in their proportions of fatal injuries from 2008 to 2009. In 2009, falls accounted for more than one-third of fatal occupational injuries in construction (34%). There were 3.3 million non-fatal injuries and illnesses reported across all industries in 2009, of which almost 95 percent were injuries, and just over 5 percent were illnesses. More than 9 percent of the non-fatal illnesses and injuries requiring days away from work were experienced by construction workers (92,540). Construction workers reported 4.3 non-fatal injuries and illnesses per 100 full-time workers (CCRT, 2013). Falls alone accounted for more than one in every five of these injuries and illnesses (22%). Construction labourers are said to experience the seventh-highest rate of non-fatal injury and illness requiring days away from work of all occupations (382 per 10,000 full-time workers). Hispanic or Latino workers experienced non-fatal injuries and illnesses disproportionately, with almost one-fifth of cases (19%) (CCRT, 2013). Falls remain the leading cause of death for workers engaged in residential construction, with an average of 40 workers suffering a fatal fall from a residential structure each year (Firl, 2012). A report from the United States (U.S. Bureau of Labour Statistics) has shown that an average of two construction workers die each day in the United States. According to Finneran and Gibb (2013:2), 'construction is one of the most hazardous industry sectors with many thousands of workers being killed and seriously injured each year all over the world'. Finneran and Gibb further asserted that H&S in construction is about using appropriate means to ensure workers are both safe and healthy. However, the situation is quite different and more challenging in a construction environment where projects differ considerably in terms of size, location and complexity.

The Ghanaian Occupational Safety and Health Act (OSHA) of 1994 was enacted on a self- regulatory basis to promote safety. The elements of this act were thought to encourage employers and employees to change their behaviour towards occupational health and safety (OHS) improvement and protect employees from occupational accidents, injuries and illnesses. It would discipline the employees as one strong team to give full support towards the achievement of the organization safety goal. A report from the European Agency for Safety and Health at Work (EASHW, 2009:5) noted that compliance with H&S is a contributory factor to the viability of economic performance. According to Windapo (2013:79), 'the compliance with statutory H&S requirements decreases with an increase in the cost of compliance, and does not increase with the degree of risk or perceived cost savings'. Windapo further indicated that complying with the H&S regulatory requirements involves upfront costs; the H&S of construction operatives should take precedence. Contractors should expect issues of non-compliance to H&S regulations when they regard some elements of the H&S regulations as unimportant. Windapo emphatically stated that owing to the contractors' cost-saving mindset, there is no way accidents will not occur on construction sites, and both public and private clients should take note. Improving H&S in the construction industry, therefore, continues to remain a priority (CIDB, 2008:1). The construction sector in developing countries plays a significant role in the physical development and employment of the largely unemployed labour force.

Studies conducted by the International Labour Organisation (ILO) and Kheni et al. (2007) showed that H&S awareness and performance was low in developing countries. Construction sites in developing countries are 10 times more dangerous than in industrialised countries. They further indicated that effective implementation of H&S programmes was absent in most construction businesses in developing countries. Kheni et al. identified weaknesses in the policy and institutional environments as an obstacle to the implementation of H&S standards on construction sites in most developing countries. Many stakeholders in the industry have long focused their attention on construction H&S. In the year 2008, the ILO reported that 2.2 million people die annually from work-related accidents and diseases, and work-related deaths appear to be on the rise. Moreover, 270 million people suffer minor injuries from work-related accidents, and an additional 160 million new people suffer from some work-related illness. Construction continues to contribute a disproportionate number of fatalities and injuries above other industrial sectors. There are high levels of non-compliance with H&S regulations. Poor coordination of the activities of the many institutions responsible for implementing H&S standards, a lack of specific H&S regulations, and an undesirable level of compliance with relevant H&S legislation are contributing factors. It is mandatory for construction industries and site operatives to have a positive change in their attitudes (Kheni & Braimah, 2014). The change of attitudes will enable OHS to take place.

1.2 SIGNIFICANCE OF THE BOOK

There is no doubt that research on construction has gone beyond safety practices and the state of H&S on construction sites. The interest of the current research is to determine the compliance with H&S and how it will help to reduce accidents on construction sites and enhance the performance of the construction industry. It is assumed that not all construction industries strictly obey the H&S regulations. H&S compliance has long been viewed from several perspectives and essentially been the subject of some researchers. Many construction workers are involved in various degrees of the accidents at their workplaces, and yet they are not properly compensated owing to the negligence of their employers. The H&S compliance model developed in this book will address all these issues. The model will predict compliance with all H&S issues about construction because an understanding of the factors is fundamental to the formulation and improvement of an H&S policy and its implementation in the construction industries and sites.

Little attention has been paid to H&S compliance in the construction industry in Ghana, in terms of either research or support for preventive initiatives. Previous works suggest that this sector has serious problems exacerbated by limited access to human, economic and technological resources. Moreover, it is now recognized that compliance methods developed specifically for developed countries are not applicable in developing countries. The compliance with H&S regulations is vital to both employees and employers. These regulations should specifically deal with construction industry workers; the stipulated policy for the prevention of accidents; and the safety of employees, employers and their properties. The development of the H&S compliance model in this book will assist to determine the ways in which employers will address the issue of H&S. It is also clear from the background of the study that H&S research conducted in Ghana and in other African countries is on practices and the state of H&S on construction sites. An obvious sign of this inadequacy is the existence of inconsistent, sometimes even conflicting, research results about the factors that shape construction H&S compliance. The discrepancies in research relate to the differences in samples, as the sample for most studies might not be representative of the population under study and the way the key variables may be defined. It may also be because of how construction research has been carried out in the global context of the studies or how the data were analysed. Hence, this book is determined to overcome these problems to achieve a better understanding of the constructs that determine compliance to H&S in small to medium-sized construction companies.

Also, this book will likewise examine the factors that influence the compliance of H&S regulations in the Ghanaian construction industry and thereafter, develop a holistic H&S compliance model for contractors in small to medium-sized companies in the construction industry. The proposed model will aid in determining and measuring H&S compliance in the Ghanaian construction industry especially, as Ghana is the study site for this book. Most of the firms do not provide adequate personal protective equipment (PPE) and sometimes operate without a safety officer. Therefore, the proper use of the integrated H&S compliance model will help to avoid mistakes that lead to accidents. The developed model will be context specific as it relates to H&S compliance in the Ghanaian construction industry.

Research on H&S compliance has gone beyond the safety practices and state of H&S on construction sites. It is assumed that not all employers make use of OHS regulations in the execution of projects. OHS compliance has long been viewed from several perspectives and essentially been the subject for some researchers in both developed and developing countries. The lack of an OHS compliance model is a significant problem for construction industries in most developing countries because this prevents them from meeting a higher level of industry performance thus stamping a better presence in the development national economy. Since most Ghanaian construction industries have failed to comply with OHS regulations, their failure has increased the number of accidents. Therefore, there is a need for this book to assist in the prevention of the prevalent occurrence of accidents and to save lives. It will also improve on the key performance indicators of the firms on construction sites.

1.3 STRUCTURE OF THE BOOK

This book is divided into thirteen chapters for guidance and ease of use. This first chapter consists of detailing the background information for the book, while the theoretical and conceptual perspectives of health and safety are discussed in Chapter 2. The third chapter of the book gives a detailed review of health and safety compliance literature with the aim of identifying gaps that can improve H&S compliance in the developing countries context. Not only are the gaps identified, ways on how the gaps can be filled are also extensively discussed. Chapter 4 explores information relating to the general construction industry literature with a focus on accident causation and how accidents can be mitigated on sites. Chapters 5 to 7 detail occupational health and safety issues in the three African countries of Nigeria, South Africa, and Ghana. The chapters explore the legislatures supporting compliance to H&S issues in the construction industries of these countries.

Thereafter, Chapter 8 gives a detailed analysis of the methodology used in conducting the research for the book. Likewise, Chapter 9 presents findings of the Delphi technique used in conducting the research and the resultant implications on small to medium-sized construction companies. In discussing the conceptual integrated health and safety compliance model for small to medium-sized construction companies, a detail explanation of the variables and their justification is presented in Chapter 10 of this book. This is followed by the presentation of the questionnaire survey findings resulting from the developed priori model (Chapter 11). Chapter 11 of the book also details the innovative statistical process that was used in the development of the model of H&S compliance for small to medium-sized construction companies. The last two chapters (Chapters 12 and 13) presents the discussion of the findings together with their contributions in the improvement and compliance of small to medium-sized construction companies to H&S issues. The last chapter of the book (Chapter 13) also provides further insight into the contributions of the book to the general H&S body of knowledge. However, the aim of this book is to substantiate whether compliance with H&S in the construction industry will reduce the rate of accidents on construction sites and enhance their performances.

Aside from the presentation of a detailed review and synthesis of the existing body of knowledge on small to medium-sized construction companies to H&S issues, the developed H&S compliance model will guide the Ghanaian and other developing economies' construction industries in their enforcement of H&S regulations. It will also streamline the use of personal protective equipment and the employment of a safety officer at all construction sites in the country. The developed model will address mistakes that occurred due to the negligence of both employees and employers. Further, the developed H&S models will help to monitor and guide future use of the model to be developed for the construction industry in Ghana and other developing countries.

1.4 SUMMARY

This chapter introduced the idea behind the conception of this research book with the emphasis on the explanation of key issues relating to the construction H&S compliance as well as the existing views and perceptions of small and medium-sized construction companies' compliance to H&S issues. The objective of the book—to promote the compliance of small and medium-sized companies in the construction industry to H&S issues, thus reducing the rate of accidents on construction sites and enhancement of the companies' performance in all aspects—was also explained while detailing the scope and areas of concern that are addressed in each of the book chapters. The concept of theoretical and conceptual perspective of H&S research will be further explained in the next chapter.

REFERENCES

Arewa, A.O. and Farrell, P. (2012). A review of compliance with health and safety regulations and economic performance in small and medium construction enterprises. In: Smith, S.D. (Ed.), *Proceedings of 28th Annual ARCOM Conference*, 3–5 September, Edinburgh, UK, Association of Researchers in Construction Management, 423–432.

Center for Construction Research and Training (CCRT). (2013). Announces re-launch of the national campaign to prevent falls. Available from: http://www.cdc.gov [Accessed 14 April 2014].

Construction Industry Development Board (CIDB). (2008). Development through partnership. Construction health and safety in South Africa. Available from: http://www.asocsa.org [Accessed 08 April 2014].

European Agency for Safety and Health at Work (EASHW). (2009). Occupational safety and health and economic performance in small and medium-sized enterprises: A review. *Journal of European Agency for Safety and Health at Work*, 12. Luxembourg: Office of the European Communities, pp. 5–13.

Finneran, A. and Gibb, A. (2013). CIB W099 safety and health in construction research. Roadmap report for consultation. CIB General Secretariat. 978-90-6363-078-2.

Firl, G. (2012). Occupational health and safety (OHS) residential construction regulations compliance. Available from: http://www.Ohsonline.com [Accessed 08 April 2014].

Fitzgerald, M.K. (2005). Safety performance improvement through culture change. *Process Safety Environmental Protection*, 83(4): 324–330.

Hoonakker, P., Loushine, T., Carayon, P., Kallman, J., Kapp, A. and Smith, M.J. (2005). The effect of safety initiatives on safety performance: A longitudinal study. *Applied Ergonomics*, 36: 461–469.

Health and Safety Executive (HSE). (2005). Promoting health and safety as a key goal of the corporate social responsibility agenda. Research Report 339. Sudbury, HSE Books.

Hughes, P. and Ferrett, E. (2008). *Introduction to health and safety in construction*. Oxford: Elsevier Butterworth-Heinemann.

Kheni, N.A. and Braimah, C. (2014). Institutional and regulatory frameworks for health and safety administration: Study of the construction industry of Ghana. *Journal of Engineering and Science* (IRJES), 3(2): 24–34. Available from: http://www.irjes.org [Accessed 04 July 2015].

Kheni, N.A., Dainty, A.R.J. and Gibb, A.G.F. (2007). Influence of political and socio-cultural environments on health and safety management within SMEs: A Ghana case study. In: Boyd, D. (Ed.), *Proceedings of the 23rd Annual Association of Researchers in Construction Management (ARCOM) Conference*, 3–5 September, Belfast, UK, 159–168.

Lee, S., Halpin, D.W. and Chang, H. (2006). Quantifying effects of accidents by fuzzy-logic and simulation-based analysis. *Canadian Journal of Civil Engineering*, 33: 219–226.

Mansingh, K.S. and Haupt, T.C. (2008). Construction accident causation: An exploratory analysis. In: Hinze, J., Boehner, S. and Lew, J. (Eds.), *Evolution of and Directions in Construction Safety and Health*, CIB W99, Rotterdam, 465–482.

Misnan, M.S., Mohammed, A.H.B., Mahmood, W.Y.W., Mahmud, H.S. and Abdullah, N.M. (2008). Development of safety culture in the construction industry: The leadership and training roles. Conference Proceedings of the 2nd International Conference on Built Environment in Developing Countries, (ICBEDC), Pulau.

Murie, F. (2007). Building safety: An international perspective. *International Journal of Occupational, Environmental and Health*, 13: 5–11.

Ofori, G. (2012). Developing the construction industry in Ghana: The case for a central agency. National University of Singapore.

Senzu, T. (2015). The proof that government of Ghana is insensitive to SMEs development. Bastiat Ghana Institute (Free Market Economic Think Tank), BGI-Scholastic Article, 78. Available from: www.modernghana.com [Accessed 22 March 2016].

Shibani, A., Saidani, M. and Alhajeri, M. (2013). Health and safety influence on the construction project performance in United Arab Emirates (UAE). *Journal of Civil Engineering and Construction Technology*, 4(2): 32–44. Available from: http://www.academicjournals.org [Accessed 17 April 2014].

Windapo, A. (2013). Relationship between degree of risk, cost and level of compliance to occupational health and safety regulations in construction. *Australian Journal of Construction Economics and Building*, 13(2): 67–82.

Windapo, A. and Oladapo, A. (2012). Determinant of construction firms' compliance with health and safety regulations in South Africa. In: Smith, S.D. (Ed.), *Proceedings of 28th Annual Association of Researchers in Construction Management (ARCOM) Conference*, 3–5 September, Edinburgh, UK, 433–444.

Section II

Theoretical and Conceptual Perspectives of Health and Safety Research

2 Theoretical and Conceptual Perspectives of Health and Safety Research

2.1 INTRODUCTION

This chapter is a research review of health and safety (H&S) compliance throughout the extant discussion of well-known existing theories. To understand the theory behind accident occurrences, there is a need to be familiar with the process and how the interacting elements within the accident causation theory work through the existing theories.

The definitions of theories have been offered by several researchers (Dublin, 1978; Kinloch, 1977; Silva, 1977) as an assertion, axiom and conceptual framework. Other various definitions for a theory are a description, maxim, model, postulation, prediction, proposition, system, theoretical model and typology. Hamilton (1997) indicated that there are many definitions of a theory within and across disciplines. These terms are used as synonyms for a theory, but have specific and often divergent meanings. It is held that a theory is essential to systematically organize and synthesise information. It also discovers relationships among variables and guides the discovery of new facts to move research forward (Creswell, 2003). Without theory-based research, a discipline chaotically moves in all directions with the lack of purpose (Mitchell & Jolley, 1992). The relationship of theory to research methods has been likened as that of a roadmap to driving. 'Without a theory, all roads look the same. A lot of time will be wasted making wrong turns and getting lost. One may not know, when he gets to his destination' (Guy, Edgley, Arafat & Allen, 1987).

Hence, advances in any field of study are unlikely without the understanding of theory and conceptual frameworks. Therefore, scientists can communicate their findings without any difficulty with sets of agreed-upon concepts. To inform the model constructs for this book on contractor H&S compliances for small to medium-sized construction companies, previous theoretical frameworks were reviewed. Lennon and Burns (2000) assert that a theory allows us to explain and predict behaviour; it also dictates which and whose behaviour is worthy of study and which should be excluded from the study. Theory development can take place through deduction or induction. In the case of deduction, the researcher moves from general to specific, from theory to the fact (Guy et al., 1987). The researcher advances a theory, collects data to test it, and reflects on the confirmation or disconfirmation of the theory by the results. The theory becomes a framework for the entire study.

2.2 ACCIDENT CAUSATION THEORIES

2.2.1 Who Should Be Blamed: The Worker or the System?

Much that has been said about the occurrence of occupational injury and illness, and assumptions have been made on their preventive measures. The terms 'blaming the victim' and 'blaming the system' were two broad sets of assumptions identified by Hopkins (2006). The first of these approaches explains occupational injury and illness in terms of characteristics of workers themselves that make them particularly susceptible (Lingard & Rowlinson, 2005). The social relations of production, such as the pressure to maintain production and bonus or piece-rate payment schemes, are seen as playing a key role in encouraging workers to ignore safe work practices. However, the physical or technological environment, which in many industries presents unusual and sometimes extreme hazardous conditions, is also recognised as a source of occupational injury and illness. According to Lingard and Rowlinson (2005:19), 'certain common features of many incidents that lead to occupational injury or illness are organisational breaches of occupational health and safety (OHS) legislation and codes'. Blaming-the-system approaches regard accidents as system failures. Construction accidents are more of a complex system interaction of plant and equipment, management systems and procedures, people and other human factor considerations than blaming of the workers or the system.

2.2.2 Accident Causation Models

Heinrich (1930) and Abdelhamid and Everett (2000) opine that accident prevention is an integral programme. It is a series of coordinated activities, directed to the control of unsafe personal performance and unsafe mechanical conditions. It is based on certain knowledge, attitudes and abilities. The multiple causation theory is related to well-known developed models (e.g. Heinrich's domino theory of 1930; Petersen, 2000). An accident causation model is not a new model to identify the root problem of safety in construction and other industries (Abdelhamid & Everett, 2000), but it provides some basis for the conceptualisation of accidents on construction sites.

2.2.2.1 Domino Theory

Heinrich's domino sequence was a classic in safety and health thinking and teaching for over 30 years in many countries around the world (Abdelhamid & Everrett 2000). Heinrich's domino theory consists of five dominoes, namely ancestry and social environment, the fault of a person, unsafe acts and condition, accident and injury. Construction work may commit unsafe acts or physical hazards that will result in an injury (Abdul Hamid, Yusuf & Singh, 2003; Hossenian & Torghabeh, 2012). The reason for the cause of most site accidents is people and management, as they are not able to handle the prevention of accidents as indicated in the theory. The majority of the accidents that take place are due to human error, and the accidents can only be prevented if management provides a conducive environment in which the employees work. The management organisational structure as enumerated by Adam in the year 1976, reflects the relationship between the causes and effects of all incidents and accidents that directly have management involvement (Heinrich, Petersen & Roos,

1980). Five elements stated in both Heinrich's domino theory and that of Adams have a similar concept, but the elements were different. The updated domino theory by Bird related the role of the mitigating measures of management in the prevention of loss. Recognising the root cause of unsafe acts or conditions is fundamental in accident detection; this makes the management of accident prevention controllable by taking into consideration the root causes of unsafe acts or conditions.

2.2.2.2 Multiple Causation Model

The root causes of accidents in the multiple causation models relate to the management system such as management policy, procedure, supervision, effectiveness and training (Abdelhamid & Everett, 2000). The contributing factors, causes and sub-causes are the main culprits in an accident scenario as inspired by the model. Questions are usually developed for injured persons, the management, supervisor and any other person involved. These questions relate to the accident and are used in the investigation to identify the root causes. Any improvement tools for inspections, supervisions, training, better definition of responsibilities and pre-job planning by supervisors are achieved from the answers provided by the employees.

2.2.2.3 Human Error Theories

The behaviour model, the human factor model and the Ferrell theory are centred on human error theory, as indicated by many researchers (Abdul Hamid et al., 2003; Taylor, Easter & Hegney, 2004; Hughes & Ferrett, 2008; Hosseinian & Torghabeh, 2012). The goals-freedom-alertness theory as developed by Kerr in 1957 and the motivation reward satisfaction model by Petersen in 1975 have been developed to describe the reason why there is repetition of accidents (Abdul Hamid et al., 2003; Taylor et al., 2004; Hosseinian & Torghabeh, 2012). Most of these theories address the human (worker) as the main problem that makes an accident happen. The permanent characteristic of humans, the combination of extreme environment and an overload of human capability and conditions tend to be the cause of humans. The overall objective of human error theory is to create a better designed workplace, tasks and tools that are suitable with human limitation. The worker is figured as the main factor of an accident in this theory's approach. Human error theory considered the design of workplace and tasks but excluded worker (human) limitation, not considering worker (human) limitation as part of the reason why accidents happen instead of blaming the worker.

2.2.2.4 Accident Root Causes Tracing Model

The accident root causes tracing model (ARCTM) has shown further advances over many of the previous accident models. The main reason for this model is to provide an investigator with an easy model for identification of root causes of construction accidents, compared to the sophisticated models of the accident's investigation (Abdul Hamid et al., 2003; Fang, Choudhry & Hinze, 2006; Jha, 2011; Hosseinian & Torghabeh, 2012). The ARCTM insists on specific issues such as worker training, worker attitude and management procedure problems being recognized and modified to avoid a reoccurrence of the accident. Cooperation between workers and management should also be encouraged to prevent the reoccurrence of accidents

(Abdelhamid & Everett, 2000; Abdul Hamid et al., 2003; Fang et al., 2006; Jha, 2011; Hosseinian & Torghabeh, 2012). The application of the ARCTM is a complement to the accident investigation process and should be able to give solutions to accident occurrence and preventive measures in construction instead of identifying who caused the accident.

A further summary of the philosophy of ARCTM is as follows:

- Avoidance of accident by new employees without any training is impossible.
- The accident will occur during the process of work for new employees unless a safety attitude is observed.
- It is the responsibility of management to clear any unsafe condition on the site and install safety values among workers.

2.2.2.5 Modified Statistical Triangle of Accident Causation

Control and management of hazards in construction sites take two main forms, namely avoidance of the occurring of hazardous events and restricting the severity potential of hazards when the hazard happens. The first step is to restrict the entrance of hazardous events into the triangle by lessening the probability of the hazard happening. The second step is designed to restrict the movement of hazardous events through the upper part of the triangle. This step lessens the risk via lessening the severity of the hazard if it happens (Carter & Smith, 2006:198). To obtain the general rationale for performing all safety risk assessments, the hazards should be considered in terms of their probability of occurrence and severity of consequence by estimating the probability of a hazard's occurrence. This has to do with its frequency, and its probable severity if it does occur, evaluating the risk associated with the hazard based upon the frequency and severity of estimations and responding to the hazard by implementing suitable control measures. However, there will be complete freedom of entry and movement within the triangle if the hazard is not identified in the first place.

2.3 MEASURING HEALTH AND SAFETY COMPLIANCE IMPROVEMENT

The level or extent of improvement depends on the defined or set goal. The goal for H&S should simply be a reduction or zero in the number of accidents from an existing record to a new one. Improvement could also be slight, moderate or significant depending on the scale that has been used. Improvement could be dramatic, or a gradual gravitation towards a set goal or vision that does not necessarily mean attaining the goal, whether clearly defined or not. The Health and Safety Executive (HSE, 2010a) further defined this in terms of the extent to which the initial state has been exceeded or what progress has been made. The Egan report for instance proposed improvement to H&S in the United Kingdom's construction industry of a 20 percent reduction in the number of reportable accidents per year. Whilst statistics released for 2010 show a fatality rate of 2.2 in the construction industry (HSE, 2010b) even though the target was 5.4 per 100,000 by the year 2010. There was a significant improvement because the rate of 2.2 is almost 4½ times higher than the national average of 0.5 as compared to the goal that had been set. Therefore, improvement relates to the targets that were initially set. The fatality rate in the previous example was 2.2 per 100,000 workers based on the fatality rate calculated from 42 people who died in 2010 in the United Kingdom as a result of workplace accidents. This shows that there has been an improvement as compared to 10 years earlier where 100 people died.

A report from the National Statistics UK and the Bureau of Labour Statistics informs that there should be a fairly acceptable measure to determine progress or improvement in this manner in that no one should die as a result of his or her work. The fatal rate of 2.2 in construction against the rate of 0.5 in all industries in the United Kingdom clearly indicates that not only benchmarking had been used to evaluate progress made on H&S performance. Benchmarking is an approach to process improvement and is defined as a systematic process of measuring and comparing an organization's performance against that of other similar organizations in key activities (Rankin, Fayek, Meade, Haas & Manseau, 2008). Benchmarking is popular in business circles because it can change the mindset of managers so that improvements in performance will be gradual as a result of incremental changes (Johnson & Scholes, 2002). Benchmarking breaks the frame within organizations and industries regarding performance standards to be achieved. It also provides the context for assessing performance. Improvement also has to do with benchmarking because it compares an organization's performance against the 'best in class', wherever that is found. Incident and injury rates are important so statistics on accidents alone may not be useful to determine and plan for improvement on occupational health and safety (OHS) compliance improvement in the construction industry.

Improvement in H&S has to do with better results than previous records expressed in terms of accidents and incidents in the construction industry. This implies that the previous record on H&S needs improvement or to be better than the existing record, which entails adding some elements or components to enhance its performance. Improvement has also been defined as something getting better or when one makes something better with a slight improvement. Improvement may be expressed in terms of statistics on injuries and fatalities with a better performance in comparison to other industries or in the way things are done about H&S compliance. Health and safety have been defined by H&S management as organized efforts and procedures for identifying workplace hazards and reducing accidents and exposure to harmful situations and substances. It also includes training of personnel in accident prevention, accident response, emergency preparedness and the use of protective clothing and equipment. The Sarbanes-Oxley Act stated that H&S compliance means conforming to or being in accord with a rule or established guidelines, such as a specification, policy, legislation, standard or law, or the process of becoming so. Therefore, the perception of stakeholders on H&S improvement in an organization or at the industry level should be reflected in the H&S compliance. If a new record is attained or ways of doing things have become better than before when compared with a previous state, record or a way of doing things in the industry, then improvement is said to be experienced or attained.

An alternative measurement that is based on evaluating management actions and worker perceptions has been proposed because of the under-reporting of statistics (Musonda, 2012). It was further indicated that evaluating management actions and worker perceptions can differentiate H&S sites based on the observed level of H&S management commitment and the perceptions of workers. Current measures of accidents and incidents are recognized as invalid measures of H&S at the workplace except in very large organizations or over a long period because they are insensitive to real changes in H&S behaviour or conditions. It will, therefore, be inadequate to measure H&S compliance improvements using only accident statistics. Suitable and appropriate measures are therefore important in addressing the issue of H&S compliance improvement. For instance, perception is one of the best measures available to truly define reality by using worker perception surveys. This is in agreement with many researchers who argue that statistics alone is not sufficient to measure performance improvement (Musonda, 2012). They argued that incident rates that involve statistics are important, but they are not always useful for process improvement. They contend that without proper investigation of causes, incident rates indicate that there is a problem, but they do not inform what the problem is if a proper investigation is not carried out. There are better methods to reflect the reality than injury statistics alone. There is the need to determine H&S compliance improvement because studies have shown that non-compliance has been found to be the major cause of accidents in the construction industry.

2.4 SUMMARY

One of the most outstanding findings in this chapter revealed that the work of Heinrich (the domino theory) and the accident root causes tracing model (ARCTM) are the most comprehensive conceptual models of construction accident causation theories. However, it is clear that they do not preclude compliance from being a useful concept, as there are limitations to all research investigations. The gaps in the theoretical framework that have been observed from the review of construction accident causation models, which the current book addresses in the preceding chapter. To overcome the limitations of other measurement techniques that have so far been used, this book adopted the use of structural equation modelling (SEM) utilizing EQS software in the measurement of the variables, even though the most widely used method in previous studies has been multiple regression. SEM is a statistical methodology that takes a confirmatory approach rather than an exploratory approach to the analysis of structural theory (Bentler, 1993; Aigbavboa, 2013).

REFERENCES

Abdelhamid, T.S. and Everett, J.G. (2000). Identifying root causes of construction accidents. *Journal of Construction Engineering and Management*, 126(1): 52–60.

Abdul Hamid, A.R., Yusuf, W.Z.W. and Singh, B. (2003). Hazards at construction sites. Proceedings of the 5th Asia-Pacific Structural Engineering and Construction Conference (APSEC 2003), 26–28 August, Johor Bahru, Malaysia.

Aigbavboa, C.O. (2013). An integrated beneficiary centred housing satisfactory model for publicly funded housing schemes in South Africa. Unpublished DPhil in Engineering Management, University of Johannesburg, South Africa. Available from: www.ujdigispace.uj.ac.za [Accessed 15 June 2014].

Bentler, P.M. (1993). Comparative fit indexes in structural models. *Psychological Bulletin*, 107(2): 238–246.

Carter, G. and Smith, S.D. (2006). Safety hazard identification on construction projects. *Journal of Construction Engineering and Management*, 132(2): 197–205.

Cresswell, J.W. (2003). *Research Design: Qualitative, quantitative and mixed methods approaches* (2nd edn.). Thousand Oaks, CA: Sage.

Dublin, R. (1978). *Theory of building: A practical guide to the construction and testing of theoretical models* (rev. edn.). New York: The Free Press.

Fang, D., Choudhry, R.M. and Hinze, J.W. (2006). *Proceedings of CIB W99 International Conference on Global Unity for Safety & Health in Construction*, 28–30 June, Beijing, China. Beijing: Tsinghua University Press.

Guy, R.F., Edgley, C.E., Arafat, I. and Allen, D.E. (1987). *Social research: Puzzles and solutions.* Boston: Allyn & Bacon.

Hamilton, J. (1997). The macro-macro interface in the construction of individual fashion forums and meanings. *Clothing and Textiles Research Journal*, 15(3): 164–171.

Health and Safety Executive (HSE). (2010a). *The Health and Safety Executive statistics, 2009/10.* Caerphilly: HSE Books.

Health and Safety Executive (HSE). (2010b). *Work-related injuries and ill health in construction.* Suffolk: HSE Books.

Heinrich, H.W. (1930). *Industrial accident prevention.* New York: McGraw-Hill.

Heinrich, H.W., Petersen, D. and Roos, N. (1980). *Industrial accident prevention* (5th edn.). New York: McGraw-Hill.

Hopkins, A. (2006). Studying organisational cultures and their effects on safety. *Safety Sciences*, 44: 875–889.

Hosseinian, S.S. and Torghabeh, Z.J. (2012). Major theories of construction accident causation models: A literature review. *International Journal of Advances in Engineering & Technology*, 4(2): 53–66.

Hughes, P. and Ferrett, E. (2008). *Introduction to health and safety in construction.* Oxford: Elsevier Butterworth-Heinemann.

Jha, K.N. (2011). *Construction project management: Theory and practice.* Delhi: Pearson Education in South Asia.

Johnson, G. and Scholes, K. (2002). *Exploring corporate strategy, text and cases* (6th edn.). Harlow: Prentice Hall.

Kinloch, G.C. (1977). *Sociology theory: Its development and major paradigms.* New York: McGraw-Hill.

Lennon, S.L. and Burns, L.D. (2000). The diversity of research in textiles, clothing, and human behavior: The relationship between what we know and how we know. *Clothing Textiles Research Journal*, 18(4): 213–226.

Lingard, H. and Rowlinson, S. (2005). *Occupational health and safety in construction project management.* Abingdon: Spon Press.

Mitchell, M. and Jolley, J. (1992). *Research design explained* (2nd edn.). Fort Worth, TX: Harcourt Brace Jovanovich College Publishers.

Musonda, I. (2012). Construction health and safety (H&S) performance improvement: A client-centred model. Unpublished DPhil in Engineering Management, University of Johannesburg, South Africa. Available from: http://www.ujdigispace.uj.ac.za [Accessed 02 June 2014].

Petersen, D. (2000). The behavioural approach to safety management. *Professional Safety*, 37–39.

Rankin, J.R.J., Fayek, A.R.F.A.R., Meade, G.M.G., Haas, C.H.C. and Manseau, A.M.A. (2008). Initial metrics and pilot program results for measuring the performance of the Canadian construction industry. *Canadian Journal of Civil Engineering*, 35(9): 894–907.

Silva, M.C. (1977). Philosophy, science, theory: Interrelationships and implications for nursing research. *Image*, 9(3): 59–63.

Taylor, G., Easter, K. and Hegney, R. (2004). *Enhancing occupational safety and health.* Burlington, MA: Elsevier Butterworth-Heinemann.

Section III

Review of the Health and Safety Compliance Literature

3 Review of the Health and Safety Compliance Literature

3.1 INTRODUCTION

This chapter addresses the gaps observed in construction accident causation theories. These gaps have not been evaluated as all-inclusive constructs in the previous models, and they form the additional new constructs for the current conceptual framework as defined in this book. The identified gaps are government support and contractor's organizational culture. This chapter is devoted to the discussion of these gaps and how to achieve them in health and safety (H&S) compliance with regards to small to medium-sized construction companies.

3.2 GAPS IN HEALTH AND SAFETY COMPLIANCE RESEARCH

The consideration of the identified gaps is based on the notion that an H&S compliance model cannot be achieved without government support and consideration of a contractor's organizational culture. This is because an H&S compliance model is not a simple, single-track factor assessment, but a combination of numerous variables. Even though the Heinrich domino theory and the accident root causes tracing model (ARCTM) are robust in nature, government support and a contractor's organization culture were missing from their conceptualisation. Hence, in order to develop a robust holistically integrated model of H&S compliance for small to medium-sized construction companies, the two identified gaps are very significant. Using the conceptual frameworks of Heinrich (1930) and ARCTM amongst others, it becomes clear that most of the research findings relating to H&S compliance were obtained in developed countries. The model developed in this book is not only based on the foundation of work laid by previous studies, but it is context specific to the developing economics construction industry.

Studies conducted on H&S in developing countries and in Ghana, for example, indicated several factors as affecting the compliance with H&S provisions in the Labour Act of Ghana. A lack of H&S training for workers, poor risk assessment and workers attitude towards H&S are commonly considered factors affecting H&S compliance. Other factors are inadequate H&S professionals, H&S policies and data collection systems. A lack of H&S education in various institutions, communication difficulties, the cost of providing and maintaining H&S on sites and accident reporting shortfalls are additional shortcomings (Dadzie, 2013). There is a low level of ratifications of International Labour Organization (ILO) Conventions that address organizational health and safety (OHS) in Ghana. This is coupled with the lack of comprehensive international OHS policy framework, inadequate resources allocated to OHS research, ineffective OHS inspection, OHS training and education, and OHS capacity building and monitoring. These barriers need serious attention to ensure effective OHS management. A lack of structures and procedures at all levels of the construction chain and a lack of strong and appropriate H&S legislation for governing construction work and site operations in construction are also major factors of non-compliance to H&S issues (Laryea, 2010).

The conceptual framework provides the perspectives from which problems are highlighted. It is most likely that there are some gaps in the Western conceptual framework that have failed to capture the factors affecting H&S in Ghana and other developing countries and in the construction H&S studies in general. This chapter of the book attempts to address the two gaps that have been identified, namely government support and contractor's organisational culture that have not been discussed in all the models. There is the need to find out why there is a contrast in H&S compliance research and whether the proposed existing theoretical framework has some gaps that need to be covered before any conclusion can be drawn.

3.2.1 Gap 1: Government Support

All employers are expected to provide a safe and healthy work environment, and are in turn expected to contribute to that safe environment through responsible behaviour. When it comes to safety, the only acceptable number is zero-zero accidents, incidents or occupational illnesses. In order to reach this goal, it is critical for all construction industries to reach excellence. Management should be accountable for the prevention of injuries and occupational illnesses. To address the gap of government support, there should be an integration of H&S policy into the management systems of all construction industries at all levels, most important at the small to medium-sized scale. Moreover, the effective implementation of government safety policies, and regular education and training is important and useful to both the government and the parties involved in all aspects of construction business. Government support for H&S compliance can be ensured by checking the H&S policy of the organisation and ensuring that it complies with the manual detailing how H&S will be implemented in every aspect of the company's work environment. Also, government support can be provided through a thorough monitoring of H&S and training in construction organisations to ensure that employees and the personnel responsible for H&S management are up to date on the latest trends of H&S issues. These points (H&S policy and the organisation, and training and information) are further discussed next.

3.2.1.1 Health and Safety Policy and the Organisation

The H&S policy is an organisation's statement detailing how it will ensure a healthy and safe work environment. Individual policies will need to be developed for specific hazards and issues, e.g. smoking in the workplace, manual handling and first aid. Policies should be supported by procedures that provide step-by-step instructions on how policies will be achieved. Section 2 of the United Kingdom's Health and Safety at Work (HSW) Act 1974 indicated that if an organization employs more than five people, it must have a written H&S policy. The key elements clearly defined in an H&S policy at an organization should include the following: a copy of a written H&S policy statement (specifying H&S aims and objectives) dated and signed by the most senior person in the organization responsible for H&S matters, and H&S responsibilities for employees at all levels (Hughes & Ferrett, 2008; Lingard & Rowlinson, 2005; Health & Safety Executive [HSE], 2009). Construction organisations with more than five employees are also required to make available a written copy of arrangements for H&S and the way of communicating these arrangements to the workforce (Yu, 2013). The adviser must be able to provide general H&S advice and also advice relating to construction H&S issues (Lingard & Rowlinson, 2005; Carpenter, 2006; Health and Safety Commission [HSC], 2007).

3.2.1.2 Training and Information

Training of employees on health and safety is important, and will improve the performance of H&S compliance over time in the organization. It will also promote a positive H&S culture (Yu, 2013). Training provides employees with knowledge and skills to perform their job safely and understand the necessary information to discharge their duties. The effectiveness of a training programme relies on the extent to which what is learnt is put into practice. An H&S programme is a vital way of ensuring that commitments made in the H&S policy are translated into effective action to prevent injury and disease. The H&S policy should be dynamic and should change in response to organisational changes that affect the management of H&S. Regular revision of the policy allows the employer to promote and maintain an organisation's H&S programme (Russell, 2012; Yu, 2013). The employer becomes legally responsible for the H&S of any employee, visitor or people nearby as well as anyone affected by his or her activities. Since the health and safety of his or her employees are his or her responsibility, the employer must take responsibility for health and safety by providing H&S training for all the employees, taking into account particular risks they face and detailing any specific H&S responsibilities in their employment contract. The employers must also provide posters or leaflets relating to H&S and use signage where necessary. He or she should also communicate the firm's policy to all employees and anyone that could be affected by the construction activities.

3.2.2 Gap 2: Contractor's Organisational Culture

This section presents a review of the literature on the concept of culture and, in particular, the organisational culture. It is alleged that sustained improvement on vital matters such as H&S would not happen without cultural change (Dingsdag, Biggs, Sheahan & Cipolla, 2006). It is insufficient to provide safe equipment, systems and procedures if the culture is not conducive to a healthy and safe working environment. It is argued that a positive culture leads to both improved H&S and as well as organisational performance (Dingsdag et al., 2006). Culture can be defined as a characteristic set of assumptions, beliefs, values, knowledge, attitudes and symbols held and shared by all members of a group and these influence behavioural patterns and perceptions. Individual and group behaviours, attitudes, norms and values, perceptions and thoughts are termed a culture by Choudhry, Fang and Mohamed (2007). Table 3.1 shows the definitions of the concept of culture from various scholars.

TABLE 3.1
Definitions of the Concept of Culture

Author	Definition of Culture
Adeogun and Okafor (2013)	Culture is 'the ways of thinking, behaving and believing that members of a social unit have in common'.
Choudhry, Fang, and Mohamed (2007)	Culture is a product of individual and group behaviours; attitudes, norms and values, perceptions and thoughts that determine the commitment to safety, style and proficiency, and the organisation's system; how workers act and react in terms of the company's ongoing safety performance.
Dingsdag, Biggs, Sheahan and Cipolla (2006)	Culture is a description of values, norms and attitudes, and collective beliefs towards H&S within an organisation. These guide behaviour by indicating to employees what will be rewarded or punished by the organisation.
Mengolini and Debarberis (2007)	A pattern of shared, taken-for-granted basic assumptions held by the members of an organisation and developed through a process of external adaptation and integration.
Fernandez-Muniz, Montes-Peon, and Vazquez-Ordas (2007)	A set of values, perceptions, attitudes and patterns of behaviour concerning safety; a set of policies, practices and steps that trigger hazard minimisation; implementation of the high level of concern and commitment to accident prevention.
ACRCCI (in Musonda 2012)	The behaviour, attitudes and values of members are dependent upon the sets of both conscious and unconscious beliefs that individual members possess and that these beliefs are seen as a key element of organizational culture.
Cooper (in Musonda 2012)	Culture is a characteristic set of assumptions, beliefs, values, knowledge, attitudes and symbols shared and held by all members of a group that influences behavioural patterns and perceptions. These can surface through observation and/or description of what goes on by those that are part of the organisation.
Gadd and Collins (2002)	The concept that describes the shared corporate values within an organisation that influences the attitudes and behaviours of its members. Safety culture is a part of the overall culture of the organisation and is seen as affecting the attitudes and beliefs of members in terms of H&S performance.
Hopkins (2006)	A product of individual or group behaviour.
Wiegmann, Zhang, Von Thaden, Sharma, and Mitchell (2002)	The enduring value and priority placed on worker and public safety by everyone in every group at every level of an organisation.
Martins and Terblanche (2003)	Organisation culture is the deeply seated (often subconscious) values and beliefs shared by personnel in an organisation.
IOSH (2004)	Safety culture consists of shared values and beliefs that interact within an organisation's structure and control systems to produce behavioural norms (the way we do things around here).
Fitzgerald (2005)	Consists of shared values (what is important) and beliefs (how things work) that interact with an organisation's structure and control systems to produce behavioural norms (the way we do things around here).
Institute of Engineering and Technology (2009)	The product of the individual or group values, attitudes, competencies and patterns of behaviour that determine the commitment to, and the style and proficiency of an organisation's H&S management.

As for the concept of culture itself, its conceptualization and definitions have been derived from the more general notion of organizational culture. Organizational culture serves as a filter through which strategies are decided, and performance results attained (Fong & Kwok, 2009). Organizational culture is an organization's values, assumptions and expectations (Hooijberg & Petrock in Fong & Kwok, 2009).

Major construction site accidents have been attributed to a contractor's poor organisation culture. The solutions of H&S should match the problems at hand because an H&S accident has its roots in the contractor's organisational culture. Culture has become popular among the various methods to improve H&S performance. Culture can be improved from one level to another. Culture creates a homogeneous set of assumptions and decision premises in which compliance occurs without surveillance (Grote, 2007). The Institution of Occupational Safety and Health (IOSH, 2004) contends that it is insufficient, for example, to provide safe equipment, systems and procedures if the culture is not conducive to a healthy and safe working environment.

In addressing the contractor's organizational culture gap, the contractor should intensify the provision of personal protective equipment (PPE) to employees and ensure their proper and continuous use. Small to medium-sized construction companies should also ensure that they do everything possible within their reach to protect the health and safety of all their employees before any work commences on site. Training on H&S programmes as well as the use of PPE should be carried out on a regular basis. Signs and notices should also be visible at all vantage points. It is the duty of the contractor to protect employees and representatives from any risk of injury at the workplace.

3.3 SUMMARY

This chapter described the gaps in the theoretical framework with emphasis on the conceptual frameworks of Heinrich and the ARCTM. The identified gaps were government support and a contractor's organisational culture. The first gap is government support that comes in the form of H&S policy, implementation of the policy and training and education. Gap 2 is the contractor's organizational culture. This takes the form of provision of personal protective equipment (PPE), training on health and safety matters, appropriate use of PPE and provision of signs and notices.

REFERENCES

Adeogun, B.K. and Okafor, C.C. (2013). Occupational health, safety and environment (HSE) trend in Nigeria. *International Journal of Environmental Science, Management and Engineering Research*, 2(1): 24–29.

Carpenter, J. (2006). Developing guidelines for the selection of designers and contractors under Construction (Design and Management) Regulations 1994. Research Report 422. Norwich: Health and Safety Executive.

Choudhry, R.M., Fang, D. and Mohamed, S. (2007). The nature of safety culture: A survey of the state-of-the-art. *Safety Sciences*, 45(10): 993–1012.

Dadzie, J. (2013). Perspectives of consultants on health and safety provisions in the Labour Act: A study into theory and practicals. *Engineering Management Research*, 2(1): 34–42.

Dingsdag, D.P., Biggs, H.C., Sheahan, V.L. and Cipolla, C.J. (2006). A construction safety competency framework: Improving OHS performance by creating and maintaining a safety culture. Brisbane: Cooperative Research Centre for Construction Innovation.

Fernandez-Muniz, B., Montes-Peon, M.J. and Vazquez-Ordas, J.C. (2007). Safety culture: Analysis of the causal relationships between its key dimensions. *Journal of Safety Research*, 38(6): 627–641.

Fitzgerald, M.K. (2005). Safety performance improvement through culture change. *Process Safety Environmental Protection*, 83(4): 324–330.

Fong, P. and Kwok, C. (2009). Organisational culture and knowledge management success at project and organisational levels in construction firms. *Journal of Construction Engineering Management*, 135(12): 1348–1356.

Gadd, S. and Collins, A.M. (2002). *Safety culture: A review of the literature.* Sheffield: Health & Safety Laboratory.

Grote, G. (2007). Understanding and assessing safety culture through the lens of organizational management of uncertainty. *Safety Science*, 45: 637–652.

Health and Safety Commission (HSC). (2007). Managing health and safety in construction: Construction (Design and Management) Regulations 2007 Approved Code of Practice. Norwich: Health and Safety Commission.

Health and Safety Executives (HSE). (2009). Safety signs and signals. The Health and Safety (Safety Signs and Signals) Guidance on Regulations. Available from: http//:www.hsebooks.co.uk [Assessed 20 September 2014].

Heinrich, H.W. (1930). *Industrial accident prevention.* New York: McGraw-Hill.

Hopkins, A. (2006). Studying organisational cultures and their effects on safety. *Safety Sciences*, 44: 875–889.

Hughes, P. and Ferrett, E. (2008). *Introduction to health and safety in construction.* Oxford: Elsevier Butterworth-Heinemann.

Institute of Engineering and Technology (IET). (2009). Safety Culture, Michael Faraday House, Six Hills Way, Stevenage, SG1, 2AY.

Institution of Occupational Safety and Health (IOSH). (2004). Promoting a positive culture: A guide to health and safety culture. Leicestershire: IOSH.

Laryea, S. (2010). Health and safety on construction sites in Ghana. In: *The Construction, Building and Real Estate Research Conference of the Royal Institute of Chartered Surveyors*, Dauphine Université, Paris, 2–3 September. Dauphine Universite, Paris, France.

Lingard, H. and Rowlinson, S. (2005). *Occupational health and safety in construction project management.* Abingdon: Spon Press.

Martins, E.C. and Terblanche, F. (2003). Building organisational culture that stimulates creativity and innovation. *European Journal of Innovation Management*, 6(1): 64–74.

Mengolini, A. and Debarberis, L. (2007). Safety culture enhancement through the implementation of IAEA guidelines. *Reliability Engineering and System Safety*, 92: 520–529.

Musonda, I. (2012). Construction health and safety (H&S) performance improvement. A client-centred model. Unpublished DPhil in Engineering Management, University of Johannesburg, South Africa. Available from: http://www.ujdigispace.uj.ac.za [Accessed 02 June 2014].

Russell, B. (2012). Workplace health and safety handbook. Available from: http://www.safework.sa.gov.au [Accessed 14 April 2014].

Wiegmann, D.A., Zhang, H., Von Thaden, T.L., Sharma, G. and Mitchell, A.A. (2002). A synthesis of safety culture and safety climate research. University of Illinois Aviation Research Lab Technical Report ARL-02-03/FAA-02-2.

Yu, W.-H. (2013). It's who you work with: Effects of workplace shares of nonstandard employees and women in Japan. *Social Forces*, 92(1): 25–57.

Section IV

Construction Industry: International Literature

4 Construction Industry
International Literature

4.1 INTRODUCTION

This chapter discusses the construction industry from global perspectives. The chapter commences by discussing the construction industry in general and the occurrence of accidents in the construction industry. This is followed by a discussion on the 'influence network on health and safety', techniques to improve health and safety compliance, and the management of a successful occupational health and safety culture in small to medium-sized construction companies. Finally, small to medium-sized enterprises' performance in terms of compliance with safety regulations is discussed.

4.2 THE CONSTRUCTION INDUSTRY

The construction industry plays an important role in the improvement of countries' economic growth. The construction industry employs about 180 million people or represents 7 percent of global employment (International Labour Office [ILO], 2005; Murie, 2007). In 2001 and 2002 about 8 percent of the total European Union (EU) workforce was employed in construction (Karjalainen, 2004; Ringen & Englund, 2006). Despite the contributions to economic growth, the construction industry has always been blamed for the high rates of accidents and fatalities. This concern has placed the construction industry among those industries with unreasonable rates of accidents, permanent and non-permanent disabilities and even fatalities. There are many pieces of evidence that portray the construction industry as a hazardous and inconsistent industry. For instance, the high rates of accidents and fatalities in this industry have placed it among the most hazardous industries (Hosseinian & Torghabeh, 2012). The costs of injuries, which are direct and indirect, workers' compensation insurance, legal liability as well as legal prosecutions have pushed parties involved to seek ways of mitigating these hazards. The world rates of occupational injuries, illnesses and fatalities are still alarming. A report from the National Safety Council (United States) in 1996 showed that 1000 construction workers died at work and 350,000 suffered disabilities. The U.S. construction workforce constitutes only 5 percent out of 20 percent of all occupational fatalities. Nine percent of all disabling occupational injuries relate to the construction industry (Hosseinian & Torghabeh, 2012). A report from the International Labour Office (ILO, 2005) indicated that construction contributes 25 to 40 percent of fatalities in the world's occupational settings. Based on fatality statistics, different countries show that the construction industry produces 30 percent of fatal industrial accidents across the European Union (EU), yet it employs only 10 percent of the working population. The construction industry in the United States accounts for 20 percent of fatal accidents. In Japan, construction fatalities account for 30 to 40 percent of fatal industrial accidents (ILO, 2005). In the developing world, the risks associated with construction work are much greater. Twenty to twenty-five percent of all fatal accidents occur in the construction industry (Karjalainen, 2004; Ringen & Englund, 2006). Globally, the construction site is the most hazardous place to work, with a high level of health and safety risks (ILO, 2005; Lingard & Rowlison, 2005; Smallwood, Haupt & Shakantu, 2008). The estimate given by ILO indicates that at least 60,000 fatal accidents occur a year on construction sites around the world. It was also asserted that the injury and fatality rates on construction projects are very high (Hinze, 2005; Lingard & Rowlison, 2005; Smallwood et al., 2008).

The construction industry in Singapore appears to have a similar record. Although the construction sector in 2005 contributed less than 10 percent to the gross domestic product, it accounted for more than 37 percent of all industrial accidents (Teo, Ling & Chong, 2005). H&S performance, even at the global level is not impressive. The ILO identified high risk in the construction industry. The construction industry is the third highest risk sector. Every year, 60,000 people are killed. There is an urgent need to improve construction H&S performance to make the industry less dangerous. According to Haslam et al. (2005:401), 'the construction sector does not have an enviable record or reputation and its H&S compliance can be described as poor'. The EU Agency for H&S at work reported that many more accidents occurred in construction per 100,000 workers than in the overall EU national workforce (Karjalainen, 2004).

Construction sites in developing countries are 10 times more dangerous than those in developed countries. A report from the South African Construction Industry Development Board (CIDB, 2008) illustrated that efforts made by several bodies to improve H&S are to no avail within the construction industry. High levels of fatalities and injuries occur in construction as compared to other industrial sectors. This development raises the level of non-compliance with H&S legislation. It is further argued that improving H&S in the construction industry remains a priority. Further causes of accidents are due to management system failure and human error (Petersen, 2000; Chua & Goh, 2004). The human error theory holds that the cause of accidents is in the organisational and management processes (Behm, 2008; Bellamy, Geyer & Wilkinson, 2008). Therefore, an effort aimed at addressing H&S should be directed more at addressing organisational and project management factors. These should include management in the industry, the project and company or organisation level as accidents are prevalent in the construction industry.

The source of hazards and unsafe behaviour has its origins not only from proximal conditions but also from distal conditions including organisational and management conditions. Data from the ILO and the World Health Organization (WHO) indicated that overall occupational accident and disease rates are slowly declining in most industrialised countries (Alli, 2008). However, the rate of accidents in developing and industrialising countries is increasing. About 5000 workers were killed in accidents at work, and about 5 million workers were victims of accidents at work leading to more than three days' absence from work in the European Union in 2004. In India and China, the rates of occupational fatalities and accidents are similar: respectively, 10.4 and 10.5 per 100,000 for fatalities, and 8,700 and 8,028 for accidents. In sub-Saharan Africa, the fatality rate per 100,000 workers is 21 and the accident rate is 16,000. It implies each year 54,000 workers die and 42 million work-related accidents take place and cause at least three days' absence from work. In Latin America and the Caribbean, about 30,000 fatalities occur each year and 22.6 million occupational accidents cause at least three days' absence from work (Alli, 2008). Construction accidents can be prevented by identifying the root causes of accidents, which is made possible by accident investigation techniques such as theories of accident causation and human errors (Hosseinian & Torghabeh, 2012).

Occupational and industrial accidents are caused by preventable factors that could be eliminated by implementing already known and available measures and methods. Arewa and Farrell (2012) posited that all firms in the United Kingdom have a legal responsibility for the H&S of anyone affected by their business, irrespective of their nature, size or volume of work. Compliance with H&S entails carrying out thorough H&S risk assessments and drawing up H&S policies for businesses with more than five employees. It also ensures that workplaces meet minimum standards of conformity and cleanliness, recording serious injuries, diseases or dangerous accidents in the accident book. One of the key challenges facing the occupational health profession in the 21st century is protecting workers' H&S in a global economy characterised by ferocious competition to reduce production costs and a marked decline in the development and enforcement of governmental workplace regulations (Brown, 2005). For occupational health to be effectively protected, there must be key components in all future trade and investment agreements. These elements include: a minimum floor of occupational health and safety regulations; an "upward harmonization" of regulatory standards and actual practice; the inclusion of employers so that they have formal responsibility and liability for violations of the standards; effective enforcement of national regulations and international standards; transparency and public participation; recognition of disparate economic conditions among trading partners; and provision of financial and technical assistance to overcome economic disincentives and lack of resources (Brown, 2005).

The opportunities for occupational health improvements presented by globalization are thus still outweighed by the shift in health costs to workers in the insecure forms of employment, particularly given the weaknesses of the protection systems in developing countries. The construction industry is a place where structures, skills, knowledge and analytical capacities are needed to coordinate and implement all of the 'building blocks' (Alli, 2008). The building blocks should be made of national occupational safety and health (OSH) systems to include both workers and the environment (Alli, 2008). OSH is generally defined as the science of anticipation, recognition, evaluation and control of hazards arising in or from the workplace that could impair the health and well-being of workers, taking into account the possible impact on the surrounding communities and the general environment. This domain is necessarily vast, encompassing a large number of disciplines and numerous workplace and environmental hazards (Suraji, Duff & Peckitt, 2001). The general perception is that construction H&S is a matter of construction management rather than the management on the part of clients and other participants in the construction process. The main problem lies with contractors and therefore H&S performance improvement can only be achieved by addressing contractor issues.

However, it is unlikely that H&S performance improvement can be achieved in the industry by only focusing on the construction stage and the contractor specifically (Suraji et al., 2001). A report from the Occupational Health and Safety Act (OHSA) of the United Kingdom indicated that any firm that employs more than 20 workers for four months or more must implement the act. The following specific obligations are required to be followed by all firms:

- Steps must be taken to eliminate or mitigate any hazard or potential hazard to the safety or health of employees before resorting to personal protective equipment.
- The arrangement must be made to ensure the safety and absence of risks to health connected to the production, processing, use, handling, storage or transport of articles or substances.
- Firms must also find out the health and safety hazards involved with any work. Consideration must be given to the precautionary measures that should be taken for such health and hazards, and precautionary measures must be implemented and communicated.
- Systems of work, plant and machinery that are safe and without risk to health must be provided and maintained.
- Employees must be trained in H&S matters (OHS Network Solutions, 2014). According to the OHS Act 85 of 1993, compliance is the minimum standard which an employer and employee need. More safety measures can be put in place as required by the nature of the work environment, but not less (Construction Regulations, 2003).

The minimum standard is then a working environment that is free from any exposure to anything that can cause harm to people, property, products, processes and environment. Neither the government nor the Department of Labour will implement day-to-day compliance with this act because the responsibility of implementing and enforcing this act is placed on the employer (Construction Regulations, 2003). Mitigation is offered by the OHS Act in Section 37 for employees to their authority. If an employee commits an act or omits to act, which leads to a violation of the OHS Act, the employee can be held liable as if he or she is the employer (Construction Regulations, 2003). Section 7 of the Construction Regulations (2003) requires that no contractor shall allow or permit any employee to enter any site, unless such a person has undergone H&S induction training pertaining to the hazards prevalent on the site at the time of entry. Many clients and contractors around the world are adopting schemes requiring site staff to have skill assessments, including H&S. For example, the Construction Skills Health and Safety Test in the United Kingdom is taken by more than 500,000 people every year and is designed to ensure everybody working in construction has a minimum level of H&S awareness at an acceptable industry standard.

Similar requirements also exist in other countries, for example, in Queensland, Australia, where H&S is an integral component of obtaining a builder or trade license. Although the Construction Regulation in South Africa requires all employees on site to 'be in possession of proof of H&S induction training', no industry-accepted standard exists. However, verification of H&S skills and awareness was incorporated into the South Africa CIDB Best Practice Construction Regulation Scheme (CIDB, 2008).

4.2.1 Complex Nature of the Construction Industry

The risk of a fatality in construction is at least five times more likely than in other manufacturing-based industries (Loughborough University and University of Manchester Institute of Technology, 2003). According to Teo et al. (2005:329), 'the complexity of the construction sector has been compounded by the extensive use of sophisticated plant, equipment, methods of construction, as well as multidisciplinary and multitasked project work force'. The following unique characteristics have contributed to the complex nature of the construction industry:

- The construction industry mostly offers temporary employment (Pellicer & Molenaar, 2009).
- Employees do often experience a change of environment from their workplaces (McDonald, Lipscomb, Bondy & Glazner, 2009).
- There is the possibility of engaging with different employees while working on temporary work sites (Misnan, Mohammed, Mahmood, Mahmud & Abdullah, 2008; Pellicer & Molenaar, 2009).
- It is an industry comprised mostly of small employers (Pellicer & Molenaar, 2009).
- The majority of the people employed in this industry have to combine a diverse range of skills to complete a project (Pellicer & Molenaar, 2009).
- A large number of sub-contractors are on the highest list off (Pellicer & Molenaar, 2009).
- Projects are mostly of short period durations (BOMEL, 2001; Dainty, Briscoe & Millet, 2001; Riley & Brown, 2001).

4.2.2 Challenges Facing Construction Workers

Construction is a hazardous occupation. For almost all key risks such as chemicals, dust, manual handling, physical hazards and psychosocial hazards, exposures are routine and excessive (Ng, Cheng & Skitmore, 2005; Murie, 2007). Construction industry work has a low social status because it is not attractive. Its problems also have low visibility and their resolution has a low priority. The construction industry all over the world offers low status, low pay and short-term employment. They are mostly unregistered, informal and hazardous jobs in a highly fragmented industry (Murie, 2007; Kulchartchai & Hadikusumo, 2010). Construction is also responsible for far more than its share of occupational accidents and work-related ill health (Hola, 2007; Kulchartchai & Hadikusumo, 2010).

4.3 ACCIDENTS IN THE CONSTRUCTION INDUSTRY

Figures from the ILO (2005) for the ten countries that joined the EU in 2004 have estimated that construction accounted for 20 percent of all work-related accidents in these regions. Within the United Kingdom, the Health and Safety Executive (HSE, 2009) has accounted for 25 percent of all fatal injuries and 16 percent of all major accidents are within the construction industry. The number and rate of fatal injuries to workers between 1999/2000 and 2008/2009 are depicted in Figure 4.1. In addition, the number and rate of major injuries to employees between 1999/2000 and 2008/2009 are depicted in Figure 4.2 (HSE, 2010).

Even though there have been significant reductions in the number and rate of injuries over the past 20 years or more, construction nevertheless remains a high-risk industry as depicted in Figure 4.2, although it accounts for only about 5 percent of employees worldwide. In Britain it accounts for 27 percent of fatal injuries to employees and 10 percent of reported major injuries. Some current results in

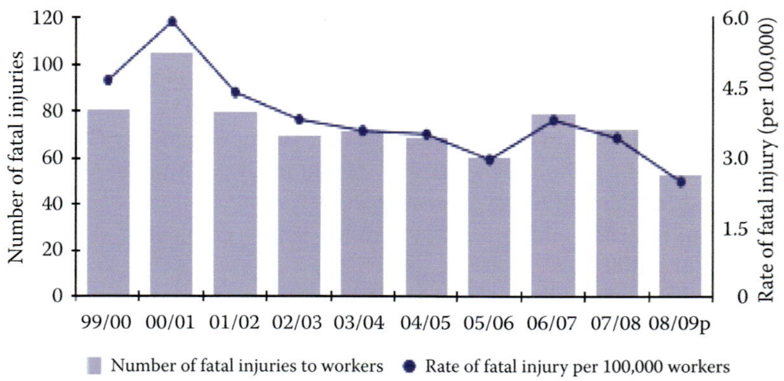

FIGURE 4.1 Number and rate of fatal injuries to workers in the United Kingdom in 1999/2000 to 2008/09. (From Health and Safety Executive (HSE), 2010, *Work-related injuries and ill health in construction*, Suffolk: HSE Books.)

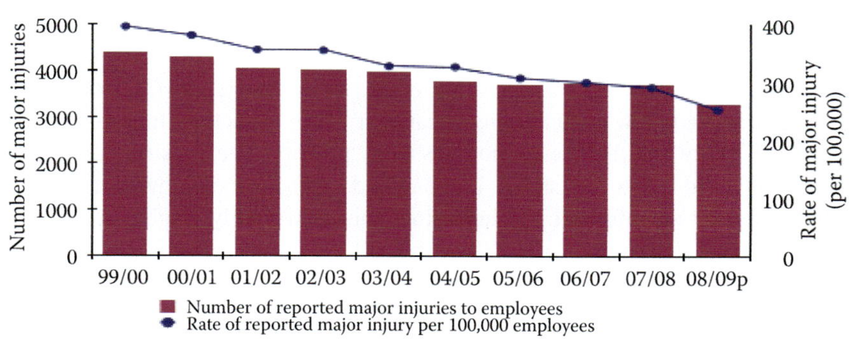

FIGURE 4.2 Number and rate of reported major injuries to employees in the United Kingdom in 1999/2000 to 2008/09. (From Health and Safety Executive (HSE), 2010, *Work-related injuries and ill health in construction*, Suffolk: HSE Books.)

construction as reported by the Great Britain Labour Force Survey (HSE, 2013) are as follows:

- There were 39 fatal injuries to workers. Twelve of these fatalities were to the self-employed. There were an average of 53 fatalities over the previous five years, including an average of 18 to the self-employed (Reporting of Injuries, Diseases and Dangerous Occurrences Regulations [RIDDOR]).
- There were an estimated 74,000 total cases and 31,000 new cases of work-related ill health.
- A total of 1.4 million working days were lost in 2011/12; 818,000 of these were due to ill health and 584,000 due to workplace injury, making a total of 0.7 days lost per worker (HSE, 2013).

4.3.1 Fatal Injuries to Workers

A quarter of fatal injuries to the workers over the previous five years were due to falls and another quarter to slips or trips as shown in Figure 4.3 (HSE, 2013). A report from HSE (2013) indicated that there were 39 fatal injuries to workers in construction in 2012/13, and 12 of these fatalities included the self-employed. The rate of fatal injury per 100,000 construction workers was 1.9 in 2012/13 compared with a five-year average of 2.3. The HSE (2013) reports (see Figure 4.3) shows that 26 percent of all fatal injuries to workers were in construction. This incidence accounts for the greatest number of fatal injuries of the industry sections. In the year 2010/11 the number of fatalities fell to the level seen in 2009/10 (HSE, 2013).

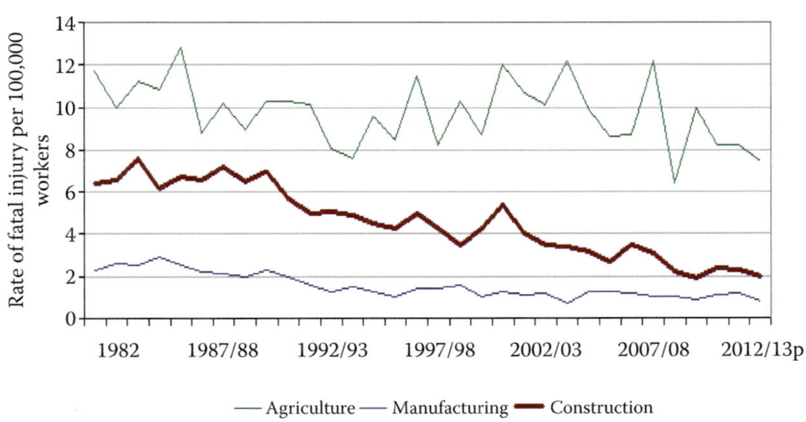

FIGURE 4.3 Long-term trends in rates of fatal injury to workers, 1981 to 2012/13 (RIDDOR). (From Health and Safety Executive (HSE), 2013, *Work-related injuries and ill health: Health and safety in construction in Great Britain*, 2013.)

4.3.2 Major Injuries

There were 1913 reported major injuries to employees in 2012/13, compared to an average of 2815 over the previous five years. The corresponding rates of major injury per 100,000 employees were 156 in 2012/13 and an average of 192 (HSE, 2013). There has been a general reduction in the rate of reported major injury since 2004/05, and the number of reported injuries has also fallen significantly over the previous five years (HSE, 2013). The fall in reported major injuries over the previous two years is similar to that for all industries, but it was significantly higher in the previous three years.

A report from the HSE (2014) indicated that 'the number of workers fatally injured in 2013/14 alone was 133 and corresponds to a rate of fatal injury of 0.44 deaths per 100,000 workers. The deaths of 133 workers in 2013/14 was 19 percent lower than the previous five years (164)'. The HSE report showed that the latest rate of fatal injury of 0.44 as compared to the five-year average rate was 0.56. The finalised figure for 2012/13 was 150 worker fatalities and corresponds to a rate of 0.51 deaths per 100,000 workers. There were 70 members of the public fatally injured in accidents connected to work in 2013/14 (excluding railways-related incidents) (HSE, 2014).

4.4 INFLUENCE NETWORK ON HEALTH AND SAFETY

The influence network for H&S in the construction industry provides a wider context for the way that Construction (design and management) Regulations 1994 (CDM) are perceived in the construction industry. The influence network process addresses issues relating to H&S in construction's five high-level objectives as shown in Figure 4.4 (BOMEL Limited, 2007). The following hierarchy was adopted by the influence network to model these influences (Table 4.1).

1. Direct performance influences—These directly influence the likelihood of an accident being caused.
2. Organisational influences—These influence direct influences and reflect the culture, procedures and behaviour promulgated by the organisation.
3. Strategy level influences—These reflect the expectations of the decision makers in the employers of those at risk and the organisations they interface with (e.g. clients, suppliers, subcontractors).
4. Environmental level influences—These cover the wider political, regulatory, market and social influences which impact the policy influences.

In terms of the construction industry, the relevant stakeholders fit into the model as shown in Figure 4.4 (BOMEL Limited, 2007). Influencing factors have been identified at each level of influence, as shown in the network in Figure 4.4. The factors have been determined based on accepted theories of human factors and safety and risk management (BOMEL Limited, 2007). The categories have been expanded further and refined through practical application to a range of scenarios. Each influence in the generic network is defined together with a scale

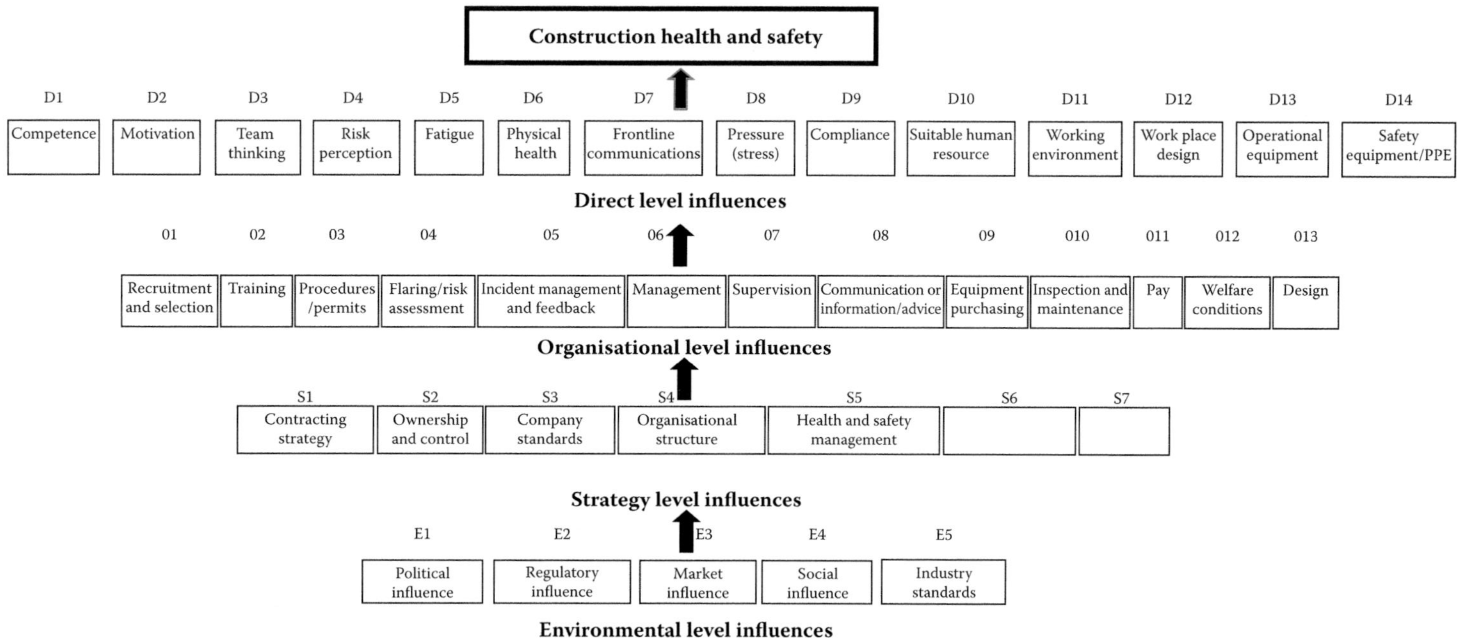

FIGURE 4.4 Influence network for health and safety in construction. (Adapted from BOMEL Limited, 2007, Improving the effectiveness of the Construction (Design and Management) Regulations 1994: Establishing views from construction stakeholders on the current effectiveness of CDM, Health and Safety Executive 2007; Manu, P., Ankrah, N., Proverbs, D. and Suresh, S., 2010, Exploring the influence of construction project features in accident causation, In: Barrett, P., Amaratunga, D., Haigh, R., Keraminiyage, K., and Pathirage, C. (Eds.), Proceedings of Council for Research and Innovation in Building and Construction (CIB) 2010 World Congress, 10–13 May, Salford, UK; Musonda, I., 2012, *Construction health and safety (H&S) performance improvement: A client-centred model*, unpublished DPhil in Engineering Management, University of Johannesburg, South Africa.)

TABLE 4.1
Construction Stakeholders Influence Network Levels

Influence Level	Definition
Director level	Applies to site operatives and technicians (i.e. people who carry out the construction work).
Organisational level	Applies to site organisation, local management and designers.
Strategy level	Applies to both client and construction company management. Contracting strategy, ownership and control and company standards apply to the client (i.e. the organisation commissioning and paying for the construction activity) and the remainder applies to contractors carrying out the work.
Environmental level	The political influence incorporates government procurement strategy and guardian of workers and public safety. Otherwise, the environmental-level influences are external to the organisations represented at the HSE as a principal regulatory influence.

Source: BOMEL Limited, 2007, Improving the effectiveness of the Construction (Design and Management) Regulations 1994: Establishing views from construction stakeholders on the current effectiveness of CDM, Health and Safety Executive 2007.

from best to worst practice. The influence in a generic network provides a basis for making judgements about the relative importance of each influence (weighting), the current quality of each influence (rating) and the potential effect on the quality of the factor by introducing risk control measures as outlined in Table 4.2 (BOMEL Limited, 2007).

4.5 TECHNIQUES TO IMPROVE HEALTH AND SAFETY COMPLIANCE

Compliance is either a state of being by established guidelines, specifications, or legislation or the process of becoming so. Compliance with a rule is not always the full test for determining the effectiveness of regulation in achieving its goals. Full compliance with a rule may not accomplish the desired outcome. For example, full compliance may be so costly that it causes more damage than it remedies (e.g. if the costs of compliance are so great, it can drive legitimate enterprises out of business). Full compliance may be possible but not adequate to achieve the desired objective (e.g. if the rule mandates a particular technology that does not accomplish the intended goal). The underlying problem to be solved may not be adequate to achieve the desired objective (e.g. if the rule mandates a particular technology that does not accomplish the intended goal). The underlying problem to be solved may not be understood well enough to identify the right solution. In evaluating the outcomes of regulations, the policymaker should consider that regulatory compliance is important but is not the only factor that determines policy effectiveness (Organisation for Economic Co-operation and Development [OECD], 2000).

TABLE 4.2
Outline of Health and Safety Regulations Influencing Construction Activities

Regulations	Relevant Activities
The Work at Height Regulations 2005	Roof working, scaffold and ladder using, etc.
Confined Spaces Regulations 1997	Working in any chamber, tank, vat, silo, pit, trench, pipe, sewer, flue, well or similar enclosed space
Provision and Use of Work Equipment Regulations (PUWER)	Any work involving work equipment (machinery, appliance, apparatus, tool or installation)
The Manual Handling Operations (HMO) Regulations 1992	Any transporting or supporting (lifting, putting down, pushing, pulling, carrying or moving) of loads
Lifting Operations and Lifting Equipment Regulations (LOLER) 1998	Any work involving equipment for lifting or lowering loads, the lifting working equipment includes cranes, forklift trucks, lifts, hoists, mobile elevating work platforms, vehicle inspection platform hoists and lifting accessories such as chains, slings, eyebolts, etc.
Electricity at Work Regulations 1998	Any work involving electrical systems or equivalent
Fire Precautions (Special Premises) Regulations 1976	Setting up temporary accommodation units such as offices, workshops or storage facilities
Fire Precautions (Workplace) Regulations 1997	Dealing with general fire precautions including means of detection and giving warning in case of fire, the provision of means of escape, means of fighting the fire, and the training of staff in fire safety
Control of Substances Hazardous to Health Regulations (COSHH) 2002	Any work involving hazardous substances (solvents, paints, adhesives, cleaners and dust) in workplaces of all types
Dangerous Substances and Explosive Atmospheres Regulations (DSEAR) 2002	Dealing with hazardous substances like petrol, liquefied petroleum gas (LPG), paints, cleaners, solvents, flammable gases and the explosive mixture in the air (dust)
Chemical (Hazard Information and Packaging for Supply) Regulations 2002	Receiving chemicals from suppliers
Control of Asbestos at Work Regulations 2002	Any work could be exposed to asbestos
Asbestos (Licensing) Regulations 1998	Applying for license for working with asbestos insulation, asbestos coating or asbestos insulation board
Ionising Radiation Regulations 1999	Any activities involving use of ionising radiation equipment (e.g. x-ray weld crack detector)
Personal Protective Equipment at Work Regulations 1992	Personal Protective Equipment at Work Regulations 1992
Construction (Head Protection) Regulations 1989	Building operations and works of engineering construction
The Control of Noise at Work Regulations 2005	Any work affected by noise

(Continued)

TABLE 4.2 (CONTINUED)
Outline of Health and Safety Regulations Influencing Construction Activities

Regulations	Relevant Activities
The Reporting of Injuries, Diseases and Dangerous Occurrences Regulations 1995	Reporting process when specific accidents occur on construction sites
Health and Safety (Safety Signs and Signals) Regulations 1996	Correctly using safety signs and signals at any workplace
Health and Safety (Display Screen Equipment) Regulations 1992	Use of display screen equipment
Health and Safety (First Aid) Regulations 1981	Providing first aid facilities

Source: Hughes, P. and Ferrett, E., 2008, *Introduction to health and safety in construction*, Oxford: Elsevier Butterworth-Heinemann.

4.5.1 Health and Safety Compliance

The phrase 'Compliance with Health and Safety' has no specific definition. It is a term used to mean orthodoxy of H&S rules and regulations. For example, in the United Kingdom, all firms have a legal responsibility for the H&S of anyone affected by their business irrespective of their nature, size or volume of work. Compliance with health and businesses entails the following:

- Carrying out thorough health and safety risk assessments.
- Drawing up a health and safety policy for businesses with more than five employees.
- Ensuring workplaces meet minimum standards of conformity and cleanliness.
- Recording serious injuries, diseases or dangerous accidents in the accident book (United Kingdom Government Business Link [UKGBL] in Arewa & Farrell, 2012).

Different factors affect compliance with H&S as stated in the HSE (2005). The main motivator for complying with H&S is the general fear of the law, that is, liability (fear of being sued by clients). It is also the threat from the local workplace if non-compliance is not remedied. Non-compliance with H&S leads to accidents, and workplace accidents have the potential to take 30 percent off company annual profits and failure to manage safety has a much larger social cost (Taylor, Easter & Hegney, 2004; Arewa & Farrell, 2012). Compliance with OHS regulations is one of the management efforts to determine whether it correlates with OSH performance; while compliance with OSH regulations brings about benefits not limited to avoiding direct and indirect costs (Smallwood et al., 2008; Okeola, 2009; Idoro, 2011). Compliance in a broader sense contributes to organisations' competitive advantages.

Factors such as technical failure and inadequate training coupled with a harsh work environment and unsafe methods of working inter alia are among the causes of non-compliance with OSH regulations in developing countries (Othman, 2012; Windapo & Oladapo, 2012). Also, a lack of adequate training is a major hindrance to OSH regulations compliance. Whilst, a safe work environment can determine how issues of compliance with OSH regulations are taken care of by construction firms.

4.5.2 Problems Raised in Health and Safety Compliance Research Study

Ineffective H&S management increases the likelihood of deficiencies in site H&S practices at hazardous waste sites. Most small and medium-sized contractor sites have not integrated key aspects of their site H&S programmes, nor has one individual been given overall H&S enforcement authority. These types of inconsistencies in site H&S practices amongst small to medium-sized contractors undermine the message that H&S requirements exist to protect the employee, and they may allow hazards to remain unabated, placing site employees at increased risk (Adenuga, Soyingbe & Ajayi, 2007; Windapo & Oladapo, 2012; Idubor & Oisamoje, 2013). Implementing H&S programmes at construction sites can be costly. Descriptions of site organisational structure and lists of emergency contacts also tend to be inaccurate. Plans that do contain adequate site exposure data often lack other site-specific details when the entire aspects are not considered as a whole. Measuring and documenting the level of employee exposures is a key element in any H&S programme and a consistent area of deficiencies at construction sites. For instance, site personal protective equipment (PPE) requirements are often developed when the site H&S plan is initially drafted. As a result, they are typically based on inadequate exposure information but are seldom modified, even when personal sampling data supporting modifications become available.

4.5.3 Reasons for Non-Compliance in Health and Safety

Reasons for non-compliance can be found at three different levels:

1. The degree to which the target group knows of and comprehends the rules
2. The degree to which the target group is willing to comply, either because of economic incentives, positive attitudes arising from a sense of good citizenship, acceptance of policy goals or pressure from enforcement activities
3. The degree to which the target group can comply with the rules

At each of those three levels, governments should employ the following mix of activities to ensure that their policy will take effect:

1. Communication with the target group to inform it about its rights and duties and to explain the rules
2. The use of many kinds of policy instruments (taxes, prohibitions and subsidies for example) to influence the behaviour of the target group, backed up with a variety of enforcement activities (such as inspections and sanctions)

3. Adequate implementation to make the policy workable in practice, which means that governments have to ensure that the necessary information is provided to the target group, and other technical facilities or mechanisms are taken (OECD, 2000:12)

4.5.4 Ineffective Government Policy Leading to Failures

Regulation that is effective at achieving objectives has usually been designed with an eye to accomplishing substantive outcomes. Policy makers need to understand and take account of the individual characteristics of the target group, including how they can reasonably be expected to respond to rules and government enforcement strategies and both their internal and external incentives to comply with regulatory objectives. The design of results-oriented policy instruments presupposes working procedures in which the policy maker carefully pays attention to both process and 'policy surroundings' (OECD, 2000).

Regulation that fails to elicit an adequate level of compliance not only fails to meet its underlying policy objective but is also due to the following:

- Creates unnecessary costs through fruitless administration and implementation.
- Postpones the achievement of the policy objective.
- Erodes general confidence in the use of regulation, the rule of law and government in general.
- Cumulatively leads to the undermining of regulations. This can lead to a vicious cycle in which more and more rules are promulgated while public confidence in government regulation lessens and compliance outcomes become worse. While many regulations have dramatically improved social welfare in many areas, failures in compliance exist.
- Failure to understand the law, collapse of belief in law, procedural injustice.
- Costs of regulatory compliance, deterrence failure, incapacitation of those regulated.
- Failure of persuasion and failure of civil society (OECD, 2000).

Companies with safety personnel have little difficulty in comprehending and using information about compliance requirements. These companies are much more likely to have effective systems for ensuring compliance than small companies without safety personnel where management usually lack the time and resources to read and understand the great volume of regulatory material on H&S standards (OECD, 2000).

4.5.5 Compliance Is Costly

Voluntary compliance is likely to be low when costs (in time, money or effort) of complying with a rule are considered to be high. Various factors contribute to what may be viewed as unreasonable compliance costs: substantive standards are too high, the transition time for coming into conformity is too short or the regulation is inflexible. If a rule seems unreasonable, instead of complying, businesses may dedicate more time and money to lobbying regulators to change it or asking for special treatment (OECD, 2000). Many OECD countries have implemented different forms of regulatory impact analysis to collect data on estimated and actual costs of regulatory compliance. When business people feel that regulators are overly legalistic in the application of rules and imposition of fines, they will tend to respond by scaling down their efforts to comply with the intent of the law, and would aim to achieve only the minimal level of compliance which the rules required. People are more inclined to see non-compliance with regulation as acceptable when they feel that the relevant regulation is too petty and restrictive. Governments rely on good drafting and enforcement practices, but they should also devote resources to adequate implementation policies, aimed at making it feasible for the target group to be able to comply with the rules of voluntary compliance (OECD, 2000).

4.5.6 Rewards and Incentives for High or Voluntary Compliance

The design of rules, and particularly for monitoring and enforcement regimes, can also encourage compliance by providing incentives or rewards for high voluntary compliance and compliance innovation. Rewards for voluntary or particularly high compliance can include reducing the burden of routine inspections, offering penalty discounts for minor incidents of non-compliance that do occur, simplifying licenses and permits, permitting the use of a label or mark certifying a high level of compliance and providing indemnities for voluntary disclosure and correction of non-fraudulent non-compliance incentives. Rewards can be an important support for voluntary compliance, along with such other alternatives to regulation as self-regulation and voluntary adoption of internal management standards. Rewards and incentives can also boost compliance with government regulation in the following ways (OECD, 2000):

- Rewards and incentives for high compliance performance can recognise the good faith efforts of enterprises that usually comply with, but occasionally inadvertently violate, a rule. For example, providing a penalty reduction when non-compliance does occur, or an indemnity to the enterprise that reports and corrects a violation avoids a situation where the regulator must take ‘unreasonable’ enforcement action against a basically compliant enterprise.
- Appropriate government and public recognition can encourage well-intentioned enterprises to become ‘compliance leaders’. This provides models for other enterprises to follow and can pull up overall compliance performance in a market sector through the dynamics of market leadership.
- In particular, rewards for compliance leaders who meet certain standards can help achieve desired policy outcomes voluntarily (for example an improvement in air quality via greater emissions reduction) without having to use unreasonable coercion with all regulated entities.

- Governments can also use rewards and incentives to encourage a small group of 'compliance leaders' to enter experimental programmes for new regulatory approaches (such as moving from a rule-based standard or process-based regulation). In this way, the enterprises receive rewards for high compliance with policy objectives under the experimental regime, and governments learn from them how to reform existing regulatory approaches in practical ways and what is feasible to expect of business.

4.6 MANAGEMENT OF HEALTH AND SAFETY AT WORK REGULATIONS 1999

The management of occupational health risks is placed on the employer under the management of H&S at Work Regulations 1999 to meet these legal requirements as well as improving the organisation's H&S performance and ultimately reduce risks and costs. The employers should have the following in place in order to adequately manage H&S at the workplace (Canadian Centre for Occupational Health and Safety [CCOHS], 2009):

- H&S policies and procedures with practical arrangements for managing occupational health risks
- Provision of employee awareness training on manual handling, control of substances hazardous to health, noise at work and hand–arm vibration
- Manual handling risk assessments and safe handling techniques for manual handling activities
- Health surveillance, sickness absence management, return to work policy and stress management strategy
- Arrangements for managing subcontractors, including procedures for managing their occupational health risks
- Employers understanding their duties under the Construction Design and Management Regulations 2007

System concepts can help employers to improve their organisations' H&S planning, policies and procedures, and to minimise risks in the construction industry (BOMEL Limited, 2007). A risk assessment scheme should also be carried out in managing occupational health risks. In practical terms, it is a thorough look at one's workplace to identify those things, situations and processes that may cause harm, particularly to people. After identification is made, the employer evaluates how likely and severe the risk is, and then decides what measures should be in place to effectively prevent or control the harm from happening. The following should be followed in order to effectively manage H&S hazards on construction sites:

- Identifying hazards
- Analysing or evaluating the risk associated with that hazard
- Determining appropriate ways to eliminate or control the hazard

Management of a successful OHS model is illustrated in Figure 4.5. The first building block of a successful OHS model is policy. It is paramount that organisations should develop a formal occupational H&S policy. The purpose of the policy is that all those involved in the programme's operation should actively participate. After it has been signed by the chief executive officer of the organisation, the policy should be placed where the employees normally report for work. Effective OHS policies set a clear direction for the organisation to follow in terms of OHS. The policy should be appropriate to the nature and scale of the organisation's OHS risks.

The second building block of an organisation OHS policy is organising. Organising is the process of creating a structure for the organisation that will enable its human resources to work together effectively towards its OHS objectives. According to Adeogun and Okafor (2013:26), 'organisations need to define the responsibilities and relationships that promote positive health and safety culture, and secure the implementation and continued development of a health and safety policy'. The defined responsibility could be such that managers take full responsibility for controlling factors that could lead to ill health, injury or loss. The arrangements start with

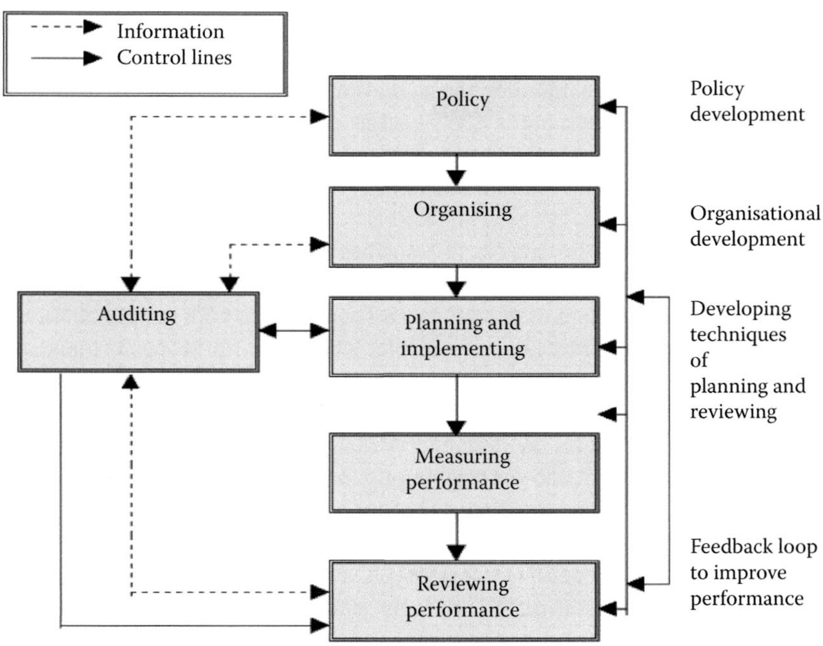

FIGURE 4.5 Key elements of successful OHS management model. (Adapted from Adeogun, B.K. and Okafor, C.C., 2013, Occupational health, safety and environment (HSE) trends in Nigeria, *International Journal of Environmental Sciences, Management and Engineering Research*, 2(1): 24–29; Ramroop, S., McCarthy, J.J. and Naidoo, K., 2004, Successful occupational health and safety: A management perspective, Proceedings of 8th World Congress on Environmental Health, Document Transformation Technologies, South Africa, pp. 22–27; and HSE, 2000).

nominating a senior person at the top of the organisation to coordinate and monitor policy implementation. OHS responsibilities are allocated to line managers, with specialists appointed to act as advisers.

Planning is essential for the implementation of OHS policies, which forms the third building block of a successful OHS model. Planning consists of an overall strategy for achieving organisational goals, and includes a comprehensive hierarchy of plans to integrate and coordinate activities. A systematic planning approach is necessary to answer the following three key questions with respect to OHS (Adeogun & Okafor, 2013):

1. Where are we now (i.e. current situation)?
2. Where do we want to be (i.e. standards stipulated by law)?
3. How do we get there?

Performance measurement is the fourth building block of a successful OHS model. Organisations need to measure what they are doing to implement their H&S policy, and to assess how effectively they are controlling risk, and how well they are developing a positive H&S culture. Measurement is essential to maintain and improve H&S performance. There are two ways to generate information on measuring performance (Adeogun & Okafor, 2013:28):

1. Active systems that monitor the achievement of plans and the extent of compliance with standards (i.e. routine procedures to monitor specific objectives, for example, quarterly or monthly)
2. Reactive systems that monitor accidents, ill health and incidents

Effective procedures for measuring performance are needed to capture both sorts of information. It is essential that personnel are given the responsibility for monitoring the achievement of objectives and compliance with standards for which they and their subordinates are responsible. There should be performance standards for managers to indicate how they will monitor performance. The auditing performance, the fifth building block, is the structured process for collecting independent information on the efficiency, effectiveness and reliability of the total H&S management system and for drawing up plans for corrective actions. To remedy deficiencies, it is necessary that there is a need to review the process of making judgements about the adequacy of performance and taking decisions about the nature and timing of the actions necessary. There should be a continuous process in the reviewing process at different levels within the organization. This should also include monthly reviews of individuals, supervisors or sections; quarterly reviews of departments; and annual reviews of sites or the organisation as a whole. The final step in the H&S management control cycle is the auditing and performance review. This constitutes the 'feedback loop' that enables an organisation to reinforce, maintain and develop its ability to reduce risks to the fullest extent. It will also ensure the continued effectiveness of the H&S management system. The use of audits and performance reviews will enable organisations to maintain and improve their ability to manage risks.

4.6.1 Safety Culture

A safety culture is necessary for the adoption of the OHS management model described in Figure 4.5 to flourish in an organisation. Integrating a safety culture into the organisation's values is a way of maintaining a safe organisation. The industrial safety transition from an unsafe to a safe organisation has been shown through the safety evolution process discussed in Figure 4.5. The ways of thinking, behaving and believing what members of a social unit have in common is known as a culture. Safety has a special place in the concerns of those who work for an organisation because a safety culture is a special case of such a culture. An organisation can be regarded as having a safety culture once there is safety in an organisation's culture.

4.6.2 Economics of Construction Health and Safety

The cost of accidents (CoA) is a financial measure that is in relation to stakeholders and can be expressed as a percentage of organisation's business volume or value of construction completed nationally. The CoA can be categorised as being either direct or indirect, and collectively these constitute the total CoA.

Direct costs are associated with the treatment of the injury and any unique compensation offered to workers as a consequence of being injured, and are covered by workmen's compensation insurance premiums. Indirect costs are borne by contractors. These include reduced productivity for the returned worker(s) and workforce, cleanup costs, replacement costs, standby costs, cost of overtime, administrative costs, replacement worker orientation, costs resulting from delays, supervision costs, costs related to rescheduling, transportation and wages paid while the injured is idle (CIDB, 2008).

The CoA is included in contractors' cost structures and it is not simple to quantify CoA. Research in South Africa estimated the total CoA to be around 5 percent of the value of completed construction. There is a cost to implementing H&S systems within a firm. This is estimated to cost between 0.5 percent and 3 percent of total project costs. This confirms other sources of research data that the total CoA usually exceeds the cost of H&S. Therefore, H&S must be seen as an enabler and catalyst for performance enhancement in relation to cost, environment, productivity, quality and schedule of work.

4.6.3 General Duties of Employers to Their Employees with Regard to OHS Issues

The general duties of employers to their employees are as follows:

- Every employer shall provide and maintain, as far as is reasonably practicable, a working environment that is safe and without risk to the health of his or her employees.
- Without derogating from the generality of an employer's duties, the matters to which those duties refer include in particular the following:
  - Providing and maintaining of systems of work, plant and machinery, so that as far as is reasonably practicable, these are safe and without risks to health.

- Taking such steps as may be reasonably practicable to eliminate or mitigate any hazard or potential hazard to the safety or health of employees, before resorting to personal protective equipment.
- Making arrangements for ensuring, as far as is reasonably practicable, the safety and absence of risks to health in connection with the production, processing, use, handling, storage or transport of articles or substances.
- Establishing, as far as is reasonably practicable, what hazards to the health or safety of persons are attached to any work which is performed, any article or substance which is produced, processed, used, handled, stored or transported and any plant or machinery which is used in his business, and he shall, as far as is reasonably practicable, further establish what precautionary measures should be taken with respect to such work, article, substance, plant or machinery in order to protect the health and safety of persons, and he or she shall provide the necessary means to apply such precautionary measures.
- Providing such information, instructions, training and supervision as may be necessary to ensure, as far as is reasonably practicable, the health and safety at work of his or her employees.
- As far as is reasonably practicable, not permitting any employee to do any work or to produce, process, use, handle, store or transport any article or substance or to operate any plant or machinery, unless precautionary measures have been taken.
- Taking all necessary measures to ensure that the entire requirements are complied with by every person in his or her employment or on premises under his or her control where plant or machinery is used.
- Enforcing such measures as may be necessary in the interest of health and safety. Ensuring that work is performed and that plant or machinery is used under the general supervision of a person trained to understand the hazards associated with it and who has the authority to ensure that precautionary measures taken by the employer are implemented.
- Causing all employees to be informed regarding the scope of their authority.

4.6.4 General Duties of Employers and Self-Employed Persons to Persons Other Than Their Employees

The following are the general duties of employers and self-employed persons to persons other than their employees:

- Every employer shall conduct his or her undertaking in such a manner as to ensure, as far as is reasonably practicable, that persons other than those in his or her employment who may be directly affected by his or her activities are not thereby exposed to hazards to their health or safety.
- Every self-employed person shall conduct his or her undertaking in such a manner as to ensure, as far as is reasonably practicable, that he or she and

other persons who may be directly affected by his or her activities are not thereby exposed to hazards to their health or safety.

- Employers shall keep the health and safety representatives designated for their workplaces or sections of the workplaces. The concerned persons should be informed of the actions taken in their respective workplaces and of the results of such actions.

4.6.5 General Duties of Employees at Work

Every employee shall at work:

- Take reasonable care for the health and safety of himself or herself and of other persons who may be affected by his or her acts or omissions.
- As regards any duty or requirement imposed on his employer or any other person by the OHSA 1993 to cooperate with such employer or person to enable that duty or requirement to be performed or complied with.
- Carry out any lawful order given to him or her, and obey the health and safety rules and procedures laid down by his or her or by anyone authorised thereto by his employer, in the interest of health or safety.
- If any situation which is unsafe or unhealthy comes to his or her attention, as soon as practicable report such situation to his or her employer or to the health and safety representative for his or her workplace or section thereof, as the case may be, who shall report it to the employer.
- If he or she is involved in any incident which may affect his or her health or which has caused an injury to himself or herself, report such incident to his or her employer or to anyone authorised thereto by the employer, or to his or her health and safety representative, as soon as practicable but not later than the end of the particular shift during which the incident occurred, unless the circumstances were such that the reporting of the incident was not possible, in which case he or she shall report the incident as soon as practicable thereafter (OHSA, 1993:10).

4.6.6 Health and Safety Representatives

Health and safety representatives are the employer representatives or his or her employees recognized by him or her or, where there are no such representatives, the employees shall consult in good faith regarding the arrangements and procedures for the nomination or election, period of office and subsequent designation of health and safety representatives to represent the employer in all H&S related matters. Employees employed in a full-time capacity at a specific workplace are acquainted with conditions and activities at that workplace or section thereof, as the case may be, shall be eligible for designation as health and safety representatives for that workplace or section. The number of health and safety representatives for a workplace or section thereof shall in the case of shops and offices be at least one health and safety representative for every 100 employees or part thereof, and in the case of all other workplaces at least one health and safety representative for every 50 employees or part thereof: Provided that those employees performing work at a workplace other than that where they ordinarily report for duty, shall be deemed to be working at the workplace where they so report for duty. If an inspector is of the opinion that the number of health and safety representatives for any workplace or section thereof, including a workplace or section with 20 or fewer employees, is inadequate, he or she may by notice in writing direct the employer to designate such number of employees as the inspector may determine as health and safety representatives for that workplace or section thereof in accordance with the arrangements and procedures referred in a typical H&S policy.

4.6.7 Functions of the Health and Safety Representatives

This section provides functions of health and safety representatives.

- A health and safety representative may perform the following functions in respect of the workplace or section of the workplace for which he or she has been designated, namely:
  - Review the effectiveness of health and safety measures.
  - Identify potential hazards and potential major incidents at the workplace.
  - In collaboration with his or her employer, examine the causes of incidents at the workplace.
  - Investigate complaints by any employee relating to that employee's health or safety at work.
  - Make representations to the employer or a health and safety committee on matters, or where such representations are unsuccessful, to an inspector.
  - Make representations to the employer on general matters affecting the health or safety of the employees at the workplace.
  - Inspect the workplace, including any article, substance, plant, machinery or health and safety equipment at that workplace, with a view to the health and safety of employees, at such intervals as may be agreed upon with the employer, provided that the health and safety representative shall give reasonable notice of his intention to carry out such an inspection to the employer, who may be present during the inspection.
  - Participate in consultations with inspectors at the workplace and accompany inspectors on inspections of the workplace.
  - Receive information from inspectors.
  - In his or her capacity as a health and safety representative attend meetings of the health and safety committee of which he or she is a member, in connection with any of the functions.
- A health and safety representative shall, in respect of the workplace or section of the workplace for which he or she has been designated be entitled to:
  - Visit the site of an incident at all reasonable times and attend any inspection in loco.

    - Attend any investigation or formal inquiry held in terms of the OHSA 1993.
    - In so far as it is reasonably necessary for performing his or her functions, inspect any document which the employer is required to keep in terms of OHSA 1993.
    - Accompany an inspector on any inspection.
    - With the approval of the employer (which approval shall not be unreasonably withheld), be accompanied by a technical adviser, on any inspection.
    - Participate in any internal health or safety audit.
- An employer shall provide such facilities, assistance and training as a health and safety representative may reasonably require and as have been agreed upon for the carrying out of his or her functions.
- A health and safety representative shall not incur any civil liability by reason of the fact that he or she failed to do anything which he or she may do or is required to do in terms of this Act (OHSA, 1993:12).

4.7 SMALL TO MEDIUM-SIZED CONSTRUCTION COMPANIES' PERFORMANCE WITH SAFETY REGULATIONS

The Confederation of British Industry (CBI) notes that small to medium-sized enterprises (SMEs) 'have special needs of their own and are all faced with the challenge of one or more of a group of special characteristics' (HSE, 2005). The following 'special characteristics' differentiate SMEs from large businesses: short track records, heavy reliance on niche markets, lack of specialist skills, low cash flow, small asset base and the need to make changes in structure and ownership at various stages of growth. SME contractors are predominantly non-unionised, which is a significant factor in H&S compliance. That is, there is no input from safety representatives or union pressure for improvements in safety in SME. Formal systems of compliance are also not common in SME. They are less likely than large businesses to have the internal expertise for dealing with regulation.

Regulation normally requires that SME contractors take some action that is likely to involve expertise, finance and management. Taking into account the characteristics of SMEs, they are likely to find regulations difficult to implement because they have limited resources that result in strict control of staffing. There are also limited personnel to monitor changing legal requirements, and interpret and implement the necessary control measures. SMEs have limited time to deal with regulatory requirements. Within SMEs, the responsibility for dealing with regulations often falls to the proprietor who may not have any specialist skills, unlike large firms that may employ an entire department of legal and technical specialists to deal with their health and safety issues. SME contractors' attitudes towards regulations and policy-making are different from those of large firms. SME contractors do not have representatives when it comes to policy review; this denies them the opportunity to implement and monitor legal requirements. Their view is not represented in policies because they do not belong to any trade body. Research indicates that SME contractors suffer from a disproportionate impact of regulations upon their business in terms of compliance costs. Work conducted by the Better Regulation Task Force (BRTF) found that smaller companies are often at a competitive disadvantage compared with larger firms because of the cost and time involved in regulatory compliance.

Although several different definitions exist for SME contractors, the most commonly accepted interpretation was that adopted by the European Commission in 1996. This definition is based upon employee numbers, annual turnover and balance sheet. SME contractors form a vital part of the economy; there are 20 million SME contractors in the European economic sector. The Department of Trade and Industry (DTI) asserted that SME contractors in the United Kingdom employed around 12.6 million people in early 2002, with an estimated annual turnover of £1.1 billion. They also account for 99.8 percent of all businesses, and 69.3 percent of all businesses do not employ any staff, for example, they are sole proprietors or self-employed owner-manager(s) (HSE, 2005). SME contractors form the bulk, by number, of companies in each industry. They offer operational flexibility to the larger firms in their role as sub-contractors. Projects undertaken by SMEs are required to have the quality enhancement of the people in any country. In addition to this, the economic growth and development of the country depend on such projects. SMEs differ significantly from large firms, and these factors affect their compliance level. The Annual Report by the Department of Occupational Safety and Health (DOSH, 2008) posited that the compliance of SME contractors to safety regulations was found to have an effect on SME contractors' behaviour and compliance levels within individual employees in the construction industry. The well-being of workers in SMEs is affected by the high rate of accidents at the workplace (DOSH, 2008). Their performance in terms of compliance with safety regulations is affected by their financial, expertise and staffing capabilities in most cases. These differences between SME contractors and large/established companies have a considerable effect when it comes to the implementation of OSH. SME contractors have limited financial resources and structure to enable them to implement an effective occupational and safety OSH system. These limited resources have contributed to their poor safety programme because they perceive it as irrelevant (McKinney in Surienty, 2012). Occurrences of accidents in the construction industry can be reduced through more sensitive or good safety behaviour of both employees and employers, whilst ineffective implementation of OSH will lead to bad safety behaviour. OSH implementation requires employers (large construction companies) to cater for the safety needs of their employers and employees (Christian, Wallace, Bradley & Burke, 2009). Hence, the development of construction SMEs in many countries depend on the support received through financial assistance schemes, initiatives which support the acquisition of equipment and training (Ofori & Toor, 2012) through government entities.

SME contractors most of the time do not satisfy the implementation role owing to a lack of enforcement from the legislative bodies overseeing the implementation of the OSH Act (Surienty, 2012). SME contractors conceive compliance in a different way to the view of enforcers (HSE, 2005). Previous studies support the contention that SME contractors want to be told exactly what to do and how to comply. Research shows that SME contractors also have major difficulties in identifying hazards and find the concept of self-assessment alien. SME contractors pose a huge challenge to enforcement resources owing to their sheer numbers.

4.7.1 Impact of the Small and Medium-Sized Enterprise on Health and Safety

Owing to the nature of their operations, management styles, financing and the markets in which they operate, they have inherent problems in complying with regulation. Fatalities are well recorded, but the levels of recording of non-fatal accidents and ill-health are known to be low. SME contractors pose a high risk to the H&S of workers and the general public. SME contractors employ fewer staff members and have the potential to impact upon fewer people compared with large businesses and yet they still operate high-risk enterprises. SME contractors contribute significantly to the overall levels of illness and injury in the construction industry when considering the number of people affected (Health and Safety Commission, 2004).

4.7.2 Factors Affecting Small and Medium-Sized Enterprises' Non-Compliance and Motivation with Occupational Health and Safety Regulations

Lack of knowledge, interest, skills, money and time have been the major factors affecting the health, safety and environmental health regulations within SME contractors leading to their non-compliance (HSE, 2005). SME companies have significantly different characteristics from large businesses in terms of their financial, expertise and staffing capabilities. These issues affect the performance of SMEs in terms of compliance with safety regulations and have generated substantial ongoing debate between practitioners about designing regulatory and enforcement strategies that optimize compliance levels (Surienty, 2012).

Research on factors affecting SME motivation found these to be fear of loss of credibility and perceived duty to comply with health and safety regulations.

Fear or loss of credibility is identified as the loss of credibility in a business as arising from some issues. Employers were seen to have a fear of adverse publicity, loss of confidence and increased regulatory attention after an accident or incident. This fear was greatest amongst high-risk businesses (e.g. chemicals and transport) of any size.

The issues of limited finance, expertise and staffing capabilities of SMEs have an effect on their performance in terms of compliance with safety regulations. This has an effect on their behaviour and attitudes towards occupational safety (Surienty, 2012). They also have a high rate of accidents at their workplaces that affect the well-being of workers and their safety (DOSH, 2008). Good safety behaviour of employees at their workplace and their approach to work can lead to a reduction in accidents in the construction industry (Christian et al., 2009). The safety needs of employees lie with the employers because it is the requirement of OSH implementation. This will lead to the effective implementation of OSH and assist in the formation of good safety behaviour in the construction industry (Surienty, 2012). Research has shown that proper implementation of OSH in SMEs has been attributed to assistance received by SMEs (Lingard & Rowlinson, 2005).

4.8 SUMMARY

The chapter focused on the issues of non-compliance to H&S issues by construction companies. The chapter detailed that the construction industry is faced with fatal accidents all over the world. The rate of accidents in the construction industry has been regarded as the highest as compared to other industrial sectors, and the fatality rate is on an annual rise, whilst the means of improving health and safety compliance in the construction industry have been provided to minimise the rampant occurrence of accidents in the construction industry. Whereas this chapter detailed that small to medium-sized construction companies are not performing very well, the next chapter of this book explores issues of H&S in developing countries.

REFERENCES

Adenuga, O.A., Soyingbe, A.A. and Ajayi, M.A. (2007). A study on selected safety measures on construction companies in Lagos, Nigeria. RICS (Cobra).

Adeogun, B.K. and Okafor, C.C. (2013). Occupational health, safety and environment (HSE) trends in Nigeria. *International Journal of Environmental Sciences, Management and Engineering Research*, 2(1): 24–29. Available from: http://www.ijesmer.com [Accessed 02 July 2014].

Alli, B.O. (2008). Fundamental principles of occupational health and safety. International Labour Organization, Geneva. Available from www.ilo.org/ [Accessed 30 April 2014].

Arewa, A.O. and Farrell, P. (2012). A review of compliance with health and safety regulations and economic performance in small and medium construction enterprises. In: Smith, S.D. (Ed.), *Proceedings of 28th Annual ARCOM Conference*, 3–5 September, Edinburgh, UK, Association of Researchers in Construction Management, 423–432.

Behm, M. (2008). Linking construction fatalities to the design for construction safety concept. *Safety Science*, 43(8): 589–611.

Bellamy, L.J., Geyer, T.A.W. and Wilkinson, J. (2008). Development of a functional model that integrates human factors, safety management systems and wider organisational issues. *Safety Science*, 46: 461–492.

BOMEL Limited. (2001). *Improving health and safety in construction, Phase 1: Data collection, review and structuring*. Norwich: HSE Books.

BOMEL Limited. (2007). Improving the effectiveness of the Construction (Design and Management) Regulations 1994: Establishing views from construction stakeholders on the current effectiveness of CDM. Health and Safety Executive 2007.

Brown, G. D. (2005). Protecting workers' health and safety in the globalizing economy through international trade treaties. *International Journal of Occupational and Environmental Health*, 11: 207–209.

Canadian Centre for Occupational Health and Safety (CCOHS). (2009). Risk assessment. Available from: http://www.ccohs.ca [Accessed 17 July 2014].

Christian, M.S., Wallace, J.C., Bradley, J.C. and Burke, M.J. (2009). Workplace safety: A meta-analysis of the roles of person and situation factors. *Journal of Applied Psychology*, 94: 1103–1127.

Chua, D.K.H. and Goh, Y.M. (2004). Incident causation model for improving feedback of safety knowledge. *Journal of Construction Engineering and Management*, 130(4): 542–551.

Construction Industry Development Board (CIDB). (2008). Development through partnership: Construction health and safety in South Africa. Available from: http://www.asocsa.org [Accessed 08 April 2014].

Construction Regulations. (2003). Safe Working Practices, Occupational Health and Safety Group. Available from: http://www.safepractice.co.za [Accessed 08 April 2014].

Dainty, A.R.J., Briscoe, G.H. and Millet, S.J. (2001). New perspectives on construction supply chain integration. *Supply Chain Management: An International Journal*, 6(4): 163–173.

Department of Occupational Safety and Health (DOSH). (2008). Annual Report: Labour and human resources statistics. Available from: http://www.mohr.gov.my/ [Accessed February 2015].

Haslam, R.A., Hide, S.A., Gibb, A.G.F., Gyi, D.E., Pavitt, T., Atkinson, S. and Duff, A.R. (2005). Contributing factors in construction accidents. *Applied Ergonomics*, 36: 401–415.

Health and Safety Commission. (2004). Health and safety statistics highlights 2003/4. Health and Safety Executive.

Health and Safety Executive (HSE). (2000). Successful health and safety management. London: HSE Books.

Health and Safety Executive (HSE). (2005). Promoting health and safety as a key goal of the corporate social responsibility agenda, Research Report 339. Sudbury: HSE Books.

Health and Safety Executive (HSE). (2009). Safety signs and signals. The Health and Safety (Safety Signs and Signals) Guidance on Regulations. Available from: http//:www.hsebooks.co.uk [Assessed 20 September 2014].

Health and Safety Executive (HSE). (2010). *Work-related injuries and ill health in construction*. Suffolk: HSE Books.

Health and Safety Executive (HSE). (2013). Work-related injuries and ill health: Health and safety in construction in Great Britain, 2013. Available from: www.hse.gov.uk/statistics/ [Accessed 11 July 2014].

Health and Safety Executive (HSE). (2014). Fatal injury statistics. Available from: http://www.hse.gov.uk/ [Accessed 11 July 2014].

Hinze, J. (2005). A paradigm shift: Leading to safety. Conference Proceedings CIB W99 on Rethinking of Construction Safety, Health Environmental and Quality, Port Elizabeth, South Africa.

Hola, B. (2007). General model of accident rate growth in the construction industry. *Journal of Civil Engineering and Management*, 13(4): 255–264.

Hosseinian, S.S. and Torghabeh, Z.J. (2012). Major theories of construction accident causation models: A literature review. *International Journal of Advances in Engineering & Technology*, 4(2): 53–66.

Hughes, P. and Ferrett, E. (2008). *Introduction to health and safety in construction*. Oxford: Elsevier Butterworth-Heinemann.

Idoro, G.I. (2011). Comparing occupational health and safety (OHS) management efforts and performance of Nigerian construction contractors. *Journal of Construction in Developing Countries*, 16(2): 151–173.

Idubor, E.E. and Oisamoje, M.D. (2013). An exploration of health and safety management issues in Nigeria's effort to industrialise. *European Scientific Journal*, 9(12): 154–169.

International Labour Office (ILO). (2005). Baseline study on labour practice on large construction sites in the United Republic of Tanzania. Working Paper 225, Geneva.

Karjalainen, A. (2004). A statistical portrait of health and safety at work in the construction industry: Actions to improve safety and health in construction. Luxembourg: European Agency for Safety and Health at Work.

Kulchartchai, O. and Hadikusumo, B.H.W. (2010). Exploratory study of obstacles in safety culture development in the construction industry: A grounded theory approach. *Journal of Construction in Developing Countries*, 15(1): 45–66.

Lingard, H. and Rowlinson, S. (2005). *Occupational health and safety in construction project management.* Abingdon: Spon Press.

Loughborough University and University of Manchester Institute of Technology (UMIST). (2003). *Causal factors in construction accidents.* Norwich: HSE Books.

Manu, P., Ankrah, N., Proverbs, D. and Suresh, S. (2010). Exploring the influence of construction project features in accident causation, In: Barrett, P., Amaratunga, D., Haigh, R., Keraminiyage, K., and Pathirage, C. (Eds.), Proceedings of Council for Research and Innovation in Building and Construction (CIB) 2010 World Congress, 10–13 May, Salford, UK.

McDonald, M.A., Lipscomb, J.H., Bondy, J. and Glazner, J. (2009). Safety is everyone's job: The key to safety on a large university construction site. *Journal of Safety Research*, 40(1): 53–61.

Misnan, M.S., Mohammed, A.H.B., Mahmood, W.Y.W., Mahmud, H.S. and Abdullah, N.M. (2008). Development of safety culture in the construction industry: The leadership and training roles. Conference Proceedings of the 2nd International Conference on Built Environment in Developing Countries, (ICBEDC), Pulau.

Murie, F. (2007). Building safety: An international perspective. *International Journal of Occupational, Environmental and Health*, 13: 5–11.

Musonda, I. (2012). Construction health and safety (H&S) performance improvement: A client-centred model. Unpublished DPhil in Engineering Management, University of Johannesburg, South Africa. Available from: http://www.ujdigispace.uj.ac.za [Accessed 02 June 2014].

Ng, S.T., Cheng, K.P. and Skitmore, M. (2005). A framework for evaluating the safety performance of construction contractors. *Building and Environment*, 40(10): 1347–1355.

OHS Network Solutions. (2014). Available from: http://www.ohsnetworksolutions.co.za [Accessed April 2014].

Occupational Health and Safety Act (OHSA). (1993). Occupational Health and Safety Amendment Act, No. 181 of 1993. South African Department of Labour. Available from: http://www.labour.gov.za [Accessed 26 January 2016].

Ofori, G. and Toor, S.R. (2012). Leadership development for construction SMEs. Proceedings of POC 2012 Conference, Working Paper Proceedings, Engineering Project Organizations Conference, 10–12 July, Rheden, The Netherlands.

Okeola, O.G. (2009). Occupational health and safety (OHS) assessments in the construction industry. 1st Annual Civil Engineering Conference, 26–28 August, University of Ilorin, Nigeria.

Organisation for Economic Co-operation and Development (OECD). (2000). Reducing the risk of policy failure: Challenges for regulatory compliance. Available from: http://www.oecd.org/regreform/regulatory-policy [Accessed 13 November 2016].

Othman, A.A.E. (2012). A study of the causes and effect of contractors' noncompliance with the health and safety regulations in the South African construction industry. *Architectural Engineering and Design Management*, 8: 180–191.

Pellicer, E. and Molenaar, K.R. (2009). Discussion of 'Developing a model of construction safety culture' by Raiq M. Choudhry, Dongping, Fang and Sherif Mohamed. *Journal of Management in Engineering*, 25(1): 44–47.

Petersen, D. (2000). The behavioural approach to safety management. *Professional Safety*, 37–39.

Ramroop, S., McCarthy, J.J. and Naidoo, K. (2004). Successful occupational health and safety: A management perspective. Proceedings of 8th World Congress on Environmental Health. Document Transformation Technologies, South Africa, pp. 22–27.

Riley, M.J. and Brown, D.C. (2001). Comparison of cultures in construction and manufacturing industries. *Journal of Management in Engineering*, 17(3): 149–158.

Ringen, K. and Englund, A. (2006). The construction industry. *Annals of the New York Academy of Sciences*, 1076(1): 388–393.

Smallwood, J., Haupt, T. and Shakantu, W. (2008). Construction health and safety in South Africa: Status and recommendation. CIDB report.

Suraji, A., Duff, A.R. and Peckitt, S.J. (2001). Development of causal model of construction accident causation. *Journal of Construction Engineering and Management*, 127(4): 337–344.

Surienty, L. (2012). Management practices and OSH implementation in SMEs in Malaysia. Available rom: http://ilera2012.wharton.upenn.edu/RefereedPapers/SurientyLilisILERA.pdf [Accessed 13 November 2016].

Taylor, G., Easter, K. and Hegney, R. (2004). *Enhancing occupational safety and health.* Burlington, MA: Elsevier Butterworth-Heinemann.

Teo, E.A.L., Ling, F.Y.Y. and Chong, A.F.W. (2005). Framework for project managers to manage construction safety. *International Journal of Project Management*, 23(4): 329–341.

Windapo, A. and Oladapo, A. (2012). Determinant of construction firms' compliance with health and safety regulations in South Africa. In: Smith, S.D. (Ed.), *Proceedings of 28th Annual Association of Researchers in Construction Management (ARCOM) Conference*, 3–5 September, Edinburgh, UK, 433–444.

Section V

Occupational Health and Safety Issues in Developing Countries

5 Occupational Health and Safety Issues in the Nigerian Construction Industry

5.1 INTRODUCTION

This chapter describes occupational health and safety (OHS) in developing countries, taking Nigeria into consideration as the first example. Other countries discussed in the proceeding and following chapters are South Africa and Ghana, as the study country. The discussions are based on OHS in developing countries and the development of human resources for OHS. Finally, a review of OHS in Nigeria is presented.

5.2 OCCUPATIONAL HEALTH AND SAFETY ACT IN DEVELOPING COUNTRIES

According to the South African OHS Act 1994, industrialization in developing countries requires effective administrative systems to control hazards and to provide decent working environments that meet international standards. A high rate of occupational accidents, particularly in construction projects, means developing countries are poor at managing the risks of hazards at workplaces. OHS is the science of anticipation, recognition, evaluation and control of hazards arising in or from the workplace that could impair the health and well-being of workers, taking into account the possible impact on the surrounding communities and the general environment (International Labour Organization [ILO], 2009). Thus, OHS is concerned with the promotion and maintenance of the highest degree of physical, mental and social well-being of workers at all occupations (ILO/WHO Committee, 1995).

There are numerous factors that hinder the performance of small to medium-sized enterprises (SMEs) in OHS performance in developing countries. For instance, the lack of resources and inadequate legal and institutional arrangements for the management of health and safety (H&S) have compounded the problem of H&S performance in developing countries. Other factors that also effect the SME to adequately comply with OHS requirements include a lack of effective mechanisms to implement laws, lack of infrastructure and equipment, rampant corruption and a lack of concerted effort by policy makers to address H&S (Cotton, Sohail & Scott, 2005; Kheni, Dainty & Gibb, 2007). It is not surprising that the H&S performance in the construction industry of developing countries is worse, because the industry is dominated by small and medium-sized contractors and most of them do not have effective systems to manage H&S. The construction industry is also characterised by a poor knowledge and awareness of H&S, apart from its labour intensiveness and utilization of old technology, which both have an impact on H&S performance. It is believed that more than 2 million people die yearly from work-related accidents and disease in the construction industries of both developed and developing countries. Unfortunately, the dearth of data and statistics makes it hard for researchers to estimate the actual number of people affected in a developing country.

5.2.1 Occupational Safety and Health Problems

Occupational safety and health can be defined as a multidisciplinary activity aiming at the

- Protection and promotion of the health of workers by eliminating occupational factors and conditions hazardous to health and safety at work
- Enhancement of physical, mental and social well-being of workers, and support for the development and maintenance of their working capacity, as well as professional and social development at work
- Development and promotion of sustainable work environments and work organizations (World Health Organization [WHO], 1999)

Hundreds of millions of people throughout the world are employed today in conditions that breed ill health or are unsafe (WHO, 1999). It is advisable to ensure the health and safety of workers within a work environment. Each year, work-related injuries and diseases kill an estimated 1.1 million people worldwide, which roughly equals the global annual number of deaths from malaria. Other problems include:

- A total of 250 million occupational accidents result in more than 300,000 fatalities annually. Many of these accidents lead to partial or complete incapacity to work and generate income.
- Annually, an estimated 160 million new cases of work-related diseases occur worldwide, including respiratory and cardiovascular diseases, cancer, hearing loss, musculoskeletal and reproductive disorders, and mental and neurological illnesses.
- An increasing number of workers in industrial countries complain about psychological stress and overwork. These psychological factors have been found to be strongly associated with insomnia, depression and fatigue, and burnout syndromes, as well as with elevated risks of cardiovascular diseases.
- Only 5 to 10 percent of workers in developing countries and 20 to 50 percent of workers in industrial countries (with few exceptions) are estimated to have access to adequate occupational health services.
- Even in advanced economies, a large proportion of work sites are not regularly inspected for occupational health and safety.

The health status of the workforce in every country has an immediate and direct impact on national and world economies. Total economic losses due to occupational illnesses and injuries are enormous (WHO, 1999). The challenges for occupational health and safety as indicated by the WHO (1999) are associated with the following:

- Occupational health problems linked to new information technologies and automation
- New chemical substances and physical energies
- Health hazards associated with new biotechnologies
- Transfer of hazardous technologies
- Aging working populations
- Special problems of vulnerable and underserved groups (e.g. chronically ill and handicapped), including migrants and the unemployed
- Problems related to growing mobility of worker populations and occurrence of new occupational diseases of various origins

The ultimate objective of occupational health is a healthy, safe and satisfactory work environment, and a healthy, active and productive worker, free from both occupational and non-occupational diseases.

Most of the prevalent hazards or accidents that occur in developing countries are due to mechanical factors, unshielded machinery, unsafe structures at the workplace and dangerous unprotected tools. These accidents affect the health of a high proportion of the workforce and could be prevented by applying appropriate measures in the work environment and working practices. Other measures are the application of appropriate safety measures and ensuring appropriate behavioural and management practices. Measures put in place will reduce accident rates, but ignorance of such precautions leads to increasing rates of occupational accidents (WHO, 1999).

Likewise, repetitive tasks and static muscular load are also found to have effect on the workforce (between 50% and 70%, respectively) in developing countries owing to exposure to a heavy physical workload, such as lifting and moving of heavy items or repetitive manual tasks. In many developing countries, musculoskeletal disorders are the main causes of both short-term and permanent work disability. This development may lead to economic losses amounting to an approximate 5 percent of the gross domestic product (GDP). Mechanization, the improvement of ergonomics and better organization of work and training will contribute to the elimination or minimization of most exposures (Hagstedt & Pieris, 2000).

Furthermore, the awareness development of occupational health and safety hazards among workers and employers and the assessment of the nature and extent of hazards can be useful in preventive approaches. Other preventive measures should be the introduction and maintaining of effective control and evaluation measures. Agencies responsible for the promulgation and enforcement of occupational health and safety regulations should be included in the development of preventive approaches in developing countries. There should also be improvement of working conditions and the location of the workplace.

On the other hand, the international labour standards (conventions and recommendations) formulated by the ILO should be employed by member countries and implemented in practice since the ILO set minimum standards of basic labour rights. The ILO conventions and recommendations on occupational safety and health define the rights of workers, and allocate duties and responsibilities to a competent authority, employers and workers in the field of occupational safety and health. The policy on occupational health and safety provides for the adoption of a national occupational safety and health policy, and describes the actions needed at the national and enterprise level to promote occupational safety and health and to improve working environments.

5.3 DEVELOPMENT OF HUMAN RESOURCES FOR OCCUPATIONAL HEALTH AND SAFETY IN DEVELOPING COUNTRIES

There is a universal shortage of both expert resources and training in developing countries due to the following three main reasons:

- A lack of effective legislation and a lack of requests from authorities and employers make the employment opportunities for such experts minimal.
- In the absence of requests, the vocational training institutions and universities have not organized and developed curricula for the training of experts in occupational health.
- In some instances, where training is available, it is oriented to clinical occupational medicine only which, though important, does not give full response to the needs of expertise in preventive workplace-oriented occupational health service.

Hence, developing countries should include in their national programmes on occupational health an element of training of sufficient numbers of experts to implement the national programme and to ensure sufficient personnel resources for OHS. The governments of developing nations should ensure that the necessary elements of occupational health are included in the basic training curricula of all who may in the future deal with occupational health issues. This should be accomplished with the training in occupational health; this should also be given in connection with vocational training and in training programmes for workers, employers and managers; and the need for a multidisciplinary approach in occupational health should be taken into consideration, ensuring involvement of occupational medicine and nursing, occupational hygiene, ergonomics and work physiology, occupational safety, work organisation and other relevant fields (Hagstedt & Pieris, 2000).

In order to further curtail the problems of H&S in developing countries, the following should be provide to strengthen research areas:

- Each government should establish or strengthen its national centre for occupational health and, if appropriate, the network of centres.
- Such a centre should be given the responsibility of carrying out research, information, training and, if appropriate, advisory and analytical and measurement services in support of occupational health practices.

- The national research programme should survey the occupational health and safety situation for developing competence and methodology in occupational health and for responding to the national occupational health and safety training programme.
- Effective international collaboration in research should be ensured.

5.4 BENEFITS OF OCCUPATIONAL HEALTH AND SAFETY FOR DEVELOPING COUNTRIES

The following are the benefits of occupational health and safety for developing countries:

- It is preventive and proactive in approach as it foresees, identifies and prevents hazards before they turn to bad occurrences with unexpected or unpredictable consequences. Prevention is effective, always better and less costly than treatment and rehabilitation (curative).
- OHS ensures workers' health is not adversely affected by their work and that work is not affected in return by poor health (mutual benefit).
- OHS fosters a healthy workforce leading to increased productivity for business owners and the economy at large (workers' well-being versus productivity). Healthy workers are essential to the success of any organization and are the best assets in any industry.
- It also reduces work-related sickness absence. This is an invaluable cost-saving benefit to employers. Sickness absence is one of the main causes of economic loss to employers owing to lost productivity/output/man-hour, and time and resources spent on sickness absence management.
- It helps businesses to attain compliance with relevant laws. It also helps businesses to save costs by reducing potential claims and litigation.
- A business organization that is successful with the health and welfare of its workers enjoys goodwill (enhances company's image), which gives a business an edge. Some companies will only award contracts to contractors with health and safety provisions.
- It is another way of caring (health protection) for the public's health. Most of the health problems from which people generally suffer can be traced to work/job tasks (e.g. low-back ache, respiratory problems, deafness, infertility and cancers) (Olusegun, 2015).

5.5 OCCUPATIONAL HEALTH AND SAFETY IN NIGERIA

This aspect of the book begins with OHS development in Nigeria and a report on the annual accident statistics from the Nigerian Federal Ministry of Labour and Productivity Inspectorate Division. This is followed by H&S issues, policy objectives and the basis for H&S compliance in Nigeria.

5.5.1 Occupational Health and Safety Development in Nigeria

The OHS regulations in Nigeria are not functioning properly and this is due to the inadequate attention received from the enforcing body (Umeokafor, Umeadi & Jones, 2014). Several researchers (Diugwu, Baba & Egila, 2012; Okolie & Okoye, 2012; Idubor & Oisamoje, 2013) have indicated that the low level of OHS regulations compliance has contributed to poor OSH in the Nigerian construction industry. There are weak statutory OSH regulations and provisions in Nigeria, and this has led to the failed OSH system. However, construction companies in Nigeria, have attributed employees' poor health and early deaths to the workers' personal habits on the job or their living conditions at home. Where little or no attention was paid to the prevention of hazards. This awareness led to the establishment of occupational health services in some Nigerian industries and the Occupational Health Legislations Act in Nigeria.

The earliest practices that can be regarded as an occupational health service in Nigeria were carried out by British companies such as the United African Company of Nigeria (UAC) and John Holt. This was followed by the establishment of some occupational health services by the Nigerian government in the Railway Corporation and coal mines. Such services included pre-employment and periodic medical examinations, and treatment of minor illnesses and accidents. In some cases, general practitioners were hired on a part-time basis to take care of sick and injured workers, especially in urban centres.

The increased industrialisation and its impact on the health, safety and welfare of workers led to the creation of an occupational health unit in the Federal Ministry of Health and the Institute of Occupation Health in Nigeria. According to Achalu (2000:25), 'the Factories Decree in 1987 was a landmark in legislation in the occupation health in Nigeria'. A substantial revision of the colonial legislation, Factories Act 1958, the 1987 decree changed the definition in providing oversight for the numerous small-scale enterprises that engaged the majority of the workforce. Enforcement of legislation is carried out by the factory inspectorate of the Ministry of Labour. This ministry produced a national policy on safety and health that details the responsibilities of employers, workers, manufacturers and government agencies in the maintenance of health and safety of workers.

5.5.2 Occupational Health and Safety in Nigeria

Occupational health, as defined by the joint committee of ILO/WHO (1995), is the promotion and maintenance of the highest degree of physical, mental and social well-being of workers in all occupation. The OHS involves the protection of workers' health from any hazard to which they may be exposed in their work environment. It is a preventive and proactive approach that involves risk assessment, hazard identification, hazard mitigation, hazard elimination and the treatment of work-related injuries. OHS activities are multifaceted and multidisciplinary. The activities of the workers have an effect on their working environment as well as on their health and safety. Hence, these influence their ability to perform their job tasks effectively (Olusegun, 2015). In the Labour Act Cap L1 under Nigerian laws, the Minister of Labour has powers to make regulations for the health, safety and welfare of workers at the workplace. The Factories Act of 2004, the Workmen's Compensation Act of 1987 and the Labour Safety, Health and Welfare Bill of 2012 are important documents aimed at protecting the H&S of the Nigerian worker. The Labour, Safety, Health and Welfare Bill of 2012 has stricter sanctions for offenders. This bill enables the state to charge corporate organizations and directors of firms for criminal offences where organizational actions or inactions result in the loss of lives and properties.

The following are the duties and responsibilities of the employer under the Nigeria OHS Act:

- Employers are duty bound under the various H&S laws to provide for all their employees at all work locations a work environment that is free from recognized hazards that are causing or that may cause death, serious injury or illness.
- Employers are liable to their employees for injuries that result from a failure to exercise due care.
- Employers are expected to comply with safety and health standards created by appropriate regulatory bodies; failure to do so provides for both civil and criminal penalties that should be vigorously enforced by the appropriate authority.
- Employers are also under a duty to furnish their employees with reasonably safe tools and appliances necessary in the performance of a particular work.

However, employers are not obliged to furnish the safest or best tools, machinery or appliances as stated in the act, which is a weakness of the Nigeria OHS Act. Some conditions handle the high level of violation of OHS regulations. These are discussed later. The Human Rights Impact Resource Centre (HRIRC) of 2009 posited a lack of strict judicial references for OHS issues in Nigeria, meaning that human rights compliance is not sufficiently embedded in the national law. The legal structures in place in Nigeria are also weak in terms of the interpretation and applications of the law, as most law edits are derivations of old colonial laws that have become obsolete. The penalty for violations of the labour laws by corporations is liberal.

5.5.3 Annual Accident Statistics from the Federal Ministry of Labour and Productivity Inspectorate Division

Table 5.1 shows that a total of 40 accidents, 46 deaths, 4 near misses and 93 injuries were reported within the 11-year period of study on the Nigeria construction industry by the Nigerian Federal Ministry of Labour and Productivity Inspectorate Division. There was no marked increase or decline in the case fatality rates of these injuries, as it indicates a case fatality rate of 58 percent in 2002, 100 percent in 2003, no fatalities in 2004, 25 percent in 2007, 75 percent in 2008, 66.7 percent in 2009, 20 percent in 2010, 25 percent in 2011 and 28.6 percent in 2012. The death and near misses in the year 2002 recorded 29 deaths (63%), the highest number of deaths; 1 (25%) near miss and 1 (2.5%) accident in the year. This was followed by 6 deaths (13%), no near misses and 2 (5%) accidents in 2008 (Umeokafor et al., 2014:121).

TABLE 5.1
Annual Accident Statistics from the Federal Ministry of Labour and Productivity Inspectorate Division (FMLPID, 2002–2012)

Year	No. of Injuries	% of Injuries	No. of Deaths	% of Deaths	No. of Near Misses	% of Near Misses	No. of Accidents	% of Accidents	Case Fatality Rate
2012	14	15.1	4	8.7	1	25	6	15	28.6
2011	8	8.6	2	4.3	–	–	7	17.5	25
2010	5	5.4	1	2.2	1	25	3	7.5	20
2009	3	3.2	2	4.3	–	–	16	40	66.7
2008	8	8.6	6	1.3	–	–	2	5	75
2007	4	4.3	1	2.2	1	25	3	7.5	25
2004	–	–	–	–	–	–	1	2.5	–
2003	1	1.1	1	2.2	–	–	1	2.5	100
2002	50	53.8	29	63	1	25	1	2.5	58
Total	93	100	46	100	4	100	40	100	49.5

Source: Adapted from Umeokafor, N. et al., 2014, The pattern of occupational accidents, injuries, accident causal factors and intervention in Nigerian factories, *Developing Country Studies*, 4(15): 119–127.

TABLE 5.2
Death and Injury Pattern, and Causes of the Accident (FMLPID, 2002–2012)

Causes of Deaths and Accidents	No. of Injuries	% of Injuries	No. of Deaths	% of Deaths	Case Fatality Rate
Fire	52	55.9	29	63	59.6
Inhaling of poisonous gases	1	1.1	1	2.2	100
Fall of a heavy object during lifting	10	10.8	6	13	60
Machinery driven by power	13	14	1	2.2	7.7
Hot thermal fluid	1	1.1	1	2.2	100
Trapped by moving parts of machines	2	2.2	2	4.3	100
Explosion	5	5.4	4	8.7	80
Electrocution	1	1.1	1	2.2	100
Malfunction of a machine	2	2.2	1	2.2	50
Total	93	100	46	100	49.5

Source: Adapted from Umeokafor, N. et al., 2014, The pattern of occupational accidents, injuries, accident causal factors and intervention in Nigerian factories, *Developing Country Studies*, 4(15): 119–127.

Table 5.2 shows that the highest number of deaths occurred within the 11-year period (2002–2012). This was as a result of fire outbreaks, which led to 52 injuries (55.9%) and 29 deaths (63%) with a case fatality rate of 59.6 percent. This was followed by deaths as a result of the fall of heavy objects during lifting. This resulted in 6 deaths (13%) and 10 injuries (10.8%) with a case fatality rate of 60 percent. Finally, explosions contributed to 4 deaths (8.7%), 5 injuries (5.4%) and a case fatality rate of 80 percent. The trapping of workers by moving parts of machines led to 2 deaths (4.3%) and 2 injuries (2.2%). However, the inhaling of poisonous gases, machinery driven by power, hot thermal fluids, electrocution and malfunction of a machine all contributed to a minimum of 1 death (2.2%).

Table 5.3 shows that the period 2002 to 2012 experienced 78 percent of the accidents under unsafe conditions due to management factors. This was as a result of the usage of unsafe equipment, outdated machines, non-functioning of lifts or out-of-use lifts owing to management factors, whereas human factors accounted for 22 percent of the unsafe conditions. This was as a result of poor maintenance practices. Management factors accounted for 91.3 percent, which included inadequate training, hence low levels of awareness and a lack of supervision. However, human factors, which accounted for 8.7 percent and included employees' failure to attend training, was the highest contributory factor of the reported accidents followed by lack of supervision. The use of outdated machines accounts for about 50 percent of the management factors in terms of unsafe conditions (Umeokafor et al., 2014).

TABLE 5.3
Accident Causal Factors (2002–2012)

Accident Causal Factors		Frequency	Percentage
Unsafe acts	Management factor	8	40
	Human factor	12	60
Unsafe conditions	Management factor	39	78
	Human factor	11	22
Remote or contributory factors	Management factor	42	91.3
	Human factor	4	8.7

Source: Adapted from Umeokafor, N. et al., 2014, The pattern of occupational accidents, injuries, accident causal factors and intervention in Nigerian factories, *Developing Country Studies*, 4(15): 119–127.

5.5.4 Measures to Ensure Occupational Health and Safety Operation in Nigeria

There must be relevant laws (separate OHS regulations or laws different from the labour laws or Compensation Act) that will adequately measure OHS in the Nigeria construction industry. Standard and comprehensive OHS regulations/guides should be drawn such as South Africa's Occupational Health and Safety Act No. 85 of 1993 (OHSSA), the Occupational Safety and Health Administration (OSHA) in the United States, and the Health Safety Executive (HSE) regulations in the United Kingdom. Bills for such laws should be raised, read and passed into law with inputs from stakeholders. This is the first step, as OHS enforcement is legislation-driven.

Likewise, an active and continuous awareness and information campaign (e.g. stakeholders' lectures/seminars and public lectures) must be embarked upon in order to get all the stakeholders and the public on board. This is to ensure that they can have OHS consciousness and embrace it. While the proper implementation and strict enforcement of the OHS legislature should be enforced, appropriate agencies (government and non-government) must be set up and made to roll into action once the laws are made. One of the problems with Nigeria is not just the absence of relevant laws but the non-enforcement of existing ones (though some of them may be old or inadequate). The training of personnel in various OHS activities, both in government and private agencies, should be encouraged. This should include training of OH nurses, OH physicians, ergonomists, safety engineers, industrial hygienists, microbiologists and the like. There must be a continuous appraisal and review of OHS regulations and standards in order to ensure they are still functional in achieving the purposes for which they were set up. In addition, the provision of OHS should be one of the criteria to be fulfilled before some businesses (especially those with OHS hazard risks and the medium-/large-scale industries) are registered by the Corporate Affairs Commission (CAC) or enlisted into and kept on the Nigerian Stock Exchange (SEC). Such measures will ensure compliance with OHS regulations by employers. Together with the aforementioned points, the 'collective will'/political will must be present. It is about people (the government, employers and employees) wanting to and doing the right thing always.

5.5.5 Benefits or Gains of Occupational Health and Safety to the Nigerian Construction Industry

The following provides information on the benefits or gains of OHS to Nigeria (achieving the economic vision 20-20) (Olusegun, 2015):

- Reduction in negative trends—Occupational diseases/injuries have negative effects on workers, families and the country at large. There is a negative ripple effect that occurs when a family's breadwinner loses his or her life in the course of work (OHS death or injury). Wives/husbands are widowed, children are orphaned and such families face untold hardship. These negative effects in turn affect the country.
- Improved overall health index—OHS is another way of caring for the public's health. It is a major aspect of public health protection and promotion. People's health and safety will be better protected at work.
- Proper compliance with OHS will add Nigeria to the league of countries practising global standards with regard to OHS. This will give international recognition to Nigeria and enable it to assume the true leadership position for which it is known in Africa.
- Another important benefit of OHS implementation to Nigeria is massive job creation and employment, thereby boosting the economy.

5.6 OBJECTIVES OF HEALTH AND SAFETY POLICY IN NIGERIA

The following are the objectives of the H&S policy in Nigeria:

- To create a general framework for the improvement of working conditions and working environments
- To prevent accidents and departures from health arising out of or in the course of work
- To ensure the provision of OSH services to workers in all sectors of economic activity

5.7 BASIS FOR HEALTH AND SAFETY COMPLIANCE IN THE NIGERIAN CONSTRUCTION INDUSTRY

According to Haupt and Smallwood (in Windapo & Oladapo, 2012) the most common issues with regard to non-compliance with regulations are that workers were never consulted about H&S by management, there is no reference made to H&S when an instruction is given to performing a task and the workers are seldom provided with personal protective equipment (PPE). In addition, there are no programmes, policies or rules; H&S representatives are not employed full time on sites; there are no inspections and meetings held; and there is a perception by workers that supervisors do not prioritise H&S though it has significant impact on safety outcomes.

A lack of training is also a major contributing factor to the cause of non-compliance. According to Idubor and Oisamoje (2013:167), 'many of the H&S violations take place in an industrializing economy, such as that of Nigeria, because they feel it is cheaper to prevent occupational accidents and diseases than deal with their effects'. They further postulate that weak legal structure and an abundant supply of cheap labour perpetuate some of these violations and attract foreign firms to operate in Nigeria.

5.8 SUMMARY

Nigeria is facing issues with H&S compliance. The non-compliance with H&S has led to injuries and accidents in the construction industry. This has been a major blow to most institutions in the developing countries. However, this calls for the proper development of an H&S policy for the construction industries in Nigeria and the developing countries with more emphasis on Africa.

REFERENCES

Achalu, E.I. (2000). *Occupational health and safety (OHS)*. Lagos: Simarch Nigeria Limited Splendid Publishers.

Cotton, A.P., Sohail, M. and Scott, R.E. (2005). Towards improved labour standards for construction of minor works in low-income countries. *Engineering Construction and Architectural Management*, 12(6): 617–632.

Diugwu, I.A., Baba, D.L. and Egila, A.E. (2012). Effective regulation and level of awareness: An expose of the Nigeria's construction industry. *Open Journal of Safety Science and Technology*, 2: 140–146.

Hagstedt, C. and Pieris, B. (2000). Occupational safety and health in developing countries: Review of strategies, case studies and a bibliography. Available from: http://www.niwl.se/arb/ [Accessed 28 February 2016].

Idubor, E.E. and Oisamoje, M.D. (2013). An exploration of health and safety management issues in Nigeria's effort to industrialise. *European Scientific Journal*, 9(12): 154–169.

ILO/WHO Committee. (1995). Definition of occupational and safety. 12th Session of Joint ILO/WHO Committee on Occupational Health. Available from: http://www.ilo.org/safework [Accessed 23 April 2014].

Kheni, N.A., Dainty, A.R.J. and Gibb, A.G.F. (2007). Influence of political and socio-cultural environments on health and safety management within SMEs: A Ghana case study. In: Boyd, D. (Ed.), *Proceedings of the 23rd Annual Association of Researchers in Construction Management (ARCOM) Conference*, 3–5 September 2007, Belfast, UK, 159–168.

Okolie, K.C. and Okoye, P.U. (2012). Assessment of national culture and construction health and safety climate in Nigeria. *Science Journal of Environmental Engineering Research*, article ID: sjeer-167.

Olusegun, A.D. (2015). Occupational health and safety in Nigeria: How the Nigerian government can create employment. Available from: https://www.linkedin.com [Accessed 27 February 2016].

Umeokafor, N., Evaggelinos, K., Lundy, S., Isaac, D., Allan, S., Igwegbe, O. and Umeokafor, K. (2014). The pattern of occupational accidents, injuries, accident causal factors and intervention in Nigerian factories. *Developing Country Studies*, 4(15): 119–127. Available from: www.iiste.org [Accessed 27 March 2016].

Umeokafor, N., Umeadi, B. and Jones, K. (2014). Compliance with occupational safety and health regulations: A review of Nigeria's construction industry. Research Gate. Available from: https://www.researchgate.net/ [7 March 2016].

Windapo, A. and Oladapo, A. (2012). Determinant of construction firms' compliance with health and safety regulations in South Africa. In: Smith, S.D. (Ed.), *Proceedings of 28th Annual Association of Researchers in Construction Management (ARCOM) Conference*, 3–5 September, Edinburgh, UK, 433–444.

World Health Organization (WHO). (1999). Fact Sheet No. 84. Geneva: World Health Organization.

6 Occupational Health and Safety in South Africa

6.1 INTRODUCTION

This chapter of the book describes occupational health and safety (OHS) issues in South Africa by looking at health and safety (H&S) policies and their objectives. This is followed by South African legislation and the construction industry, as well as the regulatory enforcement in South Africa.

6.2 OCCUPATIONAL HEALTH AND SAFETY IN SOUTH AFRICA

Occupational health and safety (OHS) is core to the successful long-term sustainability of any business, and fortunately in South Africa, health and safety (H&S) is a legislatively compliant criterion, enforced by the Occupational Health and Safety Act 85 of 1993 (OHS Act) and the Department of Labour (Action Training Academy, 2014). According to a report from the South African Construction Industry Development Board (CIDB, 2009a), the activities of the construction industry due to its poor H&S performance record are subject to various legislative and institutional frameworks. The primary objective of these is the prevention of accidents and their consequences in terms of injury, disablement, fatality and ill health within the work environment.

The CIDB report notes that South Africa is not lacking in terms of H&S legislation and that South Africa's legislative framework addresses H&S at three levels. First, in terms of the national Constitution, then in terms of acts such as the Occupational Health and Safety Act No. 85 of 1993 and the complementary Compensation for Occupational Injuries and Diseases Act No. 130 of 1993. These are followed by a range of regulations promulgated under the OHS Act, in particular, the Construction Regulations promulgated in July 2003. Although it cannot be quantified, CIDB (2009a) maintains that the Construction Regulations have had a positive impact on reducing accidents. Building contractors in South Africa do not comply fully with H&S regulations, and even though H&S issues have seen some improvement over the years, the number of people who are injured or die on construction sites in South Africa is still high according to the Department of Labour (DoL, 2007; CIDB, 2009a, 2009b; David Bettesworth Town and Regional Planners, 2011; Windapo & Oladapo, 2012).

The number and severity of H&S standard violations provide one measure of the degree to which a contractor's operations comply with OHSA standards. For instance, the level to which contractors' operations comply with OHSA regulatory requirements on construction sites in the Western Cape Region of South Africa was determined by the Master Builders Association of Western Cape (MBAWC) for years 2007, 2008, 2009 and 2010 for annual safety competitions using an audit system which was designed by the association for grading the H&S regulation compliance of construction projects. The Master Builder South Africa Audit Tool (MBSAT) used in H&S assessment was classified into 19 different elements and the associated points achievable, each element targeting different requirements of OHSA. The H&S audit undertaken by MBAWC between 2007 and 2010 when averaged and distributed by compliance to the nineteen H&S requirements revealed that the contractors attained acceptable standards in three elements: cranes, demolition, and transport and material handling. However, unacceptable standards were recorded in eleven elements and very poor standards in five elements (i.e. site plant and machinery workplace environment, health and hygiene, personal protective health and clothing, plant and storage yards, and excavation) (DoL, 2007; Weil, 2007).

Research has shown that legislation or targeted regulations can influence the H&S performance of a project, industry or a stakeholder. Another study conducted in South Africa revealed that construction regulations were perceived to have had an impact on H&S performance according to the South Africa CIDB. It was also observed from this study that the manifestations of the impact of the construction regulations in South Africa were widespread and, in particular, increased H&S awareness of and consideration for, or reference to, H&S by project managers and general contractors.

6.3 HEALTH AND SAFETY POLICY IN SOUTH AFRICA

The South Africa H&S policy proposes legislation that is enacted by the Cabinet resolution to create an integrated national OHS system. This policy applies to all agencies and government departments with the responsibility for regulating OHS.

6.3.1 Objectives of Health and Safety Policy in South Africa

A reduction in work-related accidents and diseases in South Africa is the primary objective of its H&S policy. The government expects employers and workers to adopt and implement the culture of prevention as part of the requirements. The social and economic benefits of an H&S policy are the effective prevention of work-related accidents and ill health. These include improvements in productivity, competitiveness and the quality of life of the working population. The improved levels of public policy may include the effective management of many safety hazards, while the effective control of hazardous substances at the workplace from their sources will improve levels of public health and minimize environmental pollution. The secondary objective of an H&S policy in South Africa has to do with equitable compensation to the injured in work-related accidents or those who contract occupational diseases. Medical aid, financial compensation and access to rehabilitation services are forms of compensation benefits. The contributions paid by employers as part of the compensation system, must be sensitive to an employer's OHS performance in order to serve as an incentive for improved performance.

6.3.2 Occupational Health and Safety Problems in South Africa

Occupational accidents and diseases impose an enormous cost on the South Africa economy, and much more to the performance of the construction sector. These include damage of property, loss of production time, the cost of engaging, loss of skills and retraining replacements are the cost to the employers. Employers see the expenditure of H&S on unskilled workers as a cost that must be avoided. For example, the compensation for employees with permanent injuries that reduce their earning capacity is found to be inadequate and in many cases inequitable. Rehabilitation services are only accessible to a small proportion of injured workers. Costs to workers and their families include permanent disabling injuries, debilitating disease, loss of employment and loss of breadwinners. Earlier South Africa government policy documents such as the White Paper on Social Welfare (1996) and the Green Paper on an Integrated National Disability Strategy (1998) have noted that the burden of occupational accidents and disease has been shifted disproportionately from employers to workers and their families, particularly those in rural areas.

There has been confirmation on recent research that the country has a serious occupational health problem in many sectors. This is as a result of inadequate management of occupational health risks in maintaining working environments. However, the proper implementation of effective engineering practices can eliminate these diseases. The cost of OHS preventative measures will increase owing to HIV infection that has been found to impact negatively on skills and training levels of human capital. Small to medium-sized companies have been alleged to have higher accident rates than larger firms in the same sector which exacerbate OHS problems owing to the expected growth of small to medium-sized companies in South Africa. The other problem is the widespread use of non-standard employment and sub-contracting arrangements and the growth of the informal sector have also been linked to increased OHS problems. The cost of inadequate management of OHS impacts negatively on public safety, public health and the environment.

6.4 SOUTH AFRICAN LEGISLATION AND THE CONSTRUCTION REGULATIONS

The Occupational Health and Safety Act No. 85 of 1993 (OHS Act) and the complementary Compensation for Occupational Injuries and Diseases Act No. 130 of 1993 (COID Act) are the two primary acts that impact on construction H&S in South Africa. The OHS Act replaced the previous Machinery and Occupational Safety Act No. 6 of 1983, the Machinery and Occupational Safety Amendment Act No. 40 of 1989, and the Machinery and Occupational Safety Amendment Act No. 97 of 1991, and the promulgation thereof reflected the increased emphasis on health. The Construction Regulations promulgated under the OHS Act in July 2003 have impact on construction H&S.

6.4.1 Characteristics of the Occupational Health and Safety Legislative Framework and the Construction Regulations

The following are the characteristics of the OHS legislative framework and the Construction Regulations:

- A departure from the traditionally prescriptive or 'deemed-to-comply' or 'command-and-control' approaches to a performance-based approach in terms of which no standards for compliance are set.
- The redistribution of responsibility for construction H&S away from the contractor, who was previously solely responsible, to include all participants in the construction process from the client through to the final end-user.
- Emphasis on the identification of construction hazards and the assessment of risks to eliminate, avoid or, at the very least, reduce perceived risks.
- Consideration of H&S issues not just during the construction life of the project, but from project inception through to the final demise of the facility by demolition, including the operation, utilisation and maintenance periods (CIDB, 2008).

6.5 REGULATORY ENFORCEMENT

The primary construction H&S regulatory and enforcement structures in South Africa are the OHS Inspectorate within the Department of Labour (DoL) and the Mine Health and Safety Inspectorate within the Department of Minerals and Energy (DME). Compliance with building regulations falls within the ambit of local authorities and the CIDB's act also provides for the CIDB to play a regulatory and or promotional role (Geminiani & Smallwood in Construction Handbook Report, 2009).

Construction H&S has long been the focus of attention of many stakeholders in South Africa, since the construction industry is known throughout the world to have a poor H&S record. There has been a large number of cases of non-compliance with H&S regulations in South Africa, leading to a large number of fatalities and injuries. This is as a result of not taking H&S compliance in construction seriously. Many organisations are encouraged by assessment rebate incentives and use injuries to measure their safety performance. Employee involvement in H&S management has also been raised because they are best able to see issues and to bring about real improvements. Regardless of the level of risk or company size, health and safety is always associated with the people.

A report from the Federated Employers' Mutual Assurance (FEM) of South Africa as posited by Furter (2011) stated that 'construction health and safety accidents have dropped by half in five years, from an accident frequency rate of 8.5 percent to 4.3 percent the previous year'. There was also 'a 100 percent improvement in the general South African construction accident frequency percentage'. FEM construction health and safety statistics of 'accident frequency percentage' is calculated by the number of employees injured, per hundred employees insured, over any period, usually per claim year (Furter, 2011). A company's main objective should be to have no worker injuries that will lead to an unbreakable belief that the goal of zero injuries is attainable. However, policy holders who are made up of several people have shown that it is not possible to attain the goal of zero injuries. Therefore, construction H&S should always be viewed in a positive way.

The South Africa Construction Regulations 2003 (R.1010 of 18/07/2003) of the OHS Act create minimum standard legislation that needs to be complied within the country. The Construction Regulations (CR) basically apply to any persons involved in construction work. CR 4(2) stipulates that the client shall be responsible to discuss and negotiate with the principal contractor the contents of the H&S plan and thereafter finally approve the H&S plan for implementation. In practice, a H&S plan could thus be described as a documented summary of the legal requirements to be implemented on a construction site/workplace in order to ensure a safe and healthy work environment. It describes the potential hazards of a work site, along with all company policies, controls and work practices selected to minimize those hazards. The principal contractor is responsible for the development of a plan and requires a specific plan for the task at hand from the contractor. CR 5(4) stipulates that a contractor shall provide and demonstrate to the principal contractor a suitable and sufficiently documented health and safety plan, based on relevant sections of the principal contractor's H&S specification. In the case where a sub-contractor will be used, a contractor will require a similar plan from the sub-contractor for the specific task the sub-contractor will perform. It is clear that these regulations require both principal contractors and contractors (including sub-contractors) to have a documented health and safety plan on site.

The OHS Act requires the employer to provide and maintain as far as reasonable and practical a work environment that is safe and without risk to the health of employees. Employers are obligated to have an H&S policy once the chief inspector has directed them to do so. It is compulsory for all organisations to formulate and implement a H&S policy, but the employer is still duty-bound to inform employees of work-related risks and dangers. An H&S policy could prove to be a valuable tool in this regard; in fact the importance of work-related policies and procedures cannot be overemphasised. The workplace rules and regulations bring order to the workplace. Employers need to comply with minimum standard legislation and also have the right to set a better standard for their particular business. It is imperative to communicate these standards to employees and other role players such as contractors. Policy documents also provide direction to all company activities and provide the criteria to measure and evaluate efficiency. It will not be possible to effectively control and manage employees and other role players without a policy document.

The primary objective of an H&S policy should be to prevent or reduce work-related accidents and occupational diseases. An appropriate policy could help to avoid the expense, inconvenience and other consequences of workplace accidents by making sure that employees and other role players know what is expected of them. A good policy should indicate how the organisation protects those who could be affected by its activities and be broad enough to cover all aspects of the company's activities. The policy basically demonstrates how seriously an organisation takes its H&S responsibilities. It is important to note that the employers are bound by the prescriptions of their specific policy and could be held legally liable for not complying with it.

Monitoring to ensure that the policy is still effective is also very important; a policy will only be effective if it is reviewed on a regular basis. The drafting, implementation and monitoring of an organisation's H&S policy is a clear demonstration of management commitment in this regard. Senior management must be committed to reviewing a company's policy to ensure consistency and completeness. Health and safety should enjoy the same priority as the organisation's other major objectives. Senior management must be committed to ensure that H&S policies are carried out with no exceptions. Commitment and involvement are complementary to each other and are essential elements needed to spearhead any H&S programme to success. The OHS Act provides specifications and requirements regarding the areas they govern. These regulations form part of the OHS Act. Table 6.1 gives a summary of the current OHS Act regulations in South Africa.

6.5.1 Aim and Scope of the South Africa Occupational Health and Safety Act

The aim of the South Africa OHS Act is to provide for the safety and health of persons at work in connection with the use of plant and machinery. It further provides for the protection of people other than people at work from hazards arising out of or in connection with the activities from people at work. The main objective of the act could be described as a pro-active attempt by government to prevent and avoid work-related injuries and illness. The act governs the health and safety for the diverse industry of South Africa. It regulates and controls health and safety in all organisations, from a normal office environment to more hazardous environments such as industrial plants and construction sites. The legislation requires the employer to do everything 'reasonably practicable' to protect people (construction site workers) from harm.

6.5.1.1 Severity and Scope of the Hazard and Risk Concerned

A risk assessment is basically a careful examination of what is related to the work activities of an organisation that could cause harm to people or damage to property. During the risk assessment process we identify the hazards associated with an activity (physical task or process at hand) to assess the seriousness of these hazards and to formulate systems of work, training or other methods (controls) to reduce the associated risks to a minimum or at least to an acceptable level. After evaluating these work-related hazards, risks and dangers, the employer should determine the severity of the hazards or risks concerned. This mainly refers to 'seriousness of these hazards' as previously mentioned. During the risk assessment process, the risk will be given a 'risk value' in order to determine the severity of the risk involved (example: high, medium or low risk level).

6.5.1.2 Removing or Mitigating Occupational Health and Safety Hazard or risk

The *Cambridge English Dictionary* defines *mitigate* as 'to make something less harmful'. If the identified OHS hazard or risk cannot be removed, the next option

TABLE 6.1
Summary of Current OHS Act Regulations

Health-Related Regulations	Scope of Application	Regulation
Asbestos Regulations, 2001	These regulations shall apply to every employer and self-employed person who carries out work at a workplace that may expose any person to asbestos dust at the workplace.	1. Notification of asbestos work 2. Exposure to asbestos 3. Information and training 4. Duties of persons who may be exposed 5. Assessment of potential exposure 6. Air monitoring 7. Medical surveillance 8. Respirator zone 9. Control of exposure to asbestos 10. Cleanliness of premises and plant 11. Control of exposure to asbestos of persons other than employees 12. Asbestos that forms part of structure of workplace, building, plant or premises 13. Asbestos cement sheeting and related products 14. Records 15. Personal protective equipment and facilities 16. Maintenance of control measures 17. Labelling, packaging, transportation and storage 18. Disposal of asbestos 19. Demolition 20. Prohibition
Hazardous Biological Agent (HBA) Regulations, 2001	These regulations shall apply to every employer and self-employed person at a workplace where: (a) HBA is deliberately produced, processed, used, handled, stored or transported; or (b) an incident for which an indicative list is given in Annexure A to this regulation occurs that does not involve a deliberate intention to work with an HBA but may result in persons being exposed to HBA in the performance of their work.	1. Classification of biological agents 2. Information and training 3. Duties of persons who might be exposed to HBA 4. Risk assessment by employer or self-employed person 5. Monitoring exposure at workplace 6. Medical surveillance 7. Records 8. Control of exposure to HBA 9. Personal protective equipment and facilities 10. Maintenance of control measures, equipment and facilities 11. Prohibitions 12. Labelling, packaging, transporting and storage

(*Continued*)

TABLE 6.1 (CONTINUED)
Summary of Current OHS Act Regulations

		13. Special measures for health and veterinary isolation facilities 14. Special measures for laboratories, animal rooms and industrial processes 15. Disposal of HBA
Hazardous Chemical Substances Regulations, 1995	These regulations shall apply to an employer or a self-employed person who carries out work at a workplace which may expose any person to the intake of HCS at the workplace.	1. Information and training 2. Duties of persons who may be exposed to hazardous chemical substances 3. Assessment of potential exposure 4. Air monitoring 5. Medical surveillance 6. Respirator zone 7. Records 8. Handling of hazardous chemical substances 9. Control of exposure to HCS 10. Personal protective equipment and facilities 11. Maintenance of control measures 12. Prohibitions 13. Labelling, packaging, transportation and storage 14. Disposal of hazardous chemical substances
Lead Regulations, 2001	These regulations shall apply to every employer and self-employed person at a workplace where lead is produced, processed, used, handled or stored in a form in which it can be inhaled, ingested or absorbed by any person in that workplace.	1. Exposure to airborne lead 2. Information and training 3. Duties of persons who may be exposed 4. Assessment of potential exposure 5. Air monitoring 6. Medical surveillance 7. Respirator zone 8. Records 9. Control of exposure to lead 10. Personal protective equipment and facilities 11. Cleanliness of premises and plant 12. Maintenance of control measures 13. Prohibitions 14. Labelling, packaging, transportation and storage 15. Disposal of lead waste

(*Continued*)

TABLE 6.1 (CONTINUED)
Summary of Current OHS Act Regulations

Noise Induced Hearing Loss Regulations, 2003	These regulations shall apply to an employer or self-employed person who at any workplace under his or her control, carries out work that may expose any person at that workplace to noise at or above the noise-rating limit.	1. Exposure to noise 2. Information and training 3. Duties of persons who may be exposed to noise 4. Assessment of potential noise exposure 5. Noise monitoring 6. Medical surveillance 7. Noise zone 8. Control of noise exposure 9. Record 10. Hearing protective equipment 11. Maintenance of control measures
General Regulations	**Scope of Application**	**Regulation**
Environmental Regulations for Workplaces, 1987	These regulations in general refer to the physical conditions of the work environment.	1. Thermal requirements 2. Lighting 3. Windows 4. Ventilation 5. Housekeeping 6. Precautions against flooding 7. Fire precautions and means of egress
Facilities Regulations, 1990	These regulations in general refer to sanitary facilities, toilets, bathrooms, showers, dining facilities, drinking water, certain prohibitions as well as the conditions of these facilities that form part of the work environment.	1. Sanitation 2. Facilities for safekeeping 3. Change rooms 4. Dining rooms 5. Prohibition 6. Drinking water 7. Seats 8. Condition of room and facilities
General Administrative Regulations, 2003	These regulations in general refer to sanitary facilities, toilets, bathrooms, showers, dining facilities, drinking water, certain prohibitions as well as the conditions of these facilities that form part of the work environment.	1. Access to premises 2. Exemption 3. Copy of the act 4. Health and safety committee 5. Negotiations and consultations before designation of health and safety representatives 6. Designation of health and safety representatives 7. Recording and investigation of incidents 8. Witness at inquiry 9. Returns

(Continued)

TABLE 6.1 (CONTINUED)
Summary of Current OHS Act Regulations

Draft General Health and Safety Regulations, 2005	These regulations refer to general health and safety matters or requirements set for the work environment.	1. Personal protective equipment and facilities 2. Intoxication 3. Display of substituted notices and signs 4. Admittance of persons 5. Prevention of transmission of HIV, hepatitis virus and other blood-borne diseases 6. First aid, emergency equipment, and procedures 7. Use and storage of flammable liquids 8. Work in confined spaces 9. Work in elevated positions 10. Working in danger of engulfment 11. Stacking of articles 12. Welding, flame cutting, soldering and similar operations 13. Operating trains 14. Ladders 15. Ramps
Electrical Regulations	**Scope of Application**	**Regulation**
Electrical Installation Regulations, 2009	These regulations shall apply to every user or lessor of an electrical installation as well as approved inspection authorities.	1. Responsibility for electrical installations 2. Approved inspection authorities for electrical installations 3. Functions of approved inspection authorities for electrical installations 4. Design and construction 5. Electrical contractor 6. Certificate of compliance 7. Commencement and permission to connect installation work 8. Issuing of certificate of compliance 9. Disputes 10. Application for registration as a registered person 11. Withdrawal of registration and approval 12. Substitution of lost, damaged or destroyed certificate 13. Fees payable

(*Continued*)

TABLE 6.1 (CONTINUED)
Summary of Current OHS Act Regulations

Electrical Machinery Regulations, 1988	These regulations shall apply to every employer, employee and self-employed person who carries out work whilst using electrical machinery at a workplace.	1. Safety equipment 2. Work on disconnected electrical machinery 3. Notices 4. Switch and transformer premises 5. Electrical control gear 6. Switchboards 7. Electrical machinery in hazardous locations 8. Portable electric tools 9. Portable electric lights 10. Electric fences 11. Inspection authorities 12. Earthing 13. Supports 14. Clearances of power lines 15. Protective supports 16. Insulators and fittings 17. Conductors 18. Overhead service connections and overhead service conductors 19. Crossings 20. Bare conductors on premises 21. Schemes to be submitted to the Postmaster General
Machinery Regulations	**Scope of Application**	**Regulation**
Driven Machinery Regulations, 1988	These regulations shall apply to every employer, employee and self-employed person who carries out work whilst using driven machinery at a workplace.	1. Revolving machinery 2. Circular saws 3. Band saws and band knives 4. Wood planning machines 5. Wood moulding and mortising machines 6. Sanding machines 7. Grinding machines 8. Shears, guillotines, presses 9. Slitting machines 10. Mixing, agitating and similar machines 11. Rolls and calendars 12. Washing machines, centrifugal extractors, etc. 13. Air compressors

(*Continued*)

TABLE 6.1 (CONTINUED)
Summary of Current OHS Act Regulations

		14. Refrigeration and air conditioning installations 15. Transportation plants 16. Goods hoists 17. Lifting machines and lifting tackle 18. Builder's hoists 19. Explosive powered tools
General Machinery Regulations, 1988	These regulations shall apply to every employer, employee and self-employed person who carries out work whilst using machinery at a workplace.	1. Supervision of machinery 2. Safeguarding of machinery 3. Operation of machinery 4. Working on moving or electrically alive machinery 5. Devices to start and stop machinery 6. Reporting of incidents in connection with machinery 7. Notifiable substances 8. Information regarding regulations
Lift, Escalator and Passenger Conveyor Regulations, 1994	These regulations shall apply to every employer and self-employed who installs, uses and has lifts, escalators and passenger conveyors at their workplace.	1. Permission to install and use 2. Design and construction 3. Particulars of lifts, escalators or passenger conveyors 4. Inspections and tests 5. Maintenance 6. Record keeping
Pressure Equipment Regulations, 2009	These regulations shall apply to the design, manufacture, operation, repair, modification, maintenance, inspection and testing of pressure equipment with design pressure equal to or greater than 50 kPa, in terms of the relevant health and safety standard incorporated into these regulations.	1. General requirements 2. Duties of manufactures 3. Duties of importers and sellers 4. Duties of users 5. Approved and duties of approved inspections authorities 6. Registration of steam generator 7. Pressure equipment marking 8. Pressure safety accessories 9. Inspection and test 10. Risk-based inspection 11. Repairs and modifications 12. Records 13. Access 14. Door interlocks 15. Gas reticulation equipment and systems 16. Transportable gas containers 17. Fire extinguishers

(*Continued*)

TABLE 6.1 (CONTINUED)
Summary of Current OHS Act Regulations

Specific Regulations	Scope of Application	Regulation
Regulations concerning the Certificate of Competency, 1990	Certificate of Competency: A certificate of competency as a mechanical or electrical engineer. The certificate will be issued by Chief Inspector with the recommendations of the Commission of Examiners.	1. Issue of certificates 2. Suspension or cancellation of certificates 3. Substitution of lost, damaged or destroyed certificates 4. Commission of Examiners 5. Qualifying examination 6. Acceptance as candidate
Draft Amendments to Construction Regulations, 2010	These regulations apply to all persons involved in construction work.	1. Application for a permit to perform construction work 2. Duties of client 3. Duties of principal contractor and contractor 4. Supervision of construction work 5. Risk assessment 6. Fall protection 7. Structures 8. Formwork and support work 9. Excavation work 10. Demolition work 11. Tunnelling 12. Scaffolding 13. Suspended platforms 14. Boatswain's chairs 15. Material hoists 16. Bulk mixing plant 17. Explosive powered tools 18. Cranes 19. Construction vehicles and mobile plant 20. Electrical installations and machinery on construction sites 21. Use and temporary storage of flammable liquids on construction sites 22. Water environments 23. Housekeeping and general safeguarding on construction sites 24. Stacking and storage on construction sites 25. Fire precautions on construction sites 26. Construction employees' facilities 27. Construction health and safety technical committees 28. Approved inspection authorities

(Continued)

TABLE 6.1 (CONTINUED)
Summary of Current OHS Act Regulations

Diving Regulations, 2001	These regulations basically shall apply to all diving operations and all persons engaged in diving operations in the Republic of South Africa or the territorial waters thereof.	1. Training of divers 2. Designated medical practitioners, medical examinations and medical certificates of fitness 3. Diving supervisor 4. Operations manual 5. Control of diving operations 6. Decompression 7. Compression chambers and bells 8. Plant and equipment 9. Council for diving 10. Rules, syllabi and examinations 11. Registration as learner diver 12. Registration as a diver 13. Registration as a diving supervisor 14. Applications 15. Withdrawal of certificate of registration 16. Fees payable
Explosives Regulations, 2003	These regulations shall apply to any employer, self-employed person or user who operates an explosives workplace for the purpose of manufacturing, testing, storing or using explosives.	1. Classification of explosives for manufacturing 2. Licensing of explosives workplaces 3. Non-detonatable and non-sensitised explosives 4. Danger area 5. Danger buildings 6. Safeguarding of explosives workplace 7. Design, construction and manufacture 8. Importation of explosives 9. Safety distances 10. Supervision of explosives workplace 11. Safe handling of explosives 12. Emergencies 13. Incidents 14. Closure of explosives workplaces 15. National Explosives Council 16. Approved inspection authorities 17. Standards of training

(*Continued*)

TABLE 6.1 (CONTINUED)
Summary of Current OHS Act Regulations

Major Hazard Installation Regulations, 1993	These regulations shall apply to employers, self-employed persons and users, who have on their premises, either permanently or temporarily, a major hazard installation or quantity of substance which may pose a risk that could affect the health and safety of employees and the public.	1. Notification of installation 2. Temporary installations 3. Risk assessment 4. On-site emergency plan 5. Reporting of risk and emergency occurrences 6. General duties of suppliers 7. General duties of local government 8. Closure
Regulations on Hazardous Work by Children in South Africa, 2010	These regulations are there to prohibit or place conditions upon the work that may be required, expected or permitted to be performed by child workers, and which is not prohibited in terms of any law.	1. Risk assessment 2. Respiratory hazards 3. Work in elevated position 4. Lifting of heavy weights 5. Work in cold environment 6. Work in hot environment 7. Work in noisy environment 8. Power tools and cutting or grinding equipment 9. Report to department of social development

would be to apply appropriate steps or measures to mitigate it. In the case where an employer cannot remove or eliminate a hazard or risk, steps should be taken to lessen the seriousness or extent of the hazard or risk concerned.

When dealing with this aspect, both the severity of the risk involved (risk level), as well as the 'removing or mitigating that hazard or risk' should be taken into consideration. In the case where something has to be done in order to remove or reduce the risk to an acceptable level, the employer should obtain knowledge or means in order to take the necessary steps or precautions. It implies that employers need to make informed decisions based on knowledge that is reasonably available. Existing knowledge could be gained either locally or internationally. Knowledge or means must, first, be aimed at removing and, second, at mitigating the hazard.

Likewise, employers need to take reasonable measures to safeguard employees and other persons wherever possible, but without stretching to excessive costs over H&S gains. If the risk from a hazard outweighs the cost of reducing the risk, action must be taken. If costs substantially outweigh risks, less costly action must be taken or no action if there are no alternatives. For example, spending R 5000 (US$370) to prevent the occasional bruised finger would not be reasonable, but spending R 50,000 (US$3700) on a machine guard to stop an arm amputation would be reasonable. Employers have to use their discretion in order to establish whether the amount of money that will be spent is justifiable in relation to the benefits deriving from it.

The employer still needs to take reasonable steps as follows to ensure that a workplace is safe and without risk to the health and safety of employees and others involved. In doing this, the employer should:

- Inform the employees of the decision and measures taken to protect them.
- Ascertain what type of instruction, training, supervision, personal protective equipment and the like are needed to protect them.
- Assess whether the reasonable steps actually work in order to provide a safe and healthy workplace.

The OHS Act requires the employer to provide and maintain as far as reasonable and practical a work environment that is safe and without risk to the health of employees. Section 7 of the act provides direction to employers concerning the H&S policy of the organisation. Hence, the employer is obligated to have an H&S policy and must inform employees how work-related risks and dangers could be prevented. Employers need to comply with minimum standard legislation, but must have the right to set improved standards for their particular businesses. It is imperative to communicate these standards to employees and other role players such as contractors. Policy documents provide direction to all company activities and provide the criteria to measure and evaluate efficiency. It will not be possible to effectively control and manage employees and contractors without an objective H&S policy. The primary objective of a health and safety policy should be to prevent or reduce work-related accidents and occupational diseases as far as reasonable and practical. A good policy will indicate how the organisation protects those who could be affected by its activities and be broad enough to cover all aspects of the company's activities. An employer should be able to create a policy that is suitable and sufficient in order to address the H&S needs of his or her company.

6.5.2 The Health and Safety Policy Statement

The following are some of the requirements that should be contained in a typical H&S policy statement:

- Provide a description of the organisation.
- Recognize the need to comply with minimum standard legislation of the Occupational Health and Safety Act.
- Recognize the priority of safety in relation to other organizational goals and policies.
- Acknowledge the right of every employee to work in a safe and healthy environment.
- Stipulate that management is accountable for occupational health and safety programmes and management's commitment to providing a safe and healthy work environment by eliminating or minimizing the hazards that can cause accidents and injuries.
- State the organisation's basic health and safety philosophy (statement of health and safety principles and goals).

- State the general responsibilities of all employees.
- State that the health and safety shall not be sacrificed for expediency.
- Encourage cooperation with unions and workers to involve all employees in implementing the health and safety policy into practice.
- Give an indication of the policy implementation date.
- Make provision for review date of policy.

6.5.3 Construction Health and Safety Statistics in South Africa

The section about the construction health and safety statistics gives information related to fatal, non-fatal and non-casualty incidents/accidents. Others are the analysis of FEM statistics. Table 6.2 shows that construction H&S statistics provided by the Department of Labour (DoL) covering the period 2004/05 to 2007/08 show a sharp rise in accidents up to 2007/08 to around 160 fatalities and around 400 non-fatal accidents (i.e. temporary or permanent disablements) (DoL in CIDB, 2008).

The following are the analysis of the FEM statistics:

- The dominant causes of injuries were struck by (44%), falls onto different levels (14%) and striking against (10%).
- The dominant causes of fatalities were motor vehicle accidents (MVAs) (47%), struck by (17%) and falls onto different levels (17%).
- Penetrating wounds (30%) and superficial wounds (31%) predominated in terms of the nature of injuries sustained.
- Multiple injuries caused 47 percent of fatalities.
- Injuries to hands (24%), head and neck (19%), and legs (16%) were common anatomic regions involved.
- In terms of agency, automobiles/vehicles (10%) and hand tools (6%) dominated as causes of injuries (CIDB, 2008).

TABLE 6.2
Construction Health and Safety Statistics (Excluding Motor Vehicle Accidents)

	Department of Labour: OHS			
	2004/05	**2005/06**	**2006/07**	**2007/08**
Fatal	54	81	79	162
Non-fatal	159	250	245	396
Non-casualty	11	7	10	20
Total	224	338	334	578

Source: Adapted from Construction Industry Development Board (CIDB), 2008, Development through partnership: Construction health and safety in South Africa, p. 3.

TABLE 6.3
Health and Safety Statistics

Industry	Temporary Disablement (per 100,000 Workers)	Permanent Disablement (per 100,000 Workers)	Fatalities (per 100,000 Workers)
Fishing	4088	215	473.3
Transport	1543	87	31.4
Building and construction	981	96	25.5
Mining	1746	269	23.5
Glass, bricks and tiles	1298	154	14.9
Personal services, hotels	462	23	12.5
Agriculture and forestry	772	61	12.2
Food, drink and tobacco	114	115	11.5
All industries	808	72	11.4
Chemicals	868	112	10.9
Iron and steel	1192	164	10.9
Diamonds, asbestos, bitumen	356	108	10.8
Local authorities	1096	46	9.6
Trade and commerce	494	34	6.4
Wood	1865 445	190	6.1
Educational services	595	14	5.6
Leather	381	29	2.9
Entertainment and sport	–	31	2.8
Professional services, N.O.S.	158	17	1.9
Banking and finance	114	6	1.7
Printing and paper	907	83	1.6
Charitable, religious, political and trade organisations	430	18	1.3
Medical services	268	14	0.8
Textiles	606	38	0.5

Source: Adapted from Construction Industry Development Board (CIDB), 2008, Development through partnership: Construction health and safety in South Africa, p. 7.

Table 6.3 shows that the building and construction industry has the third highest number of fatalities per 100,000 workers and the ninth highest number of permanent disabilities per 100,000 workers. The severity rate indicates to management the seriousness of the disabling injury and the amount of time that will be lost by the worker for every 1000 hours worked.

6.5.4 Compliance and Non-Compliance

Table 6.4 shows the state of construction H&S in the statistics for blitzes across the country. It is notable that 52 (5%) of the construction employers were non-compliant with the OHS Act and the Construction Regulations. Table 6.3 shows a total of 1388

TABLE 6.4
National Construction Blitz Inspection Report

Province	Total Work Places Inspected	Number Complying	Number Complying	Improvements Notices	Contravention Notices	Prohibition Notices
Eastern Cape	136	24	102	0	106	14
Free State	155	271	84	2	77	5
Gauteng: North	57	21	35	8	40	2
Gauteng: South	247	80	167	25	172	163
KwaZulu-Natal	240	126	100	7	100	3
7	75	7	68	5	57	12
Mpumalanga	237	152	85	9	50	48
North West	56	22	32	5	27	23
Western Cape	105	19	86	9	71	13
Northern Cape	107	37	70	16	315	4
Total	1415	759	829	86	1015	287
Percent (%)	100	47.5	52.5	6	73	21

Source: Adapted from Construction Industry Development Board (CIDB), 2008, Development through partnership: Construction health and safety in South Africa, p. 9.

TABLE 6.5
Construction Health and Safety Claims and Fatalities

	Number of Claims	Number of Fatalities	Number of Claims	Number of Fatalities
Province	**2006**		**2007**	
Gauteng	4257	32	5143	30
KwaZulu-Natal	1207	13	1311	10
Eastern Cape	943	7	929	7
Boland	1577	12	1629	6
Western Cape	827	3	814	1
Kimberley and Northern Cape	28	0	43	0
Free State	345	7	362	6
South Africa	9184	74	10231	60

Source: Adapted from Construction Industry Development Board (CIDB), 2008, Development through partnership: Construction health and safety in South Africa, p. 7.

notices made up of 86 (6%) improvement notices, 1015 (73%) contravention notices, and 287 (21%) prohibition notices.

Table 6.5 shows construction H&S claims and fatalities between the years 2006 and 2007 in South Africa. In the year 2006, the Gauteng Province was reported to have the highest number of claims and fatalities, as more construction activities take place in this province due it role and relevance in South Africa and in the Southern Africa region. This was followed by KwaZulu-Natal and Boland. The trend followed in 2007 in relation to claims and fatalities for Guateng and KwaZulu-Natal provinces. The fewest recorded claims and fatalities for both years were in Kimberley and Northern Cape Province for both years.

6.6 LESSONS LEARNT FROM THE SOUTH AFRICAN OCCUPATIONAL HEALTH AND SAFETY STUDIES

The following are the lessons learnt from the study of the South African OHS. The construction H&S regulatory and enforcement structures are made up of the OHS Inspectorate. This falls under the Department of Labour (DoL) and the Mine Health and Safety Inspectorate as well as under the Department of Minerals and Energy (DME). South Africa does not lack H&S legislation. The legislative framework of South Africa addresses H&S in the national constitution, the OHS Act No. 85 of 1993, and the complementary Compensation for Occupational Injuries and Diseases Act No. 130 of 1993. There is an integration of the national OHS system by Cabinet resolution and applies to agencies and government departments tasked with the responsibility for regulating OHS. Even though South Africa is faced with enormous costs due to occupational accidents, it has been able to manage these within the legislative framework.

6.7 SUMMARY

The availability of the OHS regulations in South Africa has contributed to their successful implementation, which has had a positive impact on reducing H&S accidents. Even though South Africa still experiences some OHS problems, the legislative framework has been able to contain these. This has also been achieved owing to its performance based on comparisons with the OHS legislations of Ghana and Nigeria.

REFERENCES

Construction Industry Development Board (CIDB). (2008). Development through partnership: Construction health and safety in South Africa. Available from: http://www.asocsa.org [Accessed 08 April 2014].

Construction Industry Development Board (CIDB). (2009a). Construction health and safety in South Africa: Status and recommendations. Available from: http://www.cidb.co.za [Accessed 22 April 2014].

Construction Industry Development Board (CIDB). (2009b). Construction indicators and survey results in 2008. Available from: http://www.cidb.org.za [Accessed 22 April 2014].

David Bettesworth Town and Regional Planners. (2011). The compliance debate. Available from: http://dbtownplanner.co.za [Accessed 15 April 2014].

Department of Labour (DoL). (2007). The Construction Regulations, 2003. Available from: http://www.kznhealth.gov.za/ [Accessed 26 February 2016].

Furter, E. (2011). Construction health and safety accidents have 'dropped by half in five years'. Occupational Health and Safety, Environment and Quality Management. Available from: http://www. Sheqafriac.com [Accessed 26 February 2016].

Weil, D. (2007). Assessing OSHA performance: New evidence from the construction industry. *Journal of Policy Analysis and Management*, 20(4): 651–674. Available from: http://www.jstor.org [Accessed 17 April 2014].

Windapo, A. and Oladapo, A. (2012). Determinant of construction firms' compliance with health and safety regulations in South Africa. In: Smith, S.D. (Ed.), *Proceedings of 28th Annual Association of Researchers in Construction Management (ARCOM) Conference*, 3–5 September, Edinburgh, UK, 433–444.

7 The Construction Industry and Occupational Health and Safety in Ghana

7.1 INTRODUCTION

This chapter describes the construction industry, the institutional environment, the economic contribution to the construction sector and the degree of difficulty in accessing funds. This is followed by a discussion of occupational health and safety (OHS) issues in Ghana; the regulation setting in Ghana; the Factories, Offices and Shops Act 1970 (Act 328); and health and safety (H&S) policy trends in Ghana.

7.2 THE GHANAIAN CONSTRUCTION INDUSTRY

The construction industry is considered an economic backbone and major contributor to the gross domestic product (GDP) of Ghana (Ofori, 2012). The Ghanaian construction sector has been a source of employment to both the public and private sectors (ISSER, 2008; Dadzie, 2013). There are more than 1600 building contractors in Ghana, based on the estimate provided by the Chartered Institute of Building in Ghana. Most of these construction companies fall under D4K4 and D3K4 classification. The construction sector was the third largest growing economic sector in 2004 with a constant GDP growth of about 5.8 percent from 2004 to 2005. However, according to Asamoah and Decardi-Nelson (2014), the Ghanaian construction industry contributes about 5 percent to 10 percent of the GDP to the country and employs nearly 10 percent of the working population.

7.2.1 CHALLENGES OF THE CONSTRUCTION INDUSTRY

The following are the challenges facing the construction industry in Ghana (Vibe Ghana, 2014):

- Absence of a principal development regulatory body
- Inadequate financial resources
- Lack of investment in human resource development
- Inability to embrace change
- Low technology in the industry
- Lack of appreciation of workforce in the industry
- High level of employee mobility

Most of these problems are due to weaknesses in enforcing rules, regulations and professional standards, largely due to the lack of a legal mandate, although there are professional bodies of architects, surveyors, engineers, builders and technicians to regulate the activities of their members. Construction output is usually substandard and is delivered with cost overruns and beyond timelines. Construction output lacks coordination among these construction-related institutions and bodies—both public and private (Vibe Ghana, 2014). Ghana is likely to face many negative implications for short-term economic growth and longer-term national development due to the poor performance of the industry. Countries such as Kenya, Malawi, South Africa, Tanzania and Zambia are far ahead of Ghana in their development process. To keep pace with such countries in the development of the industry, Ghana needs greater efforts (Vibe Ghana, 2014) to overcome the current limitations facing its construction industry.

7.3 INSTITUTIONAL ENVIRONMENT

The activities of many government ministries and other organisations affect the construction industry of Ghana. The activities of construction businesses and implementation of state policy in the Ghanaian construction sector are under two government ministries: the Ministry of Roads and Transport (MRT) and the Ministry of Water Resources, Works and Housing (MWRWH). The MRT handles the road sector of the economy. The Ghana Highways Authority (GHA), the Department of Urban Roads (DUR) and the Department of Feeder Roads (DFR) are under the jurisdiction of the MRT. The MWRWH handles policy implementation in respect of works, housing, water supply, sanitation and hydrology; and likewise oversees the activities of building contractors. The classification of contractors in Ghana is shown in Table 7.1. The MWRWH comprises the Public Works Department (PWD), the Department of Rural Housing, the Department of Hydrology, the Rent Control Department and agencies for implementing programmes deriving from government policies. Physical developments, particularly roads and housing, are carried out after the relevant departments

TABLE 7.1
Classification of Contractors in Ghana

Class	Required Works to Be Carried Out
A	Road works, airports and related works
B	Bridge construction, the construction of culverts and other drainage structures
C	Labour-based works
S	Structures
M	Miscellaneous road-related works

Source: Kheni, N.A., Dainty, A.R.J. and Gibb, A.G.F., 2008, Health and safety management in developing countries: A study of construction SMEs in Ghana, *Construction Management and Economics*, 26(11), 46.

are satisfied that the project meets the requirements stipulated within the planning and building regulations of Ghana. Environmental concerns have to be addressed by the client and contractor (Kheni, Dainty & Gibb, 2008).

- Class A contractors are qualified to carry out road works, airports and related works.
- Class B contractors are qualified to undertake bridge construction, culverts and other drainage structures.
- Class C contractors are qualified to carry out labour-based works.
- Class S contractors are qualified to undertake building structures.
- Class M contractors are qualified for miscellaneous road-related works.

The classification of contractors is sub-divided into categories—one to four depending on the number and qualifications of the contractor's permanent staff, equipment/machinery holding, previous experience and financial status (Kheni et al., 2008). For instance, a contractor can be designated as A1B1 or A2B2. The MWRWH classifies building contractors as belonging to one of the classes D1 through to D4, depending on the financial standing of the contractor, equipment holding and qualification and number of permanent employees. According to Frimpong and Kwasi (2013:121), 'the building contractors in Ghana are classified into four groups based on projects worth up to $75,000 (D4K4); projects ranging from $75,000–250,000 (D3K4); projects worth $250,000–500,000 (D2K2); and projects over $500,000 (D1K1)'.

7.4 ECONOMIC CONTRIBUTION OF THE GHANAIAN CONSTRUCTION SECTOR

The construction industry of Ghana contributes immensely to the economy with a high GDP. It contributes to the overall industrial development of the country and achieved a good loan portfolio for the development of the sector. The main drivers of Ghana's economy are Deposit Money Bank's credit to the private sector, Social Security and National Insurance Trust contributions, and port activities. Figures from the Bank of Ghana for the year 2014, for instance, indicate that the GDP growth rate was pegged at 4.2 percent. The inflation rate for the year ending 2014 stood at 17 percent. The Bank of Ghana's classical statistic projections only define the financial dynamism to determine the well-being of the economy and its business environment (Senzu, 2015).

7.4.1 Construction Sector's Share to Overall Gross Domestic Product

The construction sector's share in the overall GDP has improved significantly over the past year (Senzu, 2015). A report from the Ghana statistical service shows that the 2015 estimate of GDP showed a growth of 4.1 percent over the 2014 estimates. The industry sector recorded the highest growth of 9.1 percent, followed by services (4.7%) and agriculture (0.0%). Services remain the largest sector. Its share of GDP increased from 51.9 percent in 2014 to 54.1 percent in 2015. The sector's growth rate, however, decreased from 5.6 percent in 2014 to 4.7 percent in 2015. However, the construction sector had the largest activity within the industry sector with a growth of 30.6 percent and a share of 14.8 percent of nominal GDP. The share of GDP for the industry sector inched up slightly from 26.6 percent in 2014 to 26.9 percent in 2015 (Ghana Statistical Service [GSS], 2015).

7.4.2 Construction Sector's Contribution to the Overall Industrial Development

The construction sector's contribution to the overall industrial development grew from 29.8 percent in 1993 to 30.9 percent in 1995. The construction sector's share of the total industrial output picked up again from 33.9 percent in 1999 to 34.3 percent in 2000. The sector's share of the industrial sector's output reached 36.3 percent in 2005, up again from 35.6 percent registered in 2004. The construction sector's contribution further improved from 36.9 percent in 2010 to 37.4 percent in 2011, which compares favourably with the 1993 to 2011 period average of 35.9. This remarkable performance was against the backdrop of an expanded credit to the sector by the domestic money banks (Senzu, 2015).

7.4.3 Aggregate Loan Portfolio to the Construction Sector

The aggregate loan portfolio to the construction sector has been encouraging in the past years. The share of private sector credit to the construction sector continued to surge, improving by 24.9 percent to 751.64 million in 2011 from 601.82 million recorded in 2010 (Senzu, 2015). The construction sector's ability to access loans from the banks decreased significantly due to the strategic measures put in place by the banks. However, the banking sector credit facility remained tight with net loans and advances of GH¢28.1 billion as at the end of July 2016 recording a lower annual growth of 12.1 percent, compared with 24.1 percent growth in July 2015. In this regard, the growth in banks' investment portfolios (bills and securities), however, picked up by 47.3 percent to GH¢16.1 billion by the end of July 2016 compared with the 13.6 percent growth a year ago.

The changes in the banking sector credit facility have put more pressure on the construction sector's development (Bank of Ghana, 2016).

7.4.4 Construction Sector Component Output Index

The construction sector component output index with the growth dynamics within the sector and the upward-trending index suggest that the sector's performance was remarkable over the 1993 to 2011 period and would continue to remain one of the key drivers of growth in the economy, as infrastructure development remains a pivotal growth pole and paramount for promoting economic growth. The index recorded an average growth of 0.0033 percent for the 1993 to 2011 period. The index suggested 0.002 percent growth on a year-on-year basis at the end of 2012 compared with a negative growth of 0.001 registered in the corresponding period of 2011 (Senzu, 2015).

7.5 OCCUPATIONAL HEALTH AND SAFETY LEGISLATURE IN GHANA

Occupational health and safety legislation is a means by which the work environment can be controlled to ensure the safety, health and welfare of employees and persons likely to be adversely affected by the work environment. In Ghana, OHS legislation has been inherited from a British legal and institutional framework at the time when Ghana was a British dependency. For instance, the H&S of workers in the mining and wood processing industries of Ghana prior to independence was protected by the Factories Ordinance 1952. This remained the main OHS legislation in force until its repeal by the Factories, Offices, and Shops Act 1970. Regulations made under the Factories Ordinance 1952 that remained enforced include the following (Kheni & Braimah, 2014:25):

- The Factories (Woodworking) Regulations, 1959
- The Food Factories (Welfare) Regulations, 1959
- The Factories (Docks Safety) Regulations, 1960

The Ghanaian OHS legislation is influenced by the International Labour Organization (ILO). Principal ILO conventions relating to OHS which have been ratified by Ghana include the following (Kheni & Braimah, 2014:25):

- Underground Work (Women) Convention 1935 (No. 45)
- Radiation Protection Convention 1960 (No. 115)
- Guarding of Machinery Convention 1963 (No. 119)
- Hygiene (Commerce and Offices) Convention 1964
- Working Environment (Air Pollution, Noise and Vibration) Convention, 1977
- Labour Inspection Convention 1947

Existing OHS legislation in Ghana is fragmented and limited in coverage. Some key economic sectors are not covered by the country's OHS laws. A notable example is the agricultural sector, although it employs over 60 percent of the country's workforce. There is no form of OHS laws regulating the activities of the construction sector. This unfortunate situation can be traced back to colonial rule in the Gold Coast (Ghana), where the colonial government placed more emphasis on labour relations in sectors of the economy where formal employment relations existed. The mining and manufacturing sectors of the economy are examples of such economic sectors (Kheni & Braimah, 2014).

7.6 REGULATION SETTING IN GHANA

The Labour Act No. 651 of 2003 consolidates and updates the various pieces of former legislation, and introduces provisions to reflect ratified ILO conventions. The Labour Act covers all employers and employees except those in strategic positions such as the armed forces, police service, prisons service and the security intelligence agencies. Major provisions of the Labour Act include the establishment of public and private employment centres, protection of the employment relationship, general conditions of employment, employment of persons with disabilities, employment of young persons, employment of women, fair and unfair termination of employment, protection of remuneration, temporary and casual employees, unions, employers' organisations and collective agreements, strikes, establishment of a National Tripartite Committee, forced labour, occupational health and safety, labour inspection and the establishment of the National Labour Commission (Hodges & Baah, 2006).

There is also the Children's Act No. 560 of 1998, which defines a child as a person below the age of 18 years. Sections 12 and 87 prohibit engaging a child in exploitative labour, defined to mean labour depriving the child of his health, education or development. Section 91 of that text defines hazardous work. The National Vocational Training Act No. 351 of 1970 and the National Vocational Training Regulations (Executive Instrument 15) enjoin companies to introduce apprenticeship schemes when there is a technical business attached to the establishment. Employers are therefore obliged to provide training for their employees for the attainment of the level of competence required for the performance of their jobs and to enhance their careers (Hodges & Baah, 2006).

7.6.1 International Labour Organization Conventions Ratified by Ghana

Ghana joined the ILO in 1957 and immediately the Convention People's Party (CPP) government ratified many of the ILO conventions including the 'core' conventions that guarantee workers the right and freedom to form or join unions (Convention No. 87), the right to collective bargaining (Convention No. 98), abolition of forced labour (Conventions Nos. 29 and 105), and equal treatment (Conventions Nos. 100 and 111). Many other ILO conventions that sought to promote industrial harmony and welfare of workers were also ratified. These included conventions on hours of work in industry, weekly rest, minimum wage fixing, labour inspection, underground work by women, employment service, night work by women, social policy, working environment, child labour and labour administration. Ghana has so far ratified 46 ILO conventions (Hodges & Baah, 2006).

The Ministry of Manpower Development, Youth and Employment (MMDYE) is the executive body responsible for the formulation and implementation of labour laws, policies, regulations and conventions of industrial relations as well as the monitoring and evaluation of such policies and programmes. It is also responsible for the implementation of labour market programmes in collaboration with other stakeholders in the sector. The ministry used to facilitate mediation and conciliation between employees and employers in conflict situations; this role is now vested in the National Labour Congress (NLC) under the new Labour Act. The ministry is structured around four departments having separate labour administration responsibilities (Labour, Social Welfare, Cooperatives and the Factories Inspectorate). The Labour Department has 62 public employment centres throughout the country, and an Employment Information Bureau, which collates statistics on the employed and unemployed through registration, including monthly data from the employment centres.

Sections 122 to 126 of the Labour Act outline the powers and duties of labour inspection, in particular to ensure the application of the act itself (bringing violations to the notice of the Labour Department or the National Labour Commission) and providing technical assistance and advice to employers and workers in effectively complying with the act's provisions. Under the part of the Labour Act relating to occupational safety and health, section 119 entitles workers to remove themselves from exposure to imminent hazards, without risk of termination. Section 120 requires employers to report not later than seven days from the occurrence of occupational accidents or diseases occurring at the workplace. Section 124(6) lays down penalties on employers for non-compliance with a decision or order of the Minister for Labour or one of the inspectors, in the form of a fine and compensation to any person who proves that he or she suffered loss, damage or injury as a result of the non-compliance (Hodges & Baah, 2006).

7.7 HEALTH AND SAFETY LEGISLATION RELEVANT TO THE GHANAIAN CONSTRUCTION INDUSTRY

There are no H&S regulations developed specifically for the Ghanaian construction industry. Considering the high-risk nature of the sector, this limitation seriously handicaps the implementation of H&S standards on construction sites. Although the Factories, Offices, and Shops Act 1970 caters to factories, offices, shops, ports, and construction, the act provides for the Minister of Manpower, Development and Labour to make regulations in respect of construction works to address specific hazards, including imposing duties on persons in respect of the hazards. Section 57 of the act relates to building and civil engineering works. Other sections relevant to building and civil engineering works are specified in section 57(1) of the act. Under the act, construction companies are required to register their sites (sections 6–8) and to report workplace accidents and dangerous occurrences to the Factory Inspectorate Department. It also requires them to provide wholesome drinking water on their sites (20), toilet facilities on their sites (19) and personal protective equipment for their workers (25), and to take preventive measures to control or prevent specific hazards on sites. The hazards named are noise, vibrations, manual handling (26 and 27) and fire (31). The act also requires medical supervision of the health of employees where necessary (Kheni & Braimah, 2014).

Businesses are required to take measures at the workplace in respect of access and egress to the factory (site), and the construction and design of structures to ensure the safety of workers and users of facilities (33–35). Fencing and safeguards are required to be provided or constructed and maintained for the safety of persons at the factory (site) (38–40). Records of lifting machines and appliances are required to be to be kept, and they must be of sound construction, properly maintained and precautionary measures taken during their operation (37 and 43–47). Construction companies are required to take precautionary measures to prevent injury and explosions because of dust, gas and vapour present in the work environment (48 and 49). Steam boilers, receivers and containers, and air receivers are required to be of sound construction, properly maintained and precautionary measures taken to ensure their operation (50). The act provides for the training of machine operators and persons employed in processes likely to cause injury (36). The minister may make regulations to protect the health, safety, and welfare of workers (30 and 51). Other sections of the act which relate to construction works include the following:

- Sections 52 to 54 set out the authority of inspectors in ensuring health, safety and welfare of persons at workplaces and the role the courts play in such matters.
- Sections 60 to 73 set out the offences under the act and legal proceedings.
- Sections 74 to77 relate to the administration of the act.
- Sections 78 to 87 relate to general matters.

Part XV of the Labour Act 2003 (Act 651) concerns the H&S and environment of workplaces. Under this act, it is every employer's duty to ensure employees work under satisfactory, healthy and safe conditions. Other sections of the Labour Act which impact on H&S include protection of employment relationship, general conditions of employment, protection of remuneration, unions, employers' organisations and collective bargaining agreements, the National Tripartite Committee and labour inspection. The Workmen's Compensation Act 1987 imposes liability on the employer to pay compensation to employees incapacitated by accidents arising out of and in the course of their employment. Compensation payment to accident victims is independent of negligence on the part of employer or fellow worker. The employer is also required to bear the hospital expenses of the injured worker. In cases where the injured worker only requires treatment, he or she is entitled to his or her earnings while undergoing treatment for injuries he or she sustained through an accident arising out of and in the course of his or her employment. There are exceptions to employers' liability to pay compensation. These exceptions are where the injury is due to the workman having been under the influence of intoxicating liquor or drugs at the time of the accident, or where the injury was deliberately self-inflicted, or where the workman knowingly misrepresented to the employer that he was not suffering or had not previously suffered from that or a similar injury. The law applies to persons employed by both public and private organisations. The act sets out modalities for the calculation of the earnings of workers and payments of compensations to workers who sustain injuries (Kheni & Braimah, 2014:27).

7.8 HEALTH AND SAFETY POLICY TRENDS IN GHANA

Workplace H&S laws of Ghana look set to fall in line with the rest of the world. The African country by its construction industry's admission has suffered from unsafe construction sites in what is traditionally known as the most accident-prone industry. This is because they are not in touch with modern practices (IOSH, 2013). To bring Ghana in line with international standards, there should be an improvement to ensure that workplaces stick to best practices so that staff and persons who visit construction sites, factories and other working environments are not unduly exposed to injuries or accidents. Ghana has not done well in ensuring the H&S of construction sites owing to laws that are not in tune with modern practices and are hard to enforce.

7.8.1 Current Trends of Occupational Health and Safety in Ghana

A report by the Ghana News Agency (GNA) in the year 2010 indicated that there are currently two major edicts that provide guidance in the provision of occupational or industrial safety and health services, practice and management in Ghana. These include the Factories, Offices and Shops Act 1970 (Act 328) and the Mining Regulations 1970 (LI 665), but these have only been driven by the mining and the labour sectors. It was argued that their scope is limited owing to the availability of several numbers of industries operating in Ghana. The presence of the Workmen's Compensation Law 1987 (PNDC 187) has direct impacts on monitoring worker or workplace safety. The Radiation Protection Board of the Ghana Atomic Energy Commission is also proactive in monitoring companies with radiation exposure hazards for compliance. However, owing to limited resources, the effectiveness of their activities is compromised. On a proactive side, the Ghana Chamber of Mines, in collaboration with the Inspectorate Division of the Minerals Commission, formed a Technical Committee with representations from each mining company in the nation that reviews and recommends corrective actions for reported or identified unsafe acts, conditions or failures in the existing H&S system of the mining industry. This good initiative is, however, impeded by the availability of resources, and hence enforcement is challenged. Other statutes that indirectly impact on OHS include the Environmental Protection Agency Act 490 1994, the Ghana Health Service and Teaching Hospital Act 526 1999 and the National Road Safety Commission Act 567 1999 (GNA, 2010).

In the GNA report in the year 2010, it was further admitted that Ghana is among the 183 member countries of the ILO, which requires, as per the ILO convention number 155 1981, that member countries formulate, implement and periodically review a coherent policy on OHS and work environment. Ghana has not yet rectified this convention and the nation has no established authority dedicated to OHS to guide and facilitate the implementation of the 'Action at the National Level' as indicated in the R164 OHS Recommendation, 1981 (GNA, 2010). However, the Labour Act 2003, Act 651, Part XV, sections 118 to 120 apparently directs employers and employees in their roles and responsibilities in managing occupational health, safety and environment in the nation. However, it has not been stated in the acts whom to report to on issues of accidents and occupational illnesses. It is not clear what to consider as an occupational illness and who handles the implementation of corrective actions as per recommendations. Currently, accidents that occur in factories are expected to be reported to the Department of Factory Inspectorate (DFI), but companies hardly report such events to the inspectorate for investigation and correction. When these accidents are reported, it takes a long time before corrective or preventive actions are implemented. Hence, there is a little or no positive effect of the action of the DFI on the factories. The report confirms some positive safety and health practice influence among some of the companies in Ghana due to the influx of some multinational companies into the country, given their corporate expectations with specific requirements in occupational safety and health practices. This stems from their requirements for the contractors and sub-contractors, some of whom are Ghanaian, to follow their H&S standards (GNA, 2010).

Currently, the oil and gas sector has introduced its side of the approach to managing H&S. This is purely based on risks, and it is an improvement on what is in existence. In as much as this is a good effort and helps the Ghanaian construction sector to know there is more to OSH than specified in the legal framework, it tends to confuse the Ghanaian construction sector with regard to which standard to follow and what is required to make employees and employers accountable.

7.8.2 Approach to Occupational Health and Safety Management in Ghana

7.8.2.1 National Policy

- The nation has to adopt or develop a broad base OHS policy that is in line with ILO Convention 155 as a minimum. This must seek to address safety and health issues regarding all projects and operations from the design stage, through procurement, construction, and operation and decommissioning. The aim of this must seek to first protect the worker from injuries and work related ill-health, ensure standards are put in place to prevent loss of properties and damages due to accidents, and must show the commitment of the Government of Ghana.
- Achieving this means all the scattered generic OHS requirements under the different agencies of the Ghana Government such as the Environmental Protection Agency, the Department of Factory Inspectorate, the Inspectorate Division of the Ghana Minerals Commission and the Ghana Labour Commission with confusing responsibilities must be brought under a common umbrella body. Such a body must be empowered and resourced adequately to enable it to organise how the policy would be implemented nationwide under the responsibility of one body or person.
- This policy must be authenticated by the Ghana Government and form part of the nation's legal document.
- Consultation effectiveness with relevant organizations will positively impact the successful implementation of the 'Expected Ghana National Occupational Safety and Health Policy'. The relevant organizations may include but are not limited to the Ghana Minerals Commission, the Ghana Chamber of Mines, the Ghana National Petroleum Corporation, the Association of Ghana Industries, the universities, the Department of Factory Inspectorate, the Ghana Institute of Engineers, the Ghana Medical Association, the Ghana Bar Association and the Ghana Environmental Protection Agency.
- Regarding the outcomes of consultations and requirements of the OHS, the policy is salient in achieving a good control of the system. This can be done through the National Media Commission and the associated private and public media operatives.

- Competency of the human resources who would be championing the implementation and monitoring of the OHS policy must be built up. This may require introducing relevant courses such as safety engineering in our universities and polytechnics as well as occupational health in the medical schools. This training institution would need to be adequately resourced and accredited to deliver quality education in OHS for Ghanaians to be able to manage OHS issues in the nation.
- Between the industry, the public and the established body responsible for the implementation, management and monitoring of the policy is paramount.
- Control of the practices with guidelines either adopted or developed by the nation's OHS body needs specific emphasis. Specific roles and accountabilities with timelines need to be developed for planning and implementation of the OHS policy actions with clear reporting lines. This must not exclude penalties for intentional non-conformances and negligence (GNA, 2010).

7.8.2.2 Planning and Implementation

Actions that need to be put in place to ensure the achievement of the Ghanaian national OSH policy aim and objectives would need to be clearly spelt out with specific timelines, roles and responsibilities. Targets must be set by the nation for the various industries, and exceeding them should merit controls set by the government. To guide industries to achieve those targets, there have to be guidelines in areas such as emergency preparedness, hazardous material management, risk assessments, accident reporting and investigation, workplace inspections, workplace exposures monitoring, assessment and control, purchasing and supply chain policies and permit-to-work systems. There is also the need to ensure that the governing organization is empowered and resourced to proceed with continual research into workplace exposures, levels that should not be exceeded, safe ways of completing the tasks, and improvement of controls when undesired events, conditions or systems are encountered. These must tie into the roles and responsibilities as indicated under the organisation and be very specific.

7.9 LESSONS LEARNT FROM THE GHANAIAN CONSTRUCTION INDUSTRY

The following are the lessons learnt from the study of the Ghanaian construction industry and occupational health and safety. The Factories, Offices and Shops Act 1970 (Act 328) and the Mining Regulations 1970 (LI 665) are two of the major edicts that provide guidance in the provision of occupational or industrial safety and health services, practice and management in Ghana. The OHS legislation is protected by the Factories Ordinance 1952. This includes the Factories (Woodworking) Regulations 1959, the Food Factories (Welfare) Regulations 1959 and the Factories (Docks Safety) Regulations 1960. Some key International Labour Organization conventions have been ratified in Ghana, but the agriculture sector has not been covered by the existing OHS legislation whilst the construction industry lacks specific OHS regulations.

7.10 SUMMARY

The construction industry worldwide plays a significant role in the economic growth of many countries. The construction industry in Ghana faces many challenges. These include a weak regulatory and development framework, as well as financial, human resource and material constraints. Non-compliance with H&S rules and regulations in Ghana is a major issue currently facing the construction industry. The non-compliance with H&S has led to injuries and accidents. This has been a major blow to most of the institutions in Ghana because there is no standard degree meant for OHS for the construction industry in Ghana. This calls for proper development of H&S policy for the construction industry in Ghana. However, since there is no specific OHS policy for the Ghanaian construction industry, ILO Convention 155 should be adopted as the OHS policy in Ghana.

REFERENCES

Bank of Ghana. (2016). Financial stability report, 1(2), September. Available from: https://www.bog.gov.gh [Accessed 22 February 2017].

Dadzie, J. (2013). Perspectives of consultants on health and safety provisions in the Labour Act: A study into theory and practicals. *Engineering Management Research*, 2(1): 34–42.

Frimpong, S.K. and Kwasi, O.O. (2013). Analyzing the risk of investment in the construction industry of Ghana. *European Journal of Business and Management*, 5(2): 121–130.

Ghana News Agency (GNA). (2010). Occupational & industrial safety & health in Ghana. *Ghana Times*, p. 13, 22 November.

Ghana Statistical Service. (2015). Statistics for development and progress, Annual Gross Domestic Product. Available from: www.stasghana.gov.gh [Accessed 21 February 2017].

Hodges, J. and Baah, A. (2006). National labour law profile: Ghana. International Labour Organisation (ILO).

Institution of Occupational Safety and Health (IOSH). (2013). Ghana steps into health and safety. Available from: http://www.iosh.co.uk [Accessed 10 April 2014].

Institute of Statistical, Social and Economic Research (ISSER). (2008). The state of the Ghanaian economy in 2007. Legon: Institute of Statistical, Social and Economic Research.

Kheni, N.A. and Braimah, C. (2014). Institutional and regulatory frameworks for health and safety administration: Study of the construction industry of Ghana. *International Refereed Journal of Engineering and Science (IRJES)*, (3)2: 24–34. Available from: http://www.irjes.org [Accessed 04 July 2015].

Kheni, N.A., Dainty, A.R.J. and Gibb, A.G.F. (2008). Health and safety management in developing countries: A study of construction SMEs in Ghana. *Construction Management and Economics*, 26(11). Available from https://dspace.lboro.ac.uk [Accessed 10 April 2015].

Senzu, T. (2015). The proof that government of Ghana is insensitive to SMEs development. BGI-Scholastic Article 78. Available from://www.modernghana.com [Accessed 22 March 2016].

Vibe Ghana. (2014). Towards establishing the Construction Industry Development Authority in Ghana. Available from: http://vibeghana.com/2014/0 [Accessed 28 August 2014].

Section VI

Research Methodology

8 Research Methodology

8.1 INTRODUCTION

Some studies exist on health and safety (H&S) in the construction industry at the international level, but few studies exist in Ghana. Examples are perspectives of consultants on H&S provisions in the Labour Act. But little has been written on the compliance of H&S in the construction industry in general. Second, the work of Puplampu and Quartey (2012) has pointed out the low level of the ratification of International Labour Organization (ILO) conventions that address occupational health and safety (OHS) in Ghana. There is also a lack of comprehensive international OHS policy framework; inadequate resources allocated to OHS researchers; and ineffective OHS inspection, training and education; and OHS capacity building and monitoring in the Ghanaian construction industry.

However, there is the need for serious attention to be given to the identified compliance to OHS gaps to ensure effective OHS management. This is because non-compliance with H&S leads to accidents, and workplace accidents that have the potential to absorb 30 percent of company annual profits. The failure to manage safety also has a much larger social cost. The aforementioned scenario, coupled with the fact that the construction industry does not have an enviable record on H&S compliance, motivated the writing of this book. This chapter provides details about the methodological research framework for the book. The chapter consists of the following sections: research design and methodology and the quantitative and qualitative approaches used in achieving the study objective. The research design and methodology section focuses on the research procedures, the choice of research methods and the selection of participants. The use of a mixed-methods approach is included for both philosophical and practical reasons. These have been explained in detail to justify the mixed-methods approach for this book.

8.2 QUANTITATIVE VERSUS QUALITATIVE RESEARCH METHODOLOGY

Qualitative research is collecting, analysing and interpreting data by observing what people do and say, whereas, quantitative research refers to counts and measures of things. Quantitative research options are predetermined; involve a large number of respondents; and measurements must be objective, quantitative and statistically valid. Likewise, under this approach, statisticians use formulas to calculate the sample size to determine how large a sample size will be needed from a given population in order to achieve findings with an acceptable degree of accuracy (Anderson, 2006). Qualitative research is much more subjective than quantitative research and uses very different methods of collecting information, mainly individual, in-depth interviews and focus groups. The nature of this type of research is exploratory and open-ended.

A small number of people are interviewed in-depth and a relatively small number of focus groups. Participants are asked to respond to general questions. The interviewer or group moderator probes and explores the responses to identify and define people's perceptions, opinions and feelings about the topic or idea being discussed and to determine the degree of agreement that exists in the group. The quality of the finding from qualitative research is directly dependent upon the skills, experience and sensitivity of the interviewer or group moderator. This research approach is extremely effective in acquiring information about people's communication, needs and their responses to and views about the specific question.

Both quantitative and qualitative research methodologies are based on the epistemological assumptions regarding the nature of knowledge and the methods of abstracting that knowledge, as well as ontological assumptions, which relate to the nature of reality or the phenomena being investigated (Kayitsinga, 1992). The philosophical considerations that influenced the choice of the research approach for this book are discussed in the subsequent sections of this chapter.

8.3 PHILOSOPHICAL CONSIDERATIONS IN RESEARCH METHODOLOGY

As stated earlier, the choice of research methodology is usually influenced by a set of assumptions underlying each research methodology. The choice of a particular method has to be supported by the assumptions that have been brought into the research process and are reflected in the methodology (Crotty, 1998). These assumptions, though varied, tend to fall into the philosophical areas of ontology and epistemology. A brief discussion of these considerations follows.

8.3.1 Ontology

Ontology is concerned with assumptions about the variety of phenomena in the world. It refers to the claims that a particular paradigm makes about reality or truth (Hitchcock & Hughes, 1989). Ontology is about what exists, what it looks like, what components make it up and how the components interact with each other. Likewise, as with epistemology, these issues can sometimes have a major impact on methodology. Hence, any contrasting ontology of human beings in turn can sometimes demand different research methods (Burrell & Morgan, 1994).

Ontological assumptions revolve around the question of what is with the nature of reality (Crotty, 1998). In other words, it is an effort to elucidate what reality is and why things happen the way they do. In an attempt to explain reality, Kayitsinga (1992) advocates two opposite assumptions of reality: objectivity and subjectivity. Kayitsinga viewed the objectivist stance as reality existing out there, intact and tangible, but independent of individuals' appreciation and cognition (Kayitsinga, 1992; Crotty, 1998). Thus, regardless of whether individuals perceive and attach meaning to this reality, it remains unchanged (Burrell & Morgan, 1994). Hence an individual is thus 'born into and lives within the social world that has its own reality, which cannot be created by that individual' (Burrell & Morgan, 1994:4). Thus, in order to create a better understanding of reality, objectivist researchers propose the need to study the causal relationships among the elements constituting reality (Kayitsinga, 1992; Burrell & Morgan, 1994), which is advanced in the current book.

Crotty (1998) was of the belief that the objectivist view of reality is closely related to a theoretical position called positivism. Positivists postulate that the world exists as a system of observable variables waiting to be discovered (Maguire, 1987). Similarly, positivists believe that the use of scientific methods of inquiry can assist in discovering the true meaning of reality (Maguire, 1987; Crotty, 1998). The results of such investigation generate rules and theories that help to explain and sometimes provide a guide for understanding social behaviour (Maguire, 1987). The current research applied scientific methods in the development of the H&S compliance model. Also it aims to bring about the social construct that will lead to health and safety compliance in the Ghanaian construction industry. The subjectivist interpretation is that reality is not discovered, but it is constructed by human beings as they engage with the world in which they live (Crotty, 1998). In that way understanding and interpretation of reality occurs when human beings interact with their environment and others, and assign meaning to the world around them (Crotty, 1998). Thus, in research, meaning of anything is 'an expression of the manner in which the researcher as a human being has arbitrarily imposed a personal frame of reference on the world' (Kayitsinga, 1992:89). Hence, the next section of this chapter discusses how reality or knowledge is created, as an extension of the discussion of philosophical suppositions that influence researchers' choice of methodology.

8.3.2 Epistemology

Epistemology is the theory of knowledge embedded in the theoretical perspective that informs a particular study (e.g. objectivism, subjectivism). Epistemology is concerned with how phenomena can be made known to the researcher. Epistemology can sometimes also have a major impact on the data collection choices as well as on the methodology in a research process (Walker & Evers, 1988; Hitchcock & Hughes, 1995). Epistemology provides the grounds for the decision on the kind of knowledge that is considered appropriate, adequate and legitimate for the research at hand. Furthermore, research methodology is applied epistemology, thus the methodology has to be supported by an epistemology foundation. Therefore, researchers are expected to point out, explain and justify the epistemology that informs their choice of research methodology. Consequently, the choice of epistemology is widely influenced by the ontological considerations within a particular discipline. Both dimensions of ontology (objective and subjective) play an important role in the epistemology and, ultimately, the methodology chosen for this research. The next sections clearly demonstrate how ontological dimensions (objective and subjective) and epistemological considerations affect the choice of the research methodology for the current study.

8.3.3 Quantitative Methodology

Quantitative research is about quantifying the relationships between variables and finally measuring them. Statistical models are sometimes constructed to explain the observed variables. Certain characteristics are searched for and endeavoured to show something interesting about how they are distributed within a certain population. The nature of a specific research will determine the variables in which the researcher is interested. Variables are most times measured for the purpose of quantitative analysis; data are being collected concerning the variables, perhaps through a questionnaire. The variables the researcher is interested in may be dependent or independent, though there may be other features present in the problem that may be constant or confounding. Hence, the objectivist view of an integral and independent reality encourages researchers most times to adopt the epistemology of positivism.

The principles of a typical positivism view are that:

- Only phenomena that can be observed should be used to validate knowledge.
- Scientific knowledge is arrived at through the accumulation of verified facts derived from systematic observation or record keeping.
- Scientific theories are used to describe patterns of relationships between these facts to establish causal connections between them.
- The process should be neutral and judgment free. Observations should be uncontaminated by the scientist's own prediction. Thus ethical issues can be included only if they are included as part of the research.

The use of a scientifically guided research methodology where the aim is to explain and predict causal relations between elements that constitute reality are the positivist epistemology (Kayitsinga, 1992; Quattrone, 2000). The current research adopted the positivist epistemology as an approach to achieve its objective. Data were collected were quantified and analysed using mathematical formulas and other approaches displaying the success of a positivist research. Statistical rhetoric such as reliability, validity, correlation and cause-and-effect relationships were used in the current research. However, the presentation of research findings under this methodology follows an approach that emphasises explicit, exact, scientific and formal procedures, just as the use of quantitative methodology to explore and explain relationships between variables is advocated by positivist researchers. This argument makes the entire research process to be considered as totally neutral and judgment free with limited room for personal bias.

May (2001) has indicated that the researcher and the research process move together. Their perceptions, expectations, experiences and interpretations become part of the research process. May asserted that 'the relationship between the researcher and the research should be a continuous ebb and flow of information'. Therefore, the researcher's subjectivity is considered an integral part of the research process. Advocates of subjectivity suggest that it is a better option for undertaking research as opposed to objective quantitative methods (Brieschke, 1992). Sarantakos (2005) asserts that scientific research with emphasis on explicit, exact and formal procedures is appropriate for a quantitative methodology. From the given scenarios, a quantitative methodology should be supported by a qualitative methodology. Sections 8.3.5 through 8.3.7 discusses the advantages of combining qualitative and quantitative methodologies in more detail. In the following section the qualitative methodology is discussed and focuses on both the advantages and limitations of qualitative methodology when used alone.

8.3.4 Qualitative Methodology

Qualitative research allows the subjects being studied to give much 'richer' answers to questions put to them by the researcher and valuable insights that might have been missed by any other method. Not only does it provide valuable information to certain research questions in its right, but there is a strong case for using it to complement quantitative research methods. For example, if the area of interest has not been previously investigated, then qualitative research may be a vital forerunner to conducting any quantitative research. For example, it is impossible to carry out a meaningful structured questionnaire survey on health and safety compliance if the important issues to the small to medium-sized enterprises (SMEs) surrounding the provision of that service are not known. At the other extreme, qualitative research may also help the researcher to understand the findings of quantitative research. For example, it is very easy to discover that some contractors fail to keep to the compliance of H&S, but uncovering the reasons for this can be more difficult and conventional surveys may miss some of the important factors that encourage such behaviour. Hence, the subjectivist's view of reality advocates for appreciation of human involvement in the creation and shaping of knowledge (Kayitsinga, 1992). The subjectivist epistemology thus suggests that meaning or reality is not discovered but is rather imposed on the object by the subject, and in a research situation, imposed by the researcher (Crotty, 1998). In other words, with the subjectivist epistemology, the object being studied contributes less to the meaning or reality.

Thus, the researchers' input in the research process is recognised under subjectivism. The research methodology recommended by subjectivists is qualitative methodology. According to Kayitsinga (1992:92), qualitative research is a 'form of social interaction in which the researcher converses with, and learns about the phenomenon being studied'. It is part of the research process and is actively involved in creating the meaning of reality. Qualitative research is suggested as more applicable to the study of people and their environment (social sciences) than natural sciences. Consequently, advocates of qualitative research advance the use of qualitative methodology when studying people, as it enables the researcher to see through the eyes of the researched (Bryman, 2001). Constructing meaning through engagement with people involves interpretation. Interpretivism is the process by which information is extracted through interpretation. Under interpretivism, researchers seek information relating to people's views, opinions, perceptions and interpretations of the social world, which was also partly utilized in the current research. Subjectivism, constructivism and interpretivism form part of a broader list of research methods commonly employed in qualitative research. The qualitative methodology has been criticized for lacking in efficacy owing to its inability to study with a degree of accuracy the relationships between variables (Sarantakos, 2005).

In qualitative research, the researcher is the main player. He or she decides on what to concentrate on during the data collection. Views may vary according to different researchers and it is difficult to replicate and generalize the findings with ease. A small number of cases is studied as compared to large sample sizes common in quantitative ease. The population under the study may not be representative of the majority. However, advocates of qualitative research argue that generalizations are made on the assumption that the findings and inferences made during the research are supported by sound theoretical reasoning. Representation in qualitative research is in accordance with the subject of investigation, which is highly subjective and a narrow-minded view of events and what is being observed. The findings of qualitative research are difficult to subject to rigorous quality verification requirements such as reliability and validity (Creswell, 2003). It would be difficult to prove the validity of qualitative research findings through measurement. Validity requires measurement of the object of enquiry and that is not possible in qualitative research because its purpose is not to measure but to generate ideas (Stenbacka, 2001). On the other hand, reliability is concerned with producing the same result with consistency. This is not possible under qualitative research because of the involvement, influence, subjectiveness and the possibility of bias of the researcher in qualitative research. Qualitative researchers have, however, argued that quality verification using validity and reliability checks is not necessarily applicable to qualitative research because it owes its origin to scientific rhetoric and positivist paradigms common in quantitative research (Stenbacka, 2001; Creswell, 2003).

Nevertheless, both qualitative and quantitative research methodologies can be used in different situations depending on the aims and objectives of the study. Most research is centred on four primary objectives, namely 'exploration, explanation, description and prediction' (Ellram, 1996:98). Research where the objectives are either exploration or explanation, or both, would normally require qualitative research methods. This is because qualitative research has the ability to provide insight and explanation into a phenomenon that was relatively unknown (Ruyter & Scholl, 1998). It provides answers to questions, such as how or why, which are common in exploration and explanation of phenomena (Ellram, 1996). Research that is descriptive or predictive would require quantitative research methods that utilize statistical techniques to predict and describe relationships between variables. The choice between the two methodologies should be based on the aims and objectives of the study, as well as the nature of the study. In some cases, the two methods may be used jointly to cover for the weaknesses inherent in each method (Amaratunga, Baldry, Sarshar & Newton, 2002; Tashakkori & Teddlie, 2003). The combination of qualitative and quantitative methods is discussed in the next section.

8.3.5 Combined Quantitative and Qualitative Methods

Mixed-methods research involves both collecting and analysing quantitative and qualitative data. Quantitative data includes closed-ended information such as that found on attitude, behaviour or performance instruments. The collection of this kind of data might also involve using a closed-ended checklist, on which the researcher checks the behaviours seen. The analysis consists of statistically analyzing scores collected on instruments, checklists or public documents to answer research questions or to test hypotheses. In contrast, qualitative data consists of open-ended information that the researcher gathers through interviews with participants. The general, open-ended questions asked during these interviews allow the participants to supply answers in their own words. Qualitative data may also be collected by observing participants or sites of research, gathering documents from a private or public source, or collecting audio-visual materials such as videotapes or artefacts. The analysis of the qualitative data (words or text or images) typically follows the path of aggregating the words or images into categories of information and presenting the diversity of ideas gathered during data collection. The open- versus closed-ended nature of the data differentiates the two types better than the sources of the data. Researchers have theoretically supported the combination of qualitative and quantitative methods (Uysal & Crompton, 1985; Tashakkori & Teddlie, 1998; Bryman, 2001; Amaratunga et al., 2002; Creswell, 2003). Tashakkori and Teddlie (1998), for instance, asserted that quantitative and qualitative research methods are not dichotomous but rather complement one another to produce improved research findings. Mixed-methods research is commonly used as a strategic research approach that is able '(a) to demonstrate a particular variable will have a predicted relationship with another variable and (b) to answer exploratory questions about how that predicted (or some other related) relationship actually happens' (Tashakkori & Teddlie, 2003:15).

However, those advocating for the use of combined methods rejected the choice between positivism and constructivism, as none of the methods works best in isolation. The use of combined methods, often called mixed methods, has been found to alleviate the weaknesses linked with using either of the methods on their own (Tashakkori & Teddlie, 1998; Bryman, 2001; Amaratunga et al., 2002; Mangan, Lalwani & Gardner, 2004). For example, Bryman (2001:450) suggests that 'in some instances neither qualitative nor quantitative research methods may be adequate on their own, thus researchers cannot rely on just one method and have to use both to support the research process'. Quantitative and qualitative methods supplement each other by providing richness and details that are otherwise unavailable if each method were pursued separately (Jack & Raturi, 2006). Combining the methods provides a multidimensional insight into the research problem, and thus assists in gaining a broader understanding as well as a true analysis of the situation at hand (Mangan et al., 2004), which is also one of the strong points of consideration for the current research. The use of combined methods compensates for the weakness embedded in each of the research method by 'counter balancing the strengths of another' (Amaratunga et al., 2002).

The combined methodology approach improves the ability of researchers to draw conclusions from their studies, thereby resulting in more robust and generalizable research findings. Therefore, the current study, in order to achieve its objective, adopted a mixed-methods methodology in order to counterbalance the strengths and weakness embedded in each of the research methods when used separately. Further details on the justification and how quantitative and qualitative methods were used to collect data in this book are provided later in this chapter.

8.3.6 Mixed-Methods Approach

Mixed-methods research involves the use of more than one approach or method of design, data collection or data analysis within a single programme of study, with integration of the different approaches or methods occurring during the programme of study and not just at its concluding point (Johnson, Onwuegbuzie & Turner, 2007). Mixed-methods research is a research methodology with philosophical assumptions as well as methods of inquiry. As a methodology, it involves philosophical assumptions that guide the direction of the collection and analysis of data and the mixture of qualitative and quantitative approaches in many phases in the research process. As a method, it focuses on collecting, analysing and mixing both quantitative and qualitative data in a single study or series of studies. Its central premise is that the use of quantitative and qualitative approaches in combination provides a better understanding of research problems than either approach alone (Creswell, Clark, Gutmann & Hanson, 2007). The mixed-methods approach was adopted in this book based on the philosophical and practical reasons discussed earlier. A quantitative survey provides a snapshot of phenomena while qualitative methods or interviews from the Delphi experts (Delphi study), as adopted in this book, provided contextual information and human subjective information to interpret and inform the quantitative results.

Creswell et al. (2007) identified six commonly used designs in mixed-methods research. The present study uses two of them: sequential explanatory and concurrent triangulation design. The quantitative survey is the main driver of this study, complemented by the qualitative study. The use of both methods provides a richer understanding of phenomena and an explanatory account of triangulation and illuminates significant survey findings in what Teddlie and Tashakkori (2009) referred to as crossover track analysis. Although the quantitative and qualitative studies are independent, both sets of data and analyses are used in analysis. The survey (quantitative) examination of the relations and associations between the key variables and the Delphi study (qualitative) are presented in the upcoming chapters.

Both qualitative and quantitative methods were used in the current study to identify the factors that are associated with health and safety compliance practices amongst small to medium-sized construction companies. It also indicates the statistical significance of these factors in determining health and safety compliance. The current study also looked into the relationship between the identified factors and health and safety compliance to be tested (predicted). The impact of the identified independent variables on the health and safety compliance was also sought in this book. Means of exploring the identified factors by qualitative study using a Delphi technique and through literature study of health and safety compliance was employed. The quantitative research approach alone would not have provided more or detailed information on the book objective. The use of the qualitative method was employed to explore and gain a comprehensive understanding of the way the selected factors had an influence on health and safety compliance. The quantitative method was further used to verify the results in the survey. The factors that bring about health and safety compliance in small to medium-sized construction companies were obtained through the use of the mixed-methods approach. The mixed-methods approach confirms the findings of both the quantitative and qualitative approaches. This implies that the findings of the study will be useful in the establishment of the factors that should be considered for the development of health and safety regulation/compliance of the aforementioned companies. The current study adopted the qualitative method, structured (using an interview schedule) through the use of the Delphi technique, discussed later in this chapter. Findings from the Delphi survey were used to refine the survey tool (structured questionnaire) for the study and to validate the findings. The Delphi findings were used to resolve issues on health and safety compliance of small to medium-sized construction companies through consensus. The quantitative method of data collection for the study was the survey method with the use of a structured questionnaire. The analysis was done using structural equation modelling (SEM) with EQS, version 6.2, using SPSS in the development and validation of the health and safety compliance model for small to medium-sized construction companies.

8.3.7 Justification of the Mixed-Methods Approach

Both quantitative and qualitative methods have their strengths and weaknesses. Quantitative methods have been criticized for being 'sanitized and lacking in contextual realism' (Tashakkori & Teddlie, 2003). Qualitative methods are suitable for addressing questions of how and why things occur, whereas quantitative methods are more appropriate for answering what and how questions. The use of one method is not appropriate in studying the variables that predict health and safety compliance amongst small to medium-sized construction companies and is not enough to explore inputs from the various stakeholders (identified in the qualitative-Delphi study as experts). The use of the mixed-methods approach that integrated qualitative and quantitative methods was required. Tashakkori and Teddlie (2003) posited that one of the merits of a mixed-methods approach in the current study is that the techniques of the qualitative and quantitative domains are interwoven to maximize the knowledge yield of the research endeavour by adequately predicting the variables that should be given consideration by small to medium-sized construction companies in their quest to comply with H&S regulations. The mixed-methods approach was used by the researchers to discover and justify the model components within one study.

Qualitative research involves people in order to provide the realism and detail needed for the generation and building of theory (Tashakkori & Teddlie, 2003). Rich data were collected through the use of qualitative techniques in order to meet the objective of this study. The language and context of the stakeholders and the people being studied were captured during the questionnaire survey to collect the required data for the study. The qualitative data were gathered in both the first and second stage, and the data analysis approach was aligned with the positivist paradigm. The positivist paradigm sought to identify patterns and repetition within each key research issue, and also explored the level of impact, influence and agreement through the use of scales. The mixed-methods approach was adopted to answer questions that would not have been possible to answer by the qualitative or quantitative approach alone. The method was very practical because the researchers were free to use all methods possible. The generalisability of the research result was raised through the use of the mixed-methods approach, which was a major consideration in the present work. It also provided stronger evidence for a conclusion through convergence and verification of findings, thus ensuring that added insights and understanding that would have been missed if only a single method was used.

It becomes natural for individuals who intend to solve problems using both numbers and words (a combination of inductive and deductive thinking) to employ mixed-methods research as the preferred mode of understanding the thesis statement. The debate on health and safety compliance (the focus of the research) is more natural, psychological and persuasive than either words or numbers can adequately represent. This is because words, pictures and narratives can be used to add meaning to numbers, and numbers can be used to add precision to words, pictures and narrative. The Delphi technique was combined with the survey method in the current research to provide the basis for the validation of the conceptual framework for the development of a holistic health and safety compliance model for small to medium-sized construction companies.

8.4 RESEARCH DESIGN

A research design gives a detailed outline of how an investigation is to be carried out. It typically includes how data were collected, what instruments were used, how the instruments were used and the intended means for analysing data collected. The research design is a set of framework for conducting research. It assists researchers to conduct studies successfully. Usually, the research design is used to justify decisions and choices relating to the research procedure (Sekaran, 2000). This section provides the outline of this study's research design. The appropriate methodology used in this book is based on the ontological and epistemological assumptions as described earlier. The choice of research design is mostly influenced by the methodology (whether quantitative or qualitative) as well as the philosophical assumptions guiding the research process (ontology and epistemology). Objectivist ontology influences a researcher to adopt a more positivist epistemology. This gives emphasis to the use of quantitative methods in the research process (Sarantakos, 2005) such as the constructivist ontology, which culminates in qualitative methodology. The research design for this study was fixed in line with the requirements of objectivism, which is in favour of a scientific way of abstracting data when the objectivist ontology is adopted. The instruments used in collecting data were also determined by the research design. In quantitative design, the survey method is used to collect the data. The philosophical underpinnings (epistemology and ontology) provide a guide to the methodology followed in a research process. A decision on the specific research design can be hypothesised in the form of a connection beginning from the philosophical underpinnings (epistemology and ontology).

Following the decision on the methodology, researchers have to decide on the research design guided by the research questions and aims. The research design influences the choice of instruments to use in the execution of the research process. The exact justifications for a research design should follow five aspects: research purpose, theoretical framework, research questions, research methods and sampling strategy, appropriately interconnected. The current study follows these aspects of research design. Hence, the choice of research methods for the current study was influenced by the research aim, questions and objectives.

There were three considerations for the selection of the research methods to answer the predetermined set of goals in order to meet the objective of this book. The selected research methods were required

- To identify the variety of factors associated with health and safety compliance
- To be able to predict the relationship between each of the identified factors and on how they can predict health and safety compliance
- To provide in-depth information to be collected and analysed to show how contractors know the identified factors as important (influential) in determining health and safety compliance.

The current study adopted the mixed-methods research (quantitative and qualitative combined) approach as already stated, discussed and justified in the preceding sections. The mixed-methods approach was adopted to answer the research questions meeting the research objectives. Thus, a health and safety compliance model is developed for contractor health and safety compliance for small to medium-sized construction companies. The following strategies were adopted to meet the research objectives for this book:

- The first general objective was to establish the factors that determine health and safety compliance. A literature review was conducted concerning the factors that determine health and safety practices. Published articles, development reports and status reports were reviewed. The expected outcome from this objective was information and a global picture of the determinants of health and safety practices. Both international and national literature were reviewed. The review gave a general picture which was useful for the reader to understand how health and safety compliance is formed and its relevance for small to medium-sized construction companies.
- The second general objective was to establish the current theories and literature on health and safety practices, and to identify the gaps that needed consideration. The constructs that were established were included as part of the theory for the development of the holistic health and safety compliance model. A literature review was conducted from a wide source of publications including journals, conference proceedings, books and monographs. Specific theories of health and safety studies were consulted. The second general objective was also expected to provide information on the current theories on health and safety studies to determine the gaps which other scholars have not yet addressed as well as common themes, the types of methodologies that have been used in the research and how terms had been defined.
- The Delphi method was used to achieve the third and fourth objectives, which were to determine the main and sub-attribute(s) that bring about health and safety compliance and to examine the similarities in attributes that determine compliance. A further objective was to evaluate the factors and issues that affect the compliance of health and safety in small to

medium-sized construction companies since experts' views were required on the factors that determine health and safety compliance. The Delphi method was the best to use in this instance. Apart from experimental procedures that were not feasible for this study, the Delphi method or focus groups could be used.

Focus groups could have been used except that there was the challenge of assembling all identified experts to deliberate for 8 hours for a minimum of three days. The focus group would also be too expensive and beyond the budget for the fieldwork. This might have defeated the purpose of conducting a rigorous process to achieve the objective. There was no bias from the experts because they remain completely anonymous to one another and therefore, there was no undue influence from their peers. This is not the case for a focus group. A detailed explanation of the Delphi technique is given later in the chapter in order to convey how the Delphi method was conducted. The expected output was an estimation of the extent to which health and safety compliance by small to medium-sized companies is influenced by the established factors. Another expected outcome was the consensus reached on the critical factors and issues that affect the non-compliance of health and safety compliance of small to medium-sized companies based on the Ghanaian construction industry.

The conceptual model was developed for the health and safety compliance from the identified factors and their interrelationships.

- The fifth general objective of the book was achieved by drawing on the conclusions from the extensive literature review and the results and findings from the qualitative Delphi study.
- An empirical questionnaire survey was conducted and analysed using SEM to achieve the sixth objective of the research. The sixth objective was to test and validate the conceptual model developed from the fifth objective. Data obtained from the questionnaire sought to establish interrelationships between the factors that determine health and safety compliance to establish the relationship produced amongst them and which constructs have a greater influence on the determination of health and safety compliance in small to medium-sized construction companies in Ghana. The aim was to establish the core determinants of health and safety compliance in the construction companies. A detailed explanation of the survey concerning population, sampling procedure and analysis of results is presented later in this chapter. The expected output for the sixth objective was information to validate the conceptualized holistic model and based on this, to finalize the best fit model for health and safety compliance for small to medium-sized construction companies in the Ghanaian construction industry.

8.4.1 Methods

An overview of the methods used to achieve the objectives of the research in this book is given in the next sections. The methods used are literature review, the Delphi method and the questionnaire survey.

8.4.2 Literature Review

Literature is the foundation of any research. The most important aspect of developing a study is to review literature on previous works carried out by different researchers and the methodologies employed to investigate similar problems in order to establish the trends used to solve problems. For the current study, it became necessary to know the following:

- Theoretical and conceptual perspectives of health and safety research (generic literature)
- Review of the health and safety compliance literature (international)
- International literature in the construction industry
- Occupational health and safety in developing countries

The literature review on these aspects was important as it provided the broad context of the study to the reader. It also highlighted what has already been done on the subject under consideration. It further related the present research to the ongoing debate on the subject and provided a framework for comparing the results of the present research with other studies on the subject. Studies reviewed were well integrated, and adopted methods have been used in other studies. The review detailed analyses of the methods used in other studies.

Several materials were used for the literature review, such as books, reviews of articles on the subject matter (both published and unpublished), theses and dissertations. Names of leading authors and contributors on the subject matter were sourced from the references of the consulted articles. This helped to obtain their publication history and the search for information focused on research databases. The progression of research on the topic under study was made possible from articles in the mentioned sources. The methods of conducting a literature review as indicated by Boote and Beile (2005) were strictly followed.

The process of conducting a literature review specifically followed these steps (Boote & Beile, 2005):

1. Finding a broad range of high-quality, specific articles, books, dissertations and reviews directly related to the study
2. Reading and re-reading to establish progressions and trends
3. Summarizing the studies read
4. Identifying methodologies adopted in the studies
5. Relating the current study to those reviewed
6. Writing the literature

The output from the literature review was a clear perspective of the topic and an indication of where the study fits in relation to other studies on the subject matter. This also provides a framework for comparing the results of the study with other studies. A clear standpoint on the topic and an indication of where the study fits in is the product of the literature review. Research is said to be good because it advances our collective understanding. Much time and energy has been exerted in the review of literature on the current study.

Findings from the literature review have shown various factors which determine health and safety compliance. Other factors, such as key constructs in the 'bundle of factors' that should bring about health and safety compliance was likewise considered. These factors were not considered in the previously developed models. Therefore, there are missing universal factors which gives assurance on health and safety compliance of SMEs in the Ghanaian construction industry. Theories were developed about the influence of the missing factors and their interrelationships with other factors to determine health and safety compliance. A test was carried out to ascertain their influence on health and safety compliance among SME contractors. This was achieved by using the Delphi method described next.

8.4.3 The Delphi Method

The Delphi technique was first developed by two scientists, Olaf Helmer and Norman Dalkey, with the Rand Corporation in the 1950s, and named after the greatest of all Greek oracles (1963). The Delphi technique is a procedure to solicit opinions, judgement and consensus from a group of experts in a given subject field. This has to do with a long-range forecasting method of aggregating the forecasts of experts on multidisciplinary issues. The process is very interactive for soliciting and collating opinions on a particular topic through a set of designed sequential questionnaires interspersed with a summarized feedback of opinions derived from earlier responses (Dalkey & Helmer, 1963). The opinions of the experts are re-submitted a number of times (rounds) until a satisfactory consensus is reached. Each expert's opinion is revealed in the survey, but the panel members have no idea who is involved. The idea was that joint judgment of experts was a relevant measure of the outcome of the research. The Delphi technique is a qualitative methodology seeking to produce a consensus of a group of experts on an issue of concern (Miller, 1993) through a survey consisting of rounds. It is based on structural surveys and makes use of intuitively available information of the participants (experts) in their various fields. The method provides both qualitative and quantitative results, and has explorative, predictive and even normative elements (Cuhls, 2003).

The Delphi method was used during the second stage of the study to identify the main attributes that bring about health and safety compliance in SME construction companies. It was also to examine whether the attributes that determine compliance in other cultural contexts as identified from the literature are the same within the Ghanaian construction industry. The Delphi technique was used to explore the extent of the impact or influence of the main and sub-attributes on health and safety compliance in the Ghanaian construction industry.

There is agreement that Delphi is an expert survey in two or more 'rounds' in which the second and later rounds of the survey (the results) of the previous round are given as feedback. Thus, the expert's answer from the second round is based on the influence of the other experts' opinions. Thus, the Delphi method is a relatively strongly controlled group communication process, in which matters on which naturally unsure and incomplete knowledge is available are judged upon by experts (Häder & Häder, 1995). The technique requires knowledgeable expert contributors individually responding to questions and submitting the results to a central coordinator. The coordinator processes the responses, looking for central and extreme tendencies, and their validations. The results are then fed back to the input provided by the coordinator. The experts are then asked to resubmit their opinions, aided by the input provided by the coordinator. This process continues until the coordinator sees that a consensus has been formed on the questions asked. The method is intended to remove the bias that is possible when diverse groups of experts meet, which is common with other methods of decision making. With the Delphi method, the experts do not know who the other experts are.

Hence, the standard Delphi method is a survey that is directed by a coordinator as already stated and comprises several rounds with a group of experts, who are kept anonymous and for whose subjective-intuitive prognoses a consensus is aimed at (Cuhls, 2003). After each survey round, standard feedback about the statistical group judgment calculated from the median, the percentages and the interquartile range of single projections is given and if possible, the arguments and counter-arguments of the extreme answers are fed back. In the Delphi process, nobody 'loses face' because the study is done anonymously using a questionnaire. The method makes better use of group interaction whereby the questionnaire is the medium of interaction (Rowe, Wright & Bolger, 1991; Häder & Häder, 1995).

The Delphi method is especially useful for long-range forecasting, as expert opinions are the only source of information available. The Delphi technique is part of a group of decision-making (policy-making) techniques that includes the nominal group technique (NGT) and interacting group method (IGM). The Delphi technique differs in various ways from NGT and IGM primarily due to the fact that Delphi is individual based, anonymous and independent. The element of group interaction is eliminated from the process and feedback to questionnaires is in written format (Loo, 2002). Over time, the Delphi method has gained popularity and approval as a method of inquiry across many scientific disciplines. It has been used as a study instrument in the field of library and information science (Buckley, 1995) and in the medical disciplines (Linstone & Turoff, 1975). Those experienced with the Delphi technique report that the method produces valuable results which are accepted and supported by the majority of the expert community.

Similarly, in the business field, the technique has been highly rated by some as a systematic thinking tool but has been challenged in its ability to serve as an identifier of strategic issues (Schoemaker, 1993). The Delphi technique is well suited as a research approach and method for the current study. The technique has not been used in Ghana or in any other developing country to study the compliance of SME to H&S issues/regulations. The Delphi study was aimed at attracting a wide spectrum of inputs from various geographically dispersed experts in Ghana. The Delphi method was preferred to common survey methods, as the current study was addressing the 'what can' and 'what if' kinds of questions, as opposed to the 'what is' kind of questions. Delphi is more suited for these kinds of questions to explore concepts that are difficult to measure except through experimental methods. Unfortunately, an experimental survey was not feasible and appropriate for the current study. The Delphi's strength lies in the rounds used, unlike ordinary survey research which provides an opportunity for initial feedback, collation of feedback and distribution of collated feedback to participants for further review (Stitt-Gohdes & Crews, 2004). Therefore, the Delphi method is considered a robust method of rigorous query of experts. This unique process requiring group communication is central to the strength of the Delphi (Stitt-Gohdes & Crews, 2004). The Delphi process should be used when investigating policy-making or policy-evaluation strategies that will set the future direction for both the public and private sector. The Delphi method was alleged to have failed to follow accepted scientific procedures, in particular the lack of psychometric validity (Sackman, 1974). Helmer (1977) argues that the forecasting tendency, one of the major applications of the Delphi, is inevitably conducted in a domain of what might be called 'soft data' and 'soft law'. Helmer further determined that standard operations research techniques should be augmented by judgmental information and that the Delphi method cannot be legitimately criticized for using mere opinion and for violating the rules of random sampling in the 'polling of experts'. Such criticism, Helmer (1977) argued, rests on a gross misunderstanding of what the Delphi method is; it should be pointed out that a Delphi inquiry is not an 'opinion poll'.

The Delphi method was considered as a useful tool for the current study, as the book is aimed at the future direction for health and safety compliance in small to medium-sized construction companies in Ghana.

As all the preceding definitions illustrate, in no instance is reaching a majority opinion the ultimate goal in a typical Delphi study; it is rather the reaching of agreement (consensus). Buckley (1995) posited that 'Delphi is a tool for discovering agreement and identifying differences rather than forcing consensus'. In principle, agreement alone is not a sufficient condition for arguing the acceptance of the Delphi method. But as with the majority of research methods, the method of use and application has an enormous influence on the eventual success of the inquiry (Buckley, 1995). Hence, where no agreement is achieved, the Delphi still helps to clarify the issue being investigated. As one of the majority of research methods, the method of use and application have an enormous influence on the eventual success of the inquiry. Hence, where no agreement is achieved, the Delphi still helps to clarify the issue being investigated. One of the common reasons for failure in a Delphi study is ignoring and not exploring disagreement (Linstone & Turoff, 2002). The current study is not only about reaching or forcing a consensus, but recognizes disagreement and explores the reason for such.

In addition to the aforementioned criticism of the Delphi technique, different authors have pointed out their views on the weaknesses of the Delphi technique as follows:

- It has not been shown consistently that the results from the Delphi method are any better than those achieved through other structured judgmental techniques (Rowe et al., 1991).
- The Delphi study is at the mercy of the worldview and biases of the coordinating or monitor team, who chooses the experts, interprets the returned information and structures the questions. There is an enormous debate whether the experts should be chosen from within or outside the

organisation initiating the study and whether they should be experienced in the subject area of the study in question (Masini, 1993).
- Linstone (1978) disagreed with the process and how the questionnaire was structured, which Linstone believes can lead to a bias (like IQ tests), which assume a certain cultural background. Hence, the experts may give responses they think the monitoring group wants to hear, or they may not respond at all. Consequently, the cultural background of respondents will have an impact upon the results.
- Lang (1995) states that the process of choosing the experts is often not considered seriously enough. Yet, it is the calibre of the experts that determines the quality of the outcomes of the study (Lang, 1995).
- In the process of achieving consensus, extreme points of view run the risk of being suppressed, when in fact they may provide important new information or insights (Lang, 1995).
- The flexibility of the technique means it can be adapted to a whole range of situations, which in turn can make it vulnerable to misrepresentation and sloppy execution (Amara, 1975).
- Garrod (2008) found that the Delphi technique can be extremely sensitive to the level of the experts' expertise, composition of the panel clarity of the questions, the way the research or coordinator reports reasons for outliers and the administration of the questionnaire.

Despite these limitations, the Delphi is a particularly good research method for developing consensus amongst a group of entities having expertise on a particular topic where information required is subjective and where participants are separated by physical distance (Linstone & Turoff, 1975; Brill, Bishop & Walker, 2006). The Delphi method has been validated in the literature as a reliable empirical method for reaching consensus in a number of areas. Amongst these areas are journalism (Smith, 1997), visual literacy (Brill et al., 2006) and health care (Whitman, 1990). Beside these areas, the method has also been used in many other disciplines, such as in information technology (IT) research to identify and rank key issues for management attention (Delbecq, Van de Ven & Gustafson, 1975), scientific study of geographic information systems (Hatzichristos & Giaoutzi, 2005), quality management (Saizarbitoria, 2006), terrorism (Parente, Hiob, Silver, Jenkins, Poe & Mullins, 2005), banking (Beales, 2005), social sciences (Landeta, 2006), housing satisfaction studies (Aigbavboa, 2013), privatization of utilities (Critcher & Gladstone, 1998) and education (Yousuf, 2007). Based on the extensive usage of the method over time, the Delphi method in research is an accepted practice. However, it may not be appropriate for all research activities.

8.4.3.1 Epistemological Approach towards the Delphi Design

There are variances amongst the various group techniques, coupled with the definition of the Delphi method as compiled by various scholars and cognisance of the various criticisms from the epistemological foundation for defining the approach towards a typical Delphi study design. Amongst these include reducing the effects of personal bias. This is done by ensuring that all expert feedback is anonymous.

Through this, the technique captures the opinions, experience, and knowledge of each panel member. Personal knowledge is harvested and interpersonal interaction biases are stripped away. The concreteness of the framework of the Delphi design is vital in researching the overall objective of the study. The basic premises of the Delphi research design towards health and safety compliance is entrenched in some form of general agreement and consensus regarding the core ingredients and components of the subsequent framework.

Given the current status of health and safety compliance issues in Ghana and the absence of a generally agreed upon health and safety policy, the search for consensus and a point of departure in issues on health and safety policy that will better serve the construction industry is therefore justified through the use of the technique. Hence, the objective of the Delphi design for this work is to obtain the most reliable consensus of opinion of a group of experts in the specific field being studied (health and safety compliance amongst small to medium-sized construction companies). The Delphi method is best used where there is little past data available applicable to extrapolate from, and where social, economic, ethical and moral considerations are pre-eminent. Considering the outcome of the literature review of the current research, there is no structured research so far carried out that has adopted the technique with regard to health and safety compliance in small to medium-sized construction companies in Ghana regarding the health and safety definition, function and nature of the Delphi technique. It is justified that the Delphi technique is the best method to explore the subject of the research and to achieve the aim and objectives for this book.

8.4.3.2 When to Use the Delphi Technique

The Delphi method is mainly used when long-term issues have to be assessed such as the subject of the current research. This is because it is a procedure used to identify statements (topics) that are relevant for the future, it reduces the tacit and complex knowledge to a single statement, and makes it possible to judge upon (Cuhls, 2003). Hence, the use in combination with other methodologies such as survey design can be interesting.

One or more of the following properties could lead to the need for the use of the Delphi technique (Linstone & Turoff, 2002):

- When the problem of inquiry does not lend itself to precise analytical techniques but can benefit from subjective judgments on a collective basis (Buckley, 1994).
- The research needs to contribute to the examination of a broad or complex problem with no history of adequate communication and may represent diverse backgrounds with respect to experience or expertise, which is a major premise of the research.
- More individuals are needed than can effectively interact in a face-to-face exchange.
- Time and cost to make frequent group meetings is limited.
- The efficiency of face-to-face meetings can be increased by a supplemental group communication process.

- Disagreements among individuals are so severe or politically unpalatable that the communication process must be refereed and anonymity assured.
- The heterogeneity of the participants must be preserved to ensure validity of the results, such as the avoidance of domination by quantity or by strength of personality called the 'bandwagon effect'.

The Delphi method as a foresight tool possesses certain degrees of invariance to survive in the changing challenges of the past 50 years (Cuhls, 2003). Hence the process could serve different understandings of predicting or premonition, and is probably understood by the users as being relevant for covering technical, organisational and personal perspectives. What the users of the Delphi technique especially like are the sets of data about the future that are collected. Writing down future topics seems to have an immense psychological effect because it transfers implicit to tacit knowledge to the more visible explicit and therefore transferable knowledge.

8.4.3.3 Components of the Delphi Technique

The main components of the Delphi technique consist of five major characteristics, which are also adopted in the study (Loo, 2002). The study should consist of a panel of carefully selected experts representing a broad spectrum of opinion on the topic or issue being examined. The process is as follows:

- The participants are usually anonymous.
- The coordinator constructs structured questionnaires and feedback reports for the panel over the course of the Delphi.
- It is an iterative process often involving three to four iterations called 'rounds' of questionnaires and feedback reports.
- There is an output, typically in form of a research report containing the Delphi results, the forecasts, policy and programme options (with their strengths and weaknesses), recommendations to senior management and possibly an action plan for developing and implementing the policy programmes.

Likewise, Hasson, Keeney and McKenna (2000) recommended that the following research guidelines for using the Delphi technique be addressed in designing a Delphi approach: research problem identification, understanding the process, selection of experts, informing or invitation to experts, data analysis, and presentation and interpretation (these are discussed shortly). Therefore, given the nature of the current study, it was further believed that the Delphi technique was well suited to obtain credible inputs from experts in industry, academics and government to serve as key input towards the research objectives.

1. Research problem identification—There are four objectives that call for the use of the Delphi technique. One of those objectives is to relate informed judgments on a topic that spans a wide range of disciplines. The decision to use the Delphi technique must centre upon the appropriateness of the

available alternatives. The use of experts in a field of study is a perfectly suited technique (Reid, 1988):

- The technique has not been utilized in past studies about health and safety compliance in small to medium-sized construction companies in Ghana.
- It offers the opportunity to check the validity of the cross-disciplinary (social, psychological, ethical, managerial, cultural, anthropological) nature of the issue.

2. Understanding the process—The Delphi technique is a multistage process designed to combine opinions into group consensus (McKenna, 1994). The process is as follows:
   - Pilot testing of a small group.
   - Initial questionnaire (qualitative) comments solicited (not in all cases).
   - Initial feedback (quantitative) after statistical analysis of the initial opinions.
   - Subsequent questionnaire (qualitative) comments solicited again.
   - Subsequent feedback (quantitative) after statistical analysis. This provides an opportunity for participants to change their opinions if necessary.
3. Selection of experts—It is important to select panel members who are impartial and are interested in the topic. Some studies have over 60 experts, some as few as 15. Selection of people knowledgeable in the field and their commitment to multiple rounds of questions on the same topic is essential. In the section that provides more detail regarding the practical design and execution of the Delphi study for this work, how the experts were chosen for the study will be presented.
4. Informing or invitation to experts—It is imperative to explain what is required of them, how much time it will require, what they will be required to provide, what the objective of the study is and what will be done with the information.
5. Data analysis—This is the process where opinions of the experts are solicited. Two or three rounds are preferred (Green, Jones, Hughes & Williams, 1999). An 80 percent consensus should be the goal for any Delphi study (Green et al., 1999). Likewise, percentages should not be used possible as informed by some scholars, but rather the process should stop when stability of the data occurs (Crisp, Pelletier, Duffield, Adams & Nagy, 1997). Contrary to the view of Crisp et al. (1997), percentage estimation was found suitable to this study as one of the means to achieve consensus, hence a 60 percent consensus goal was set for the three-round Delphi study. Also, analytical software can be utilised to analyse the responses, and provide feedback to the experts on the central tendencies (median and interquartile range) and on the levels of dispersion (standard deviation). Hence, the criteria for qualitative studies, such as the Delphi technique, should be credibility (truthfulness), fittingness (applicability), audit ability (consistency) and confirmatory ability (Lincola & Guba, 1985).
6. Presentation and interpretation—There are a number of methods for presenting the data from a typical Delphi study, with two methods being

graphical and statistical. These two methods have been used in the current research. Therefore, given the nature of the current research, it is further believed that the Delphi technique is well-suited to obtain credible inputs from experts in industry, academics and government to serve as key input in the development of a health and safety compliance model for small to medium-sized construction companies in Ghana. The next section provides an overview of how the Delphi technique was used in this study meeting the objective of the book.

8.4.3.4 Designing, Constructing and Executing the Delphi Technique Study

Given the rationale behind the Delphi technique and the main features explained earlier, the design, construction and execution of the Delphi study for the current research will follow a sequential process (Loo, 2002). The four vital planning and execution activities followed as pointed out by Loo (2002) are as follows:

1. Problem definition
2. Panel selection
3. Determining the panel size
4. Conduction of the Delphi iterations

Supporting Loo's (2002) approach, Delbecq et al. (1975) suggest a basic Delphi methodology that includes distinct stages such as Delphi question development (objective), expert panel selection, sample size, first questionnaire, first questionnaire analysis and follow-up questionnaires. This methodology forms the basis of the current Delphi research study and is explained in the subsequent sections. Table 8.1 gives a summary of the Delphi design, construction and execution.

TABLE 8.1
Key Delphi Questions and Phrasing for the Study

Key Delphi Questions	Phrasing for This Study
Why are you interested in this study?	This study was initiated because of the belief that not all contractors provide the requisite health and safety requirement for their workers or receive government support. This assumption is concrete because there is a lack of understanding of the diverse features that determine health and safety compliance.
What do you need to know that you do not know now?	This is despite the knowledge about the features that bring about health and safety compliance. These have not been put together as a model to inform policy and predict health and safety compliance in small to medium construction companies. The attributes that determine health and safety compliance will come out clearly at the end of this study.
How will the results from the Delphi study influence health and safety compliance?	The result of the Delphi study will enable the development of a conceptual framework for the health and safety compliance model to be developed. The attributes that would collectively predict and ensure health and safety compliance in the Ghanaian construction industry will be established.

8.4.3.4.1 Phase 1: Delphi Questionnaire Development

The formulation of the Delphi question is vital to the whole process. It is paramount that the panel of experts understand the broad context within which the questionnaire is designed. For instance, the current work studied the concept of what determines health and safety compliance. Hence, the concept had to be broadly clarified. Key questions were asked to achieve the objectives of the Delphi study. The basis of constructing the questions for this current study was based on the earlier discussions, with corresponding wording and phrasing given for this study.

8.4.3.4.2 Phase 2: Delphi Expert Panel Selection

A critical part of conducting a Delphi interview technique is the selection of the right experts (panellists, participants or respondents). This is the critical part of conducting a Delphi interview technique. Their role is also crucial to the success of the research (Hasson et al., 2000).

Ensuring a high commitment response rate in the subject under examination depends on the selection of the experts, their interest and involvement. Hasson et al. (2000) posited that 'controversial debate occurs when a professional becomes an expert'. The claim that a group represents valid expert opinion has been criticized as scientifically untenable. McKenna (1994) defined 'an expert' as a panel of informed individuals (otherwise called 'experts' hereafter). The definition of 'an expert' by McKenna (1994) was supported by Goodman (1987) who stated that the Delphi technique 'tends not to advocate a random sample of panellist … instead the use of experts or at least of informed advocates is recommended'. Delphi inquiry is not an opinion poll. Relying or drawing a random sample from the population of experts is not the best approach, rather once a set of experts has been selected (regardless of how, but following a predetermined qualifying criteria), it provides a communicative device for them that uses the conductor of the exercise as a filter in order to preserve anonymity of responses, which is the core of the Delphi technique (Helmer, 1977).

Therefore, the most significant danger in selecting the panel of experts lies in the path of 'least resistance' through the selection of a group of cosy friends or like-minded individuals, which thus negates the strength of the process (Linstone & Turoff, 2002). Since experts form the cornerstone of the Delphi technique, clear inclusion criteria should be applied and outlined as a means of evaluating the results and establishing the study's potential relevance to other settings and populations.

The selection of panellists for the study was based on criterion sampling. Panellists were selected for a purpose to apply their knowledge to a concept raised in the study based on the criteria that were developed from the research objective under investigation. A Delphi study does not depend on a statistical sample that attempts to be representative of any population. It is a group decision mechanism requiring qualified experts who have deep understanding of the issues (Okoli & Pawlowski, 2004). Therefore, one of the most critical requirements is the selection of qualified experts as it is the most important step in the entire Delphi process because it directly relates to the quality of the results generated (Hsu & Sandford, 2007). The careful selection of the panel of experts is the keystone to a successful Delphi study (Stitt-Gohdes & Crews, 2004). There are detailed criteria for the selection of panel experts. In a typical Delphi study, experts should meet two recommendations (Dalkey & Helmer, 1963; Rogers & Lopez, 2002):

1. The experts should exhibit a high degree of knowledge of the subject matter.
2. They should be representatives of the profession so that their suggestions may be adaptable or transferable to the population.

Delphi participants in any study should also meet four 'expertise' requirements (Adler & Ziglio, 1996): knowledge and experience of the issues under investigation, capacity and willingness to participate, sufficient time to participate in the Delphi studies and effective communication skills.

In choosing panellists for this study, each expert was required to meet at least five of the following minimum criteria.

- Residency—Have lived in any of the metropolitan/municipal/districts of Ghana and in a related developing country at least more than one year.
- Knowledge—Has knowledge of health and safety in the construction industry.
- Academic qualification—Has been presented an earned degree (national diploma, bachelor's degree, master's degree, PhD) related to any field, certification of employment/experience focusing on construction development/ sustainable issues.
- Experience—Has a history of or is currently performing consultation services for the government of Ghana, individuals, businesses, agencies, companies and/or organizations relating to construction or other sustainable development. The experts must exhibit a high degree of knowledge of experience in the subject matter and have extensive theoretical knowledge.
- Employment—Currently serves (or has previously served) in a professional or voluntary capacity (e.g. at place of employment – institution, business, agency, department, company) as supervisor or manager of an establishment that is involved with small to medium-sized construction companies in Ghana.
- Influence and recognition—Has served or is currently serving as a peer reviewer for one or more manuscripts received from a journal editor prior to its publication in the primary literature, with focus of the manuscript(s) on construction H&S.
- Authorship—Is an author or co-author of peer-reviewed publications in the field of construction with emphasis on H&S in Ghana. Has prepared and presented papers at conferences, workshop or professional meetings focusing on construction H&S.
- Research—Has submitted one or more proposals to or has received research funds (grant/contract) from national, local government, regional and/or private sources that support construction, sustainable development and studies related to health and safety.
- Teaching—Has organised, prepared and successfully presented one or more health and safety or construction industry training workshops focusing on

the group for which expertise is sought. The workshop or course must have been on health and safety practices. Or has served as an individual or as a collaborative instructor in the teaching of one or more polytechnics or university courses focusing on construction, H&S, or construction industry development or related field.

- Membership—Member in a professional body (as listed on the expert questionnaire). Should be the representative of a professional body so that their opinions may be adaptable or transferable to the population.
- Willingness—Panel members must be willing to fully participate in the entire Delphi studies.

The selected participants should represent a wide variety of backgrounds to guarantee a wide base of knowledge (Rowe et al., 1991). The recommendations of Rowe et al. (1991) were adopted for the current study. The number of respondents was large enough to ensure that all perspectives are represented but not so large as to make the analysis of the results unmanageable by the researcher (Linstone & Turoff, 1975). The adoption of five of these criteria was considered more stringent than the recommended number of at least two criteria (Dalkey & Helmer, 1963; Rogers & Lopez, 2002). The five minimum criteria were framed after the four recommendations made by Adler and Ziglio (1996), with the inclusion of experts' residency status, which was considered to be compulsory for all selected experts. This was considered significant because experts were required to have a wide-ranging understanding of health and safety practices within their locality. A minimum number of five criteria was set because the method may be undermined if panellists to be recruited lack specialist knowledge, qualifications and proven track records in the field (Keeney, Hasson & McKenna, 2001) amongst others. The expertise comes in many guises and may include those who are 'experts by experience' (Hardy, O'Brien & Gaskin, 2004).

Panel members were identified from three sources: professionals in the building and civil engineering firms, academicians, and consultants with quantity surveying, building and civil engineering background. The involvement in the Delphi process was a key consideration. The panellists were recruited via e-mail, with a brief overview of the study objective included therein. Thereafter, those who consented to the preliminary invitation were sent a detailed description of the Delphi study. Experts were asked to send their curriculum vitae in order to confirm their areas of expertise and to ascertain whether they meet the qualifying criteria. The five criteria set for the study were met by all experts selected. The first-round Delphi survey for the current study was sent to the selected experts after a verification exercise. The survey questionnaires contained closed-ended questions. Experts were judged whether they qualified to be included in the study based on their curriculum vitae. Twenty invitations were sent out based on the set criteria. Thirteen responded that they were willing to participate and completed the first round, but only nine remained as panellists throughout the study. This number of panellists was considered adequate on literature recommendations from scholars who have previously employed the technique. Ten to fifteen experts as suggested by Delbecq et al. (1975) could be sufficient if the backgrounds of the experts are homogenous, which was achieved in the current study. Rowe et al. (1991) indicated that the size of a Delphi panel ranges from three to eighty in peer-reviewed studies. Okoli and Pawlowski (2004) and Skulmoski, Hartman and Krahn (2007) also indicated a panel size of about ten to eighteen members, whilst Hallowell and Gambatese (2010) suggested that since most studies incorporate between eight and sixteen panellists, a minimum of eight is suggested. This was beyond the given limit in the current study. The size of a panel should be dictated by the study characteristics, number of available experts, the desired geographical representation and the capacity of the facilitator (Hallowell & Gambatese, 2010).

The panel of nine experts was considered adequate based on the fact that the Delphi method does not depend on statistical power for arriving at consensus amongst experts but rather on group dynamics. Most of the potential experts who agreed to participate during the initial stage withdrew from the study, probably owing to the rigorous nature and time required in the Delphi method. The eventual withdrawal of four experts during the first round of the study led to the completion of the study with nine experts. This indicates the quality of the study and its nature in the current Ghanaian environment. All the panel members were Ghanaians.

Four of the experts are currently at the Cape Coast, two of the experts each are in Accra and Kumasi. One of the experts is in Koforidua (Table 8.2). A total of 88.9 percent of experts were males, while the remaining 11.1 percent were females.

The highest qualifications held by the experts are tabulated in Table 8.3. Three of the experts had a doctor of philosophy (PhD) and six experts had a master of science (MSc). All experts were from various fields, but they were all involved in construction projects. Their curriculum vitae analysis shows that they were involved in both building and civil engineering works and related fields.

TABLE 8.2
Residential Location of Experts

Regional Capitals/Cities	Number of Experts
Cape Coast	4
Koforidua	1
Kumasi	2
Accra	2
Total	9

TABLE 8.3
Qualification of Panel of Experts

Highest Qualification	Number of Experts
Doctor of philosophy (PhD)	3
Master of science (MSc)	6
Total	9

In terms of the experts' field of specialization, one of the selected experts was a specialist in construction health and safety, two were procurement corruption/ management specialists, one was a geotechnical engineer, one was a structural engineer, two were quantity surveyors, and two were procurement/construction managers, as shown in Table 8.4

Table 8.5 shows that two experts had 1 to 5 years of experience, six had 6 to 10 years of experience, while only one expert had 21 to 30 years of experience. All experts were professionally registered at the highest level with various professional regulating bodies such as the Ghana Institute of Construction, the Ghana Institute of Engineers and the Ghana Institute of Surveyors. Others are affiliated with the Institute of Engineers and Technology, the Royal Institute of Chartered Surveyors, the Chartered Institute of Marketing, and the International Association of Valuators, Consultants and Analysts. Two experts each were fellows in the Ghana Institute of Surveyors, the Institute of Engineers and Technology and the Ghana Institute of Surveyors. Three experts were members of the Ghana Institute of Construction and

TABLE 8.4
Panel of Experts' Field of Specialization

Field of Specialization	Number of Experts
Health and safety	1
Procurement corruption/management	2
Geotechnical engineering	1
Structural engineering	1
Quantity surveying	2
Procurement/construction management	2
Total	9

TABLE 8.5
Panel of Experts' Years of Experience

Years of Experience	Number of Experts
1–5	2
6–10	6
11–15	–
16–20	–
21–30	1
Mean	9
Median	6.5
Mode	10
Range	27
Cumulative total years of experience	90

one expert each was a Member of the Royal Institute of Chartered Surveyors, the Chartered Institute of Marketing, and the International Association of Valuators, Consultants and Analysts.

8.4.3.4.3 Phase 3: Determining the Panel Size

The Delphi technique requires a qualitative approach rather than a quantitative approach. The number of participants or experts in the Delphi technique is expected to be lower than in normal quantitative surveys. Various scholars have recommended different sample sizes for determining the minimum number of experts to participate in a typical Delphi survey. Dalkey and Helmer used a panel of seven experts in their original Delphi experiment in 1953 (Helmer, 1983). According to Linstone (1978:296), 'a suitable minimum panel size is seven'. Linstone justified this by indicating that the research runs the risk of accuracy and deteriorates quickly when the numbers increase. The justification given by Linstone was supported by Cavalli-Sforza and Ortolano (1984) who postulated that a 'typical Delphi panel has about eight to twelve members', while Phillips (2000) stated that the optimum number of participation should be between seven and twelve. Miller (1993) was of the view that any additional responses beyond the first thirty responses would not generate new information. Similarly, Dunn (1994) suggested ten to thirty participants, apprising that as the complexity of the policy issue increases, the sample size needs to be larger to include the entire range of participants both for and against the policy issue area. Dunn (1994) further emphasised the inclusion of both formal and informal stakeholders who have absolute interest in the policy issue in the study. They should also have varying degrees of influence, hold a variety of positions and be affiliated with different groups.

The following are the factors highlighted by Skulmoski et al. (2007) to be considered to determine the sample size for a Delphi technique:

- Heterogeneous or homogeneous sample—Where the group is homogeneous, then a smaller sample of between ten to fifteen people should yield sufficient results. Nevertheless, if an unrelated group is involved, for instance in an international study, then a larger sample will likely be required and several hundred people might participate (Delbeq et al., 1975). However, the Delphi coordinator needs to exercise caution because heterogeneous groups can greatly increase the complexity and difficulty of collecting data, reaching consensus, conducting analysis and verifying results.
- Decision quality/Delphi manageability trade-off—There is a reduction in group error (or an increase in decision quality) as sample size increases (Linstone & Turoff, 2002). However, above a certain threshold, managing the Delphi process and analysing the data becomes cumbersome in return for marginal benefits.
- Internal or external verification—The larger the group, the more credibly the results can be said to be verified. However, a smaller sample might be used, with result verification conducted through follow-up research. The current research adopted a smaller sample premise and is verified through a follow-up questionnaire survey.

However, the selection of an initial respondent panel for the Delphi study varies. It was concluded from the literature review that a typical sample size varies between seven to fifty panellists since there is no agreement on the desired 'typical' number of panellists to be adopted in a Delphi studies. Rather, the method can be modified to suit the circumstances and the research question. Owing to time constraints and conflicting schedule of the experts, the current study did not involve a large number of experts. A sample size of nine experts was adopted based on the following premises; this was also in conjunction with the qualifying criteria as established in phase two of the Delphi study:

- Experts should be fairly and practically split between academics and practitioners. The two categories may provide input for various perspectives and balance the theoretical and practical considerations.
- Panellists in both categories should have extensive experience relating to general construction industry issues, or in other specific aspects such as in H&S studies. Regardless of the aforementioned criteria, the current study also adopted the recommendation of Rowe et al. (1991) that the selected participants should represent a wide variety of backgrounds to guarantee a wide base of knowledge and experience. The number of panel experts also depends on the topic area, as well as the time and resources at the researcher's disposal. The adopted number of panel of experts of nine seems appropriate, considering the data required and subsequent analyses each panel of expert will generate.

8.4.3.4.4 Phase 4: Conducting the Delphi Iterations

Sequences of questionnaire rounds are used to obtain iterative responses to issues in a Delphi study (Masser & Foley, 1987). The decision to use 'two to ten rounds depends on repetition of judgment and group pressure for conformity owing to accuracy' (Woudenberg, 1991). Critcher and Gladstone (1998) suggested between two to five rounds. Three rounds of iterative process were used in the current study for the Delphi method. This was done to achieve consensus between the panel members on the influence and impact of health and safety. The designed Delphi questionnaire was sent out electronically to all members of the panel of experts to respond to the questions according to their ability and expertise. The Delphi questionnaire was developed based on the findings from the literature review and was specifically designed to address and achieve the Delphi-specific objectives defined for the study. The Delphi study for the current research consists of three rounds. The experts took a minimum of one month to complete the questionnaires. A questionnaire was designed for each round based on the responses to the previous one.

The round one questionnaire was designed based on a summary of the comprehensive review of literature highlighting sets of attributes and sub-attributes that are potentially relevant to health and safety compliance decisions among small to medium-sized construction companies. Issues relevant to health and safety practices were also extracted from the reviewed literature. These were structurally and constructively put together to frame the first round of the Delphi survey. Therefore, round one of the Delphi study was intended to be a brainstorming exercise used to produce a list of empirical attributes that determine health and safety compliance as well as other issues relating to health and safety practices in Ghana and other subsequent issues relating to the objective of the Delphi study. Closed-ended questions were used in this round. The responses were analysed and formed the basis of round two and round three of the study. Frequencies were obtained to measure the degree of consensus reached amongst participants regarding the attributes that determine health and safety compliance in the Ghanaian construction industry and for other related questions.

The purpose of the second round of the study was to allow experts to review and comment on the attributes that determine health and safety compliance and other issues relating to health and safety practices in Ghana, which were proposed by the panel of experts in round one. Closed-ended questions were used in this round to investigate participant comments expressing agreement, disagreement or clarification concerning proposed attributes that determine health and safety compliance in Ghana. The specific nature of the closed-ended questions stimulated participants' reactions. Frequencies were likewise obtained to measure the degree of consensus reached amongst participants regarding the attributes that determine health and safety compliance and for other related questions.

The final round three was specifically aimed at

- Informing the experts of the findings of the analysis of responses to the questionnaire of round two
- Requesting their final affirmation or comments on attributes and issues that did not receive any consensus in round two

The round three questionnaire was designed based on the measures of frequency responses to the questionnaire of round two. Closed-ended questions were used and frequencies were obtained to indicate consensus reached amongst experts regarding attributes that determine health and safety compliance and health and safety issues as presented in the study. Consensus was reached regarding most of the attributes that determine health and safety compliance in the Ghanaian construction industry over the three rounds of the Delphi survey. Based on the findings of the analyses of responses to the Delphi rounds, a list of attributes that determine health and safety compliance was prepared, which informs the conceptual framework for the broader study, while issues surrounding health and safety practices were highlighted. These responded accordingly to the set objective of the Delphi study. The Delphi survey was conducted via electronic mail, and follow-up e-mails were used to encourage prompt responses to the questionnaires. Using mail provides a free and faster means of communication.

With regard to the Delphi questionnaire, the experts were requested to rate the likelihood of an attribute influencing health and safety compliance and the impact of sub-factors in predicting health and safety compliance in the Ghanaian construction industry, if they were present. The probability scale ranged from 1 to 10 representing 0 to 100 percent. Interval ranges were set at 10 (Table 8.6). The panel of experts

TABLE 8.6
Influence or Likelihood Scale

0%–10%	10%–20%	21%–30%	31%–40%	41%–50%	51%–60%	61%–70%	71%–80%	81%–90%	91%–100%
1	2	3	4	5	6	7	8	9	10
			X				X		

TABLE 8.7
Impact Scale

No Impact		Low Impact		Medium Impact		High Impact		Very High Impact	
1	2	3	4	5	6	7	8	9	10
								x	

were asked to rate the negative impact that would result if a particular health and safety attribute was absent. This was based on a 10-point ordinal scale ranging from negligible to very high impact. This aspect indicated the importance of the health and safety compliance as shown on Table 8.7.

The experts were also required to state their level of agreement using a 5-point Likert scale with certain statements and to support their choices where necessary, with regard to health and safety compliance, policy issues, and health and safety practices in order to arrive at a consensus. Group medians were calculated as a measure of the central tendency for each response on each element. Thus, an indicator of whether consensus had been reached on the questions for each element was determined. The median was deemed to be more suitable for the type of information that was being collected. This is because the median eliminates bias and takes into consideration outlier responses and makes consensus more reasonable. The mean, on the other hand, may not reflect a reasonable central tendency as it considers only the outlier responses.

Group medians from the Delphi first round were computed for each element. These were then sent back to the expert panel members so that responses in the second round could be made taking into account the group median. The experts in the second round were asked to either maintain their original responses made in round one, or they could change their initial responses to be more in line with the group median. Panel members who had responses to units above or below the group median in the second round were requested to state their reasons for sticking to a response that does not agree with the group median. Experts were again requested to consider reasons for the outliers in making their decisions in the third round. Group medians and the absolute deviations were again computed for the third round.

From these calculations and after three rounds of the Delphi process, it was determined that consensus had been reached. Reasons for other experts' sticking to their ratings have, however, been taken into consideration. After the third round, group medians and the absolute deviations were again computed for the third round. Calculations for the third round of the Delphi process indicated that there was no need to proceed to the fourth round, as there was no further value that could be added to the degree of consensus attained at that level. Throughout the entire Delphi process, anonymity of panel members was maintained to avoid undue influence on other members. The aspect of anonymity is crucial to the credibility of the Delphi technique. There was a rigorous establishment of the complex 'what would happen if' kind of questions that ideally should be established from an experimental study. This was in fact extremely difficult and not feasible to do.

8.4.3.5 Specific Objectives of the Delphi Study

There are various characteristics and factors which determine health and safety compliance from literature, as measured from different health and safety theories, and categories, such as health and safety practices. It is not clear from literature what was specifically the level or extent to which the identified factors contribute to the health and safety compliance of small to medium-sized contractors. Attempts have been made through various studies to determine the influence of these factors in health and safety studies, but none has been specifically related to health and safety compliance. Various factors determine health and safety compliance, albeit varied in different cultural and health and safety backgrounds and typological settings. Also, previous models have not been adequately organized into a model to form a holistic attribute which determines health and safety compliance. Based on the foregoing, a more reliable measure of the determinants of health and safety compliance was therefore necessary in order to establish not only whether these factors have influence on health and safety compliance, but also the extent to which their influence had on the identified gap. Based on the context of the study, this kind of investigation would ordinarily call for an experimental kind of research. However, the experimental method of research was not feasible and practical considering the time frame and ethical issues and the willingness of would-be participants.

Hence, the Delphi method was considered the most suitable method to achieve the general objective of determining the influence and impact of the identified factors on health and safety compliance of small to medium-sized construction companies. The broader research aim was to develop a health and safety compliance on SME contractors' in developing countries: a case study of Ghana. The Delphi method was therefore chosen at the first stage to formulate the conceptual model. This had to be validated later from responses obtained from the questionnaire survey analysed using the structural equation modelling software EQS, version 6.2. At the Delphi stage, factors that had been identified from literature that defined and determined health and safety compliance were formulated into questions. Experts were asked to give their rating as being influential or having an impact on health and safety compliance. The output from the Delphi process was a set of attributes which determine the health and safety compliance that would be implemented. Consideration was given to these in order to achieve better health and safety compliance among SME contractors in Ghana. This is because a number of studies have identified and shown that different attributes determine health and safety compliance (Adenuga, Soyingbe & Ajayi, 2007; Othman, 2012; Windapo & Oladapo, 2012; Idubor & Oisamoje, 2013).

The current study extends the aforementioned studies by looking at health and safety practices holistically with the addition of new constructs to develop a model to predict health and safety compliance of SME contractors' and to what extent. The following specific objectives were raised, considering the various attributes that determine health and safety compliance.

The following are the specific objectives for achieving the Delphi survey based on the research objectives:

- To identify the main and sub-attributes that determine H&S compliance in small to medium-sized construction companies in Ghana
- To determine the factors that enable small to medium-sized construction companies to comply with H&S
- To identify the factors that affect small to medium-sized construction companies in non-compliance with H&S regulations
- To identify the effects of H&S non-compliance on small to medium-sized construction companies
- To evaluate the management issues that affect the government in the implementation of H&S policies in small to medium-sized construction companies

The philosophy behind the preceding objectives is to do away with the tendency of a non-coherent dialogue on health and safety compliance in Ghana. The following are the outcomes in achieving the objectives:

- Determining the key factors and constructs that are of critical significance (influence) to determine health and safety compliance among small to medium-sized construction companies
- Determining a holistic conceptual model on health and safety compliance in the Ghanaian construction industry

8.4.3.6 Computation of Data from Delphi Study

Computation of data from the Delphi study was conducted using Microsoft Excel, a spreadsheet software program. The first stage involved analysis to determine consensus on responses to the predetermined criteria. This involved determining the group median responses for each question. After the third round of the Delphi, absolute deviations (*Di*) of the group medians (m(*X*)) of each rating for the relevant questions as pre-determined were calculated using Equation 8.1:

$$Di = [x\ i - \mathrm{m}(X)] \tag{8.1}$$

where

Di = Absolute deviation

$x\ i$ = Panel rating

$\mathrm{m}(X)$ = Measure of central tendency

A computation of each and every question element was completed for the likelihood and impact of the attributes in predicting health and safety compliance and health and safety improvement of small to medium-sized construction companies in Ghana. The influence or impact of the absence or presence of a particular health and safety practice element on the overall health and safety compliance of the other elements is presented as well as issues relating to health and safety in Ghana. For every round of responses from the experts, besides the group median value computation, their respective interquartile deviation (IQD) was also computed as a measure of the central tendencies to determine consensus. The median value was adopted as a measure of central tendency because of its effect to minimize the effects of potentially biased individuals. While the IQD scores were used to summarize the variability in the data, the IQD helped to identify which measures were most appropriate to influence health and safety compliance amongst small to medium-sized construction companies. A robust picture of the overall data set was provided as the IQD removes or ignores outlying values through the use of the IQD. The interquartile range is a measure that indicates the extent to which the central 50 percent of values within the data set are dispersed. This is based upon, and related to, the median. The median is adopted as a measure of central tendency for studies of this nature, as opposed to the mean and IQD. To compute the variation of the median from the responses for each question in each round, the absolute deviation given in Equation 8.1 was done. This is the absolute difference between a response within a data set and a given point. The point from which the deviation is measured is a measure of central tendency, which is the median. The results from the Delphi analysis are presented as numbers and percentages in tables in the next chapter of this book. The outcomes show the predictions of the influence of health and safety compliance factors and other issues in health and safety practices in Ghana.

8.4.3.7 Determination of Consensus from the Delphi Process

It is required that consensus should be reached on all objectives as set for the Delphi process. Consensus was determined by measuring the central tendency of the various responses on all questions. The group median and the IQD were computed for all responses. In order to achieve consensus, the deviation of all responses about the group median was determined to be not more than one unit, likewise for the IQD. This is considered to be suitable, as the scale that was used for both probability (influence) and impact was 1 to 10. The deviation of all responses was calculated using the absolute median (Equation 8.1), while the IQD was calculated based on the recommended statistical process of the absolute value of the difference between the 75 and 25 percentiles. A percentile (or centile) is the value of a variable below which a certain percent of observations fall. The percentile is often used in the reporting of scores from norm-referenced tests, as in the present situation. The 25 percentile is also known as the first quartile (Q1), the 50th percentile as the median or second quartile (Q2), and the 75 percentile as the third quartile (Q3).

Hence, the deviation between the 75 and 25 percentiles give an absolute value referred to as the interquartile range or deviation. The interquartile range deviation is a robust statistic, having a breakdown point of 25 percent, and is thus often preferred to the total range, with smaller values indicating higher degrees of agreement (consensus). Smaller values in the interquartile range would then indicate higher degrees of consensus. However, consensus is difficult to measure in Delphi studies. The foregoing has been established from literature, namely that there is no consensus on how to determine consensus regarding a set of opinions. Holey, Feeley, Dixon and

Whittaker (2007:2) suggest that consensus is the same as agreement and that agreement can be determined by the following:

- The aggregate of judgments
- A move to a subjective level of central tendency
- Alternatively, by confirming stability in responses with the consistency of answers between successive rounds of the study

Some other researchers have used frequency distribution to assess agreement and the criterion of at least 51 percent responding to any given response category being used to determine consensus (McKenna, 1994). Other studies (Rayens & Hahn, 2000) have used means and standard deviations with a decrease in standard deviations between rounds indicating an increase in agreement. IQD has also been used to determine consensus (Rayens & Hahn, 2000), and this has also been adopted for the current study. Studies conducted by Rayens and Hahn (2000) have included another criterion to determine consensus in addition to the IQD to achieve stability. The criterion to achieve consensus was that the IQD should equal one unit for which more than 60 percent of respondents should have answered either generally positive or generally negative. Items which had an IQD ≠1 for which the percentage of generally positive or generally negative responses was between 40 percent and 60 percent were determined to indicate a lack of consensus or agreement. An IQD of 1.00 or less is an indicator of consensus, however, a change of more than 1.00 IQD point in each successive stage is the criterion for convergence of opinion. How to use or interpret IQD as a method of data analysis for the Delphi process to achieve consensus has not been identified in a literature review.

The potential range of IQD values depends on the number of response choices, with larger IQDs expected as the number of response choices increases. The use of a particular IQD as a cut-off for consensus requires consideration of the number of response choices. The following criteria for determining consensus was used by Holey et al. (2007):

- Percentage response
- Percentages for each level of agreement for each question to compensate for varying response rates
- Computation of median, standard deviation and their associated group rankings
- Computation of the means, standard deviation and their associated group rankings using the importance ratings
- Computation of the weighted kappa (k) values to compare the chance eliminated agreement between rounds

Consensus can also be reached when the following are present (Holey et al., 2007):

- An increase in percentage agreements
- Convergence of importance ratings

- An increase in kappa values
- A decrease in comments as rounds progressed
- A smaller range of responses
- Smaller values of standard deviations

There is little agreement on how to measure consensus in a Delphi study. It is, however, agreed that for consensus to have been achieved, there has to be a convergence of ideas and reasoning towards a subjective central tendency measure. Hence, in the current study, consensus was determined to have been reached if the following were achieved:

- More than 60 percent of responses are generally positive or negative with certain questions.
- The average of the absolute deviation was not more than one unit. The absolute deviation is calculated from Equation 8.1.
- The IQD was less than 1.00, meaning that items with IQD = 0.00 were considered to have reflected a high degree of consensus.

Therefore, the scales of consensus adapted for this study are as follows:

- Strong consensus: median 9–10, mean 8–10, IQD ≤1 and ≥80% (8–10)
- Good consensus: median 7–8.99, mean 6–7.99, IQD ≥1.1≤2 and ≥60%≤79% (6–7.99)
- Weak consensus: median ≤6.99, mean ≤5.99, and IQD ≥2.1≤3 and ≤59% (5.99)

8.4.3.8 Reliability and Validity of the Delphi Process

'Reliability is the extent to which a procedure produces similar results under constant conditions at all times' (Els & Delarey, 2006:52). This kind of statistical reliability is not possible in a Delphi study because another panel may reach a different conclusion depending on its knowledge of the subject area and interest. Care was taken that credibility was shown in truthfulness, fittingness was exhibited in applicability, audit ability was shown in response consistency and conformability was exhibited in the responses from all participants to reach reliability. Credibility was also ensured during the selection of the panel. All members of the panel of experts had distinguished themselves based on the set criteria for the selection of a panel of experts, the depth of their knowledge and experience as presented previously. Validity was boosted by the removal of preconception or influence from other members by keeping all members completely anonymous from each other and hence, eliminating the 'bandwagon' effect, which is one of the strengths of the Delphi method.

Furthermore, the number of iterations that were implemented in the Delphi study also enhanced the internal validity. Thus, expert panellists were given a chance to change their opinion or maintain it with a written explanation or argument for dissenting views. Feedback to the researcher and constant communication between the researcher and the panellists individually was another way of ensuring internal validity of the study. The external validity of a study deals with the extent the results from the study can be generalised to a larger population. This is usually determined by how participants are selected to be part of the study. This process was however not necessary, as the validation process of the conceptual model has been done using the questionnaire survey. The selection of participants for the Delphi study guaranteed external validity as a scientific criterion. The panel comprised members from varied sectors, all with in-depth knowledge on the construction industry. All the members of the panel of experts reside in one of the major cities in Ghana. They were highly experienced with sound working experience. The study therefore fulfilled the requirements for external validity in line with standard research ethics.

8.4.4 Questionnaire Survey

Phase three of the study involved collecting data from the field through the use of questionnaires. Phase three formed the pinnacle of the study. The Delphi study findings were that health and safety compliance is a multidimensional construct, which consists of a safe environment, safe acts of workers, safe working conditions, reaction of workers to safe conditions, government support and contractors' organisational culture. These factors have been collectively considered for the development of a holistic integrated health and safety compliance model for small to medium-sized construction companies. Four of the factors have been previously considered in the development of an H&S compliance model in other cultural contexts, but not for small to medium-sized construction companies. None of the existing models to date have included both government support and contractor's organisational culture as factors to develop a model that will help small to medium-sized construction companies in minimising accidents on site.

The following specific objectives of the questionnaire survey were used to validate the findings from the Delphi study:

- Identify the factors that are more influential on H&S compliance
- Establish the influence of the identified factors on the construction industry H&S compliance
- Determine the influence of H&S compliance on the construction industry
- Determine the goodness of fit of the hypothesized integrated H&S compliance model to the data sampled

Previous models of H&S compliance established in developed countries cannot be relied on in developing countries, hence, the determination of H&S compliance is less likely much more for small to medium-sized construction companies. The findings of what determines H&S compliance in developing countries are rarely known from the previously conducted research. There is a lack of research into the overall impact and influence of the direct and holistic active involvement of health and safety compliance constructs. In the absence of a health and safety compliance model, the achievement of health and safety compliance by the small to medium-sized construction companies is unlikely.

The integrated conceptual model shown in Figure 10.1 (see Chapter 10) was made up of the following interrelationships:

- Safe environmental features have an impact on H&S compliance and greatly influence its determination.
- Safe act of workers features have an impact on H&S compliance and greatly influence its determination.
- Safe working condition features have an impact on H&S compliance and greatly influence its determination.
- Reaction of workers to safe condition has an impact on H&S compliance and greatly influences its determination.
- Government support has an impact on H&S compliance and greatly influences its determination.
- Contractor's organisational culture has an impact of H&S compliances and greatly influences its determination.
- The integrated holistic Health and Safety compliance model describes the determinant (constructs) which determines H&S compliance in small to medium-sized construction companies.

As a result of the objectives of the study, it was obvious that collecting facts would involve contractors, professionals and employees in the building construction industry. A field survey was considered the most suitable method of collecting the required data. A complete biographical section was required that related to the building construction industry. Data relating to safe environment, safe act of workers, safe working condition, reaction of workers to safe condition, government support, contractor's organisational culture, and H&S compliance were required. Other data required were factors that have an influence on H&S compliance and factors that have influence on the H&S compliance of small to medium-sized construction companies. These types of information could not be obtained through other means of data collection except through a field survey. This is because other means of data collection would not give an adequate representation of the aforementioned relationships. Also, the decision to choose a survey method was based on a number of factors. The factors included the sampling technique to be adopted, the type of population, the question form, the question content, the response rate and the duration of data collection. The most appropriate survey method for this research was a personally administered questionnaire. This method was chosen for the following reasons:

- The questions could be answered by indicating the proper response with a cross. Respondents could seek clarity on any question so as to meet consistent question objectives (Sekaran, 2000; Aaker, Kumar & George, 2009).
- A higher response rate of almost 100 percent can be ensured since the questionnaires were not left with the respondents, but collected once they were completed (Malhotra, 1999; Sekaran, 2000).
- Anonymity of respondents was ensured because respondents were not required to disclose their identities (Burns & Bush, 2002; Sekaran, 2002).

This type of survey can be very time consuming if a wide geographic region is involved, but the respondents were located in the major cities of Ghana (Accra-Tema, Kumasi and Takoradi). Data were collected from small to medium-sized construction companies that were identified.

Apart from the aforementioned reason for adopting a personally administered questionnaire survey method, the following reasons justify the use of the questionnaire survey method:

- The philosophy underpinning the research is based on the positivist theory as discussed previously, which uses quantitative methods and collection of data by use of questionnaires.
- Validation of the conceptual model developed at phase one (literature review) and phase two (Delphi) entailed using an alternative method to validate the findings. This therefore eliminated the use of methods similar to the Delphi and its derivatives. Hence, this called for a collection of data on the health and safety compliance in small to medium-sized construction companies through a questionnaire survey.
- The field survey was considered to be more descriptive in that most of the small to medium-sized construction companies' contractors are located in the major cities of Ghana.
- Likewise, the interpretation and presentation of the data can easily be done and understood by various readers when adopting a positivist philosophy of research, as it follows a logical explanation of the method.
- A large number of research questions can be asked in a questionnaire to target many respondents within a stipulated time frame.
- A questionnaire requires minimal investment and is relatively easy to obtain generalizations from (Bell, 1996).
- Specific information about views, attitudes and perceptions of a group of respondents, which are difficult to measure using an observational technique, can be easily elicited via a questionnaire (McIntyre, 1999; Yuen, 2007).
- Data collected through a questionnaire can be easily analysed.
- Data entry and analysis for the questionnaire can be easily done using computer software packages, such as the SPSS and EQS (Bell, 1996; Hishamuddin, 2007; Yuen, 2007).

8.4.4.1 Questionnaire Survey Instrument

A questionnaire is a method used to gather information related to the opinions of a large group of people. A standardised questionnaire exposes each respondent to the same set of questions (Pinsonneault & Kraemer, 1993; Brace, 2008). This study applies a formal standardised questionnaire in order to achieve the research objectives. A well-structured questionnaire was used to collect data during the field survey. The questionnaire was based on the literature review conducted in the first stage of the research, as well as the findings from the Delphi study in stage two. The questionnaire consists of two sections. The first section was designed to collect information about the profile of the respondent and firm. This information included biographical, socio-economic and other information deemed necessary to meet the research objectives.

Section two of the questionnaire included questions on health and safety compliance as related to a safe environment, safe act of workers, safe working condition, reaction of workers to safe condition, government support, contractor's organisational culture, and health and safety compliance. This section was meant to collect information on the extent of the H&S compliance for each sub-attribute variable as provided. The questionnaire was designed to assess the influence of the identified constructs on health and safety compliance. The first section had fourteen questions relating to the profile of the respondents and firms. Section two had nine subheadings with a different number of questions in each section. The respondents were asked to rate each of the items on a 5-point Likert scale regarding the extent to which they agreed or disagreed with factors that determine health and safety compliance. The length of the questionnaire was ten pages, including the cover letter. This was in line with the recommended length of between five to twelve pages.

To avoid bias resulting from questionnaire design, the questions were constructed in such a way that they were direct, short, comprehensible, avoided ambiguity, not vague, avoided generalisations, not leading, not double-barrelled or presumptuous, simple and familiar to the respondents. Instructions of the questionnaire were kept simple with no technical or specialized words being used. However, it was recognised that this type of questionnaire has a few weaknesses in that

- There is an absence of probing beyond the answer given.
- There is a lack of control over who answers the questionnaire.
- They can be characterized by a low response rate because of cost.

These weaknesses were addressed by refining the questions and keeping them simple, but care was taken not to deviate from the objectives of the instrument, keeping the overall questionnaire within the recommended limits and ensuring that only the right person(s) completed the questionnaire. The absence of further probing is characteristic of this type of questionnaire. This aspect was not a major concern as the data to be collected was meant to validate the integrated conceptual model initially developed in the previous phase of the study. One type of response format was used, namely close-ended. To obtain the respondents' extent of agreement and disagreement towards the model-identified constructs, a labelled scale response format was used. Apart from the simplicity of administering and coding in further statistical analysis (Burns & Bush, 2002), the labelled scale response format is appropriate for health and safety compliance, as it allows the respondent to respond to questions to varying degrees that describe the dimensions being studied (Aaker et al., 2009).

Labelled Likert scales were appropriate to measure the responses for this study. This scale was adopted based on the following reasons:

- It yields higher reliability coefficients with fewer items than the scales developed using other methods (Hayes, 1998).
- This scale is widely used in health and safety compliance research and has been extensively tested in both marketing and social science studies (Garland, 1991).

- It offers a high likelihood of responses that accurately reflect respondent opinion under study (Zikmund, 2000; Burns & Bush, 2002).
- It helps to increase the spread of variance of responses, which in turn provide stronger measures of association (Wong, 1999; Aaker et al., 2009).

8.4.4.2 Variables

The research instrument was designed to measure the exogenous variables, namely safe environment (SE), safe act of workers (SAW), safe working condition (SWC), reaction to workers to safe condition (RWSC), government support (GS), contractor's organisational culture (COC); and the endogenous variable health and safety compliance. These variables were hypothesised to be characterised by indicator variables, which collectively constituted the questionnaire items apart from the influence on SME contractors' H&S compliance and influence on H&S compliance of SME contractors, which were also measured by the questionnaire. Table 8.8 gives a comprehensive summation of the latent and indicator variables.

8.4.4.3 Population

A study population refers to the entire group of items in which the researcher has an interest (Cooper & Schindler, 2006; Neuman, 2006). The population for this study is made up of all SME contractors in Ghana. The population of the study was based on SME contractors in the major cities in Ghana (Accra, Tema, Kumasi and Takoradi). The cities are located in the Greater Accra, Ashanti and Western Regions of Ghana. All small to medium-sized contractors had ongoing projects at the time of the study.

8.4.4.4 Sample Frame

The sample frame was established by obtaining a list of registered small to medium-sized construction companies in good standing within the last 10 years from the Ghana Ministry of Works, Housing and Water Resources. The selection of the number of contractors in the list was made possible based on a probability sampling technique (discussed in the next section). The contractors selected from the list were instructed on the cover letter of the questionnaire that their responses were of utmost importance to the researcher.

8.4.4.5 Sampling Method

The sample size was based on the number of respondents from 250 to 500 or more using the required number under structural equation modelling (SEM). A sample is a subset of items a researcher selects from a specific population (Neuman, 2006). Sampling is the process of selecting a sample consisting of units, such as people from a population of interest. By studying the sample we may answer the questions posed regarding some aspects of the population from which they were chosen (Trochim & Donnelly, 2007). The two general sampling methods are probability and non-probability sampling, which are usually differentiated by their randomness. A non-probability sample is also known as a non-random sample where samples are selected in some way not suggested by probability theory, but, sampling elements are selected using something other than a mathematically random process

TABLE 8.8
Conceptual Model Indicator Variable

Latent Variable Construct	Measurement Variables
Safe environment (SE)	Safe and healthy work environment
	Safe storage of equipment
	Safe storage of building materials
	Safe storage of formwork and false work
	Safe transportation of building materials
	Safe transportation of formwork and false work
	Safe transportation of equipment
	Provision of warning system
Safe act of workers (SAW)	Inspect workplace before commencing any activity
	Tidy up workplace at the end of any activity
	Use appropriate tools and equipment
	Do not work under the influence of alcohol and other drugs
	Do not smoke in flammable materials stores
	Ensure equipment or tools are in good condition before usage
	Ensure proper lifting, handling or moving of objects.
	Ensure proper stacking of objects or materials in safe locations
	Avoid annoyance and horseplay at the workplace
	Ensure the use of personal protective equipment
	Do not remove safety guards from the workplace or equipment
	Do not throw or accidentally drop objects from high levels
	Ensure proper positioning of tasks
	Do not work under the effects of alcohol and other drugs
	Do not service equipment which is in operation
	Concentrate on the task at hand
	Work in good physical conditions
Safe working condition (SWC)	Provision of training
	Good inspection programme
	Provision of incentives to workers
	Provision of safety regulation of equipment
	Good company safety policies
	Good salaries
	Payment of Social Security and National Insurance Trust
	Provision of sufficient lighting system for enclosed areas
	Safe movement around workplace
	Provision of guidance on the recommended illumination level for various types of tasks
	Workers should have adequate ventilation
	Provision of adequate facilities (toilet, drinking water, washing and canteen)
	Availability of facilities within a reasonable distance from the work area
	Provision of change room for workers
	Facilities must be available to both day and night workers
	Provision of safe means to facilities all the time
	Provision of break periods for workers to access facilities

(*Continued*)

TABLE 8.8 (CONTINUED)
Conceptual Model Indicator Variable

Latent Variable Construct	Measurement Variables
Reaction of workers to safe condition (RWSC)	Attend safety education programme
	Attend safety training programme
	Adhere to warning signs and notices
	Follow safety regulations
	Adhere to company safety policies
	Adhere to guidance on recommended illumination level for various tasks
	Put to proper use available facilities (toilet, drinking water, washing and canteen)
	Adhere to regular use of provided change room
Government support (GS)	Formulate H&S policy for construction
	Implementation of H&S policy by government representatives
	Monitoring of H&S policy implementation by government representatives
	Provision of H&S policy update by government representatives
	Provide H&S training by government representatives
Contractor's organisational culture (COC)	Provision of personal protective equipment
	Provision of signs and notices on sites
	Training of workers on health and safety
	Involve workers in H&S programmes
	H&S staffing
	H&S inspection
	Company H&S policy
	Management commitment to H&S
	Assessment of hazard identification and risk
	Consultation on H&S information to workers
	Update on H&S information to workers
Health and safety compliance (HSC)	Accidents on sites will be minimized
	Compensations paid to accident victims will be reduced
	Reduce cost of training on H&S
	Limited number of H&S education by government representatives
	Limited number of H&S monitoring by government representatives
	Improvement in H&S performance
	Increase in productivity

(Neuman, 2006). On the other hand, probability sampling allows each segment of the population to be represented in the sample. Probability sampling is also known as random sampling. Cooper and Schindler (2006) state that a probability sample is one based on the concept of random selection, a controlled procedure that ensures each population element is given a known non-zero or allows each member of the population to have an equal chance of being selected. In this case, the samples are chosen from a larger population by a process known as random selection.

The various sampling techniques employed in the selection of a probability sample are simple random, stratified random, systematic and cluster sampling. Simple random sampling allows the sample to be chosen by a simple random selection where every member of the population has an equal chance of being selected, while the stratified random sampling occurs in populations that consist of different strata or groups. In order to have equal representation in a stratified sample, the researcher selects samples equally from each one of the strata or group, whereas cluster sampling sub-divides an expansive area into smaller units, for instance, a country could be sub-divided into regions and further into towns. The clusters must be as similar to one another as possible, with each cluster containing an equally heterogeneous mix of individuals and a subset of the identified clusters is randomly selected.

The current study used the probability sampling method, which allows all segments of small to medium-sized construction companies as defined earlier to be represented in the sample, making sure that a representative sample of contractors were selected for this study. Therefore, a simple random and cluster sampling techniques were used, which allows each member of the population to have an equal chance of being selected (Kerlinger & Lee, 2000), whilst a cluster sampling technique divides the population into an expansive area with each cluster containing an equally heterogeneous mix of individuals. The rationale for selecting this method of sampling was based on the nature and composition of the contractors in the major cities of Ghana. Cluster random sampling was used to ensure representativeness. The selection of a representative sample for this study was based on the justification by Smith (2004) who informed that random sampling must be used for a study of this nature, hence it was adopted. Respondents were randomly selected and questioned (questionnaire survey) based on the knowledge that they have regarding health and safety compliance. All contractors from each city location had an equal chance to be drawn and to be part of the sample. Each category was classified as a cluster. Since the clusters differ in size, a proportional representation of each cluster was calculated.

8.4.4.6 Sample Size

Sample size is the number of observations or replicates to be included in a statistical sample. The sample size is an important feature of any empirical study in which the goal is to make inferences about a population from a sample, just like the current study. The sample size used in the study was determined based on the expanse of data collection, and the need to have sufficient statistical power to validate the conceptual model. The sample size for the current study was not based on the entire population of the selected SME contractors in the major cities of Ghana; therefore, the sample size was not equal to the population size. The question of how large a sample should be, depends on the following: the kind of data analysis the researcher plans to use; how accurate the sample has to be for the researcher's purposes and the population characteristics (Neuman, 2006).

The determination of sample size depends on the proposed data analysis techniques, finance and access to sampling frame (Malhotra, 1999). The proposed data analysis technique for this research is SEM utilizing EQS software, which is very sensitive to sample size and less stable when estimated from small samples (Tabachnick & Fidell, 2001). As a general rule of thumb, at least 300 cases are deemed comfortable, 500 are very good and 1000 are excellent (Comrey & Lee, 1992; Tabachnick & Fidell, 2001). Thus it was decided to target a sample size of 700 respondents from the major cities of Ghana. A large sample size alone does not guarantee a representative sample; this is because a large sample size without random sampling or with a poor sampling frame is less representative than a smaller one with random sampling and an excellent sampling frame (Neuman, 2006). However, the larger the size of the sample, the more likely its mean and standard deviation will be representative of the population's mean and standard deviation. A larger sample also makes it less likely that the researcher will obtain negative results or fail to determine the truth.

Hence, researchers should endeavour to maximize the sample size using the following guidelines for selecting a sample size (Leedy & Ormrod, 2005):

- For small populations with fewer than 100 people or other units, there is little point in sampling and surveying the entire population.
- If the population size is around 500, 50 percent of the population should be sampled.
- If the population size is around 1500, 20 percent should be sampled.
- Beyond a certain point (at about 5000 units or more), the population size is almost irrelevant and a sample size of 400 should be adequate.

For small populations, a researcher needs a large sampling size, and for moderately large populations, a smaller sample size of about 10 percent is needed to be equally accurate (Neuman, 2006). However, there is general disagreement on the 10 percent sample size recommendation for smaller populations. A sample size does not have to be large for it to be representative of the population. They state that the absolute size of a sample is much more important than its size compared to the population, and how large a sample should be is a function of the variation in the population parameters under study and the estimating precision needed by the researcher. It is suggested that a sample of 400 may be appropriate sometimes, while more than 2000 are required in yet other circumstances; and in another case, perhaps only 40 are called for (Cooper & Emory, 1995; Cooper & Schindler, 2006).

Furthermore, Smith (2004) simplified the process of sampling size by recommending that one may use 20 cases or 5 percent, whichever is greater for the population. The sample size should vary with the type of study. For instance, a routine review study would require 5 percent or 30 observations; a query review study would require 10 percent or 40 cases, whichever is greater; an intensive review study would require a sample size of 15 percent or 60 cases; and a sentinel event would require 100 percent of the observations. Hence, because of the kind of data analysis method (SEM) to be used in this study and the avoidance of negative results, which will jeopardize the model goodness of fit thus failing to establish the truth with regard to the constructs which predict health and safety compliance, a large sample size of 556 was considered. This is because the role of sample size is crucial in SEM analysis.

Therefore, the sample size requirement in this study was a function of the model framework development consideration. Harris and Schaubroeck (1990) proposed a sample size of 200 at least to guarantee robust SEM. Kline (2010) suggested that a very complicated path model needs a sample size of 200 or more, whereas Bagozzi and Yi (2012) proposed that the sample size should be above 200. Also, based on Smith's (2004) research classification, the study is both a query and an intensive review; hence the selection of the sample size demanded 10 percent or 40 cases and 15 percent or 60 cases, whichever is greater. The study sample size also agrees with Neuman's (2006) recommendation of a 10 percent sampling size. Therefore, the total sample size of the respondents from the major cities in Ghana was 558, which aligned perfectly with the analysis of covariance structure estimation requirement for SEM.

8.4.4.7 Sample Selection

The sample selection step required a detailed specification of all the steps discussed earlier (Malhotra, 1999). A total of 700 respondents were chosen from all localities for the research, which was equivalent to the sample size. Each contractor was divided into different clusters using the company name. A systematic random sampling was then applied through the selection of every fifth contractor in each cluster. For ease of identification of the fifth contractor, company names were used to calculate the number of the next fifth contractor. This process was essential in obtaining true representativeness of the entire sample.

8.4.4.8 Fieldworkers

Fieldworkers were recruited prior to the actual survey to assist with the administering of the questionnaires. A fieldworker is defined as an objective collector of data. He or she may or may not have formal qualifications but is perceived to have access to a particular community. A fieldworker primarily mediates or facilitates learning of individuals and groups to create an environment in which people can participate (Maart & Soal, 1996). Tamblyn and Shelton (1996) defined the data collection skills in the comprehensive market research manual that fieldworkers should possess, and thus recommended that fieldworkers must be selected with great care and be trained for a stipulated minimum of four hours before undertaking quantitative data collection. They concluded that it is essential for fieldworkers to have a good understanding of the area and the respondents and to be trained in the skills necessary for relating to people, analysing situations and designing strategies.

For this study, fieldworkers were recruited from bachelor of technology students of the Department of Building Technology at Cape Coast Technical University. They were selected based on the researchers' working knowledge of their ability and competence in H&S in the construction industry. Their resident status in the survey areas was also considered. The fieldworkers were trained by the researchers on the use (administering) of the questionnaire. A five-hour intensive training workshop was organised for the fieldworkers. A day's technical training (different days for respective fieldworkers) was also organised for the fieldworkers on the selected sites according to which areas they were assigned to collect the data. Fieldworkers were trained to identify specific contractors on the identified cluster and the study sample size; to gain the consent of respondents to be surveyed; to conduct interviews using the standard questionnaire; and to maintain standard procedures in conducting the survey and recording the answers. During the collection of data, a total of four fieldworkers were used. This was based on the sample size (700) and number of days (60) available for the data collection. The researchers identified the clusters to be surveyed in relation to work of the fieldworkers and supervised the first few days of data collection in the various study areas. The research coordinator conducted five to ten interviews to ensure that the fieldworkers followed instructions as stipulated on the survey instrument. The coordinator controlled the data quality by checking for errors during the survey and after each survey, and checked whether the questionnaires were completed fully and correctly. It was established that all the respondents answered the questions and identified problem areas, which were adequately resolved.

8.4.4.9 Data Collection

The selection and training of the fieldworkers took place after determining the sample size for the study. The questionnaire was personally administered to the contractors by the fieldworkers selected by the study coordinator. A personally administered or face-to-face structured questionnaire for data collection was the preferable option used for the current study. The questionnaire was given to respondents to be completed by themselves and, where necessary, was given clarity. It took approximately 35 minutes to complete each questionnaire, although respondents were informed in the cover letter that it would take 30 minutes to complete. The process of data collection took two months, from the middle of the month of June 2015 to the middle of August 2015. Most of the questionnaires were completed by the fieldworkers. The data collection process took a long period owing to this process. All 557 questionnaires collected were sent for data capturing after completion.

8.4.4.10 Data Analysis from the Questionnaire Survey

Coding the responses, data cleaning, screening the data and selecting the appropriate data analysis strategy are steps involved in data analysis. Coding of the questionnaire involved identifying, classifying and assigning a numeric or character symbol to data, which may be done in two ways: pre-coded and post-coded (Wong, 1999). The aspect of data analysis from the questionnaire survey will be discussed later in the book. Taken from the list of responses, a number corresponding to a particular selection was given. This process was applied to every question that needed this treatment. Upon completion, the data was then entered into a statistical analysis software package (SPSS) for the next analysis steps. In choosing the appropriate statistical analysis technique, the research elements were considered, namely the research problem, objectives, characteristics of data and the underlying properties of the statistical techniques (Malhotra, 1999). To meet the purposes of this study, descriptive and inferential analyses, and the goodness-of-fit measures of the model were applied where necessary. The data analysis involved the use of multiple analytical techniques to facilitate ease of communicating the results, while at the same time improving its validity. Hence, the use of SEM utilizing EQS software. Raw data from the questionnaire were entered into the SPSS software and were later exported to the SEM software EQS, version 6.2, for analysis. The motivation for the choice of the SEM and particularly the use of the software EQS is explained in the next section.

Inferential analysis refers to the cause-effect relationships between variables, which the current study hopes to establish between the identified model constructs. Inferential statistics use the results obtained from samples to generalise about a population (Forzano, 2008). Inferential statistics used for this research were correlations, exploratory factor analysis (EFA), confirmatory factor analysis (CFA) and structural equation modelling (SEM). SEM was used for the development and validation of the H&S compliance model. The statistical significance of the constructs was evaluated. The result's statistical significance was expressed by a *p*-value (Forzano, 2008). When the *p*-value is high, there is less possibility of an association between two variables (McClave, Benson & Sincich, 2008), whereas a smaller *p*-value gives a better likelihood of association. The *p*-value chosen in the present study is 0.05, which implies a 95 percent chance that the population mean is within a listed range of values (McClave et al., 2008).

SEM is currently the most inclusive statistical procedure in social and scientific research, catering for all operations of the general linear modelling (GLM) group of statistics such as analysis of variance (ANOVA), multivariate analysis of variance (MANOVA) and multiple regression (Kline, 2005:14). Though there are many ways to describe SEM, it is most commonly thought of as a hybrid between some form of ANOVA or regression and some form of factor analysis. In general, SEM allows one to perform some type of multilevel regression/ANOVA on factors. SEM is conceptually used to answer any research question involving the indirect or direct observation of one or more independent variables or one or more dependent variables. The primary goal of SEM is to determine and validate a proposed causal process or model. In the current study, the conceptualized holistically integrated H&S compliance model for small to medium-sized construction companies is being validated. SEM takes a confirmatory approach to the analysis of a structural theory bearing on some phenomenon (Byrne, 2010). SEM simultaneously estimates all coefficients in the model, and, therefore, it is able to assess the significance and strength of a relationship in the context of the entire postulated model (Dion, 2008). Hence, considering the conceptualized model in this study of unobserved (exogenous) variables, which had to be estimated from the observable variables, methods of analysis such as ANOVA could not be used as they lack a direct way of distinguishing between observed measures and the underlying constructs (Kline, 2005). A clear distinction is made in SEM between true variance and error variance, which implies that model parameters are estimated by taking measurement error into consideration. Likewise, a CFA was carried out on each exogenous variable to determine best fit for the model before the SEM was performed.

The choice of the software EQS for analysis was enhanced by the benefit of utilizing the Satorra-Bentler scaled statistics ($S - B\chi^2$), which provides an adjusted, more robust measure of fit for non-normal data. This approach is more accurate than the normal chi-square test statistics (χ^2) (Byrne, 2006). Likewise, EQS offers several different estimation methods for non-normal data as well, including the robust maximum likelihood (RML). EQS, version 6.2, a software package, was used for SEM as it is a user-friendly software that provides a graphical user interface that is easy to understand. EQS also enables data to be imported directly from SPSS. EQS was seldom used by previous researchers to enhance conceptual understanding of health and safety compliance research as compared to other techniques, such as AMOS and LISREL (Tong, 2007). Being a user-friendly graphically modelling interface,

EQS offers a wider variety of goodness-of-fit measures (Kline, 2005; Tong, 2007; Musonda, 2012; Aigbavboa, 2013; Mustapha, Aigbavboa & Thwala, 2016).

8.4.4.11 Data Screening and Preliminary Analysis

The use of the Statistical Package for the Social Sciences (SPSS) program, version 22, was to ensure consistency in data and provide meaningful interpretation of results. Screening of data was carried out during the preliminary analyses. This includes data cleaning, the handling of missing data, the normality test and the outliers. The screened data were further analysed using more complex analyses including exploratory and confirmatory factor analysis and SEM.

8.4.4.12 Data Cleaning

To ensure the accuracy of the data being coded and entered into the data file, a verification procedure was carried out to ensure the accuracy of the data being coded and entered into the data file. In this process, data were examined using descriptive statistics and graphic representations of the variables (Tabachnick & Fidell, 2007). Data cleaning was achieved by frequency tables, histograms, bar stem-and-leaf displays and box plots (Pallant, 2007). A summary of the values for the respondents' profiles and firms were obtained in frequency tables. Percentages and graphic displays were used for the descriptive methods to simplify and characterize the data.

8.4.4.13 Missing Data

Few questionnaires from the 558 questionnaires returned had missing information. Missing data in the current study were handled by direct maximum likelihood robust using the full information of the maximum likelihood in the EQS software program when confirmatory factor analysis and SEM were conducted.

8.4.4.14 Normality

A sample size can affect a study's finding where the outcome of smaller samples have too little statistical power for the test to realistically identify significant results (Hair, Anderson, Tatham & Black, 1998). Sample size can also be easily 'over-fitting' to the data in that they fit the sample very well but yet have no generalizability (Hair et al., 1998). Large sample sizes of more than 200 to 400 respondents have disadvantages due to making the statistical tests overly sensitive as a result of the increased statistical power from the sample size which the data can incur non-normality (Hair et al., 1998). Therefore, the data obtained for the current study were analysed for normality to ensure its suitability using standard multivariate analysis. Normality of data can be examined through statistical approaches such as skewness and kurtosis, the Kolmogorov-Smirnov test and graphical approaches, for example, histograms and box plots (Pallant, 2007). The variable's frequency value distribution should approximate the bell-shaped curve or a straight diagonal line to attain normality of the data (Hair et al., 1998; Pallant, 2007). The skewness and kurtosis were used for this study and they established that the data was slightly non-normal. To overcome the non-normality of the data, maximum likelihood estimation with robust standard errors and chi-square (MLR) and the EQS program, version 6.2, were used in this study when the CFA and SEM was analysed. This estimator method rectifies non-normal data.

8.4.4.15 Treatment of Outliers

Outliers are cases with extreme values on a single variable (univariate) or on a combination of variables (multivariate) (Pallant, 2007). Some causes of outliers are data entry errors, unusual events, unexplainable observations, and unusual or unique combined patterns (Hair et al., 1998). Univariate outliers were identified by examining the box plot of each variable (Tabachnick & Fidell, 2001). Few outliers were identified in the current study and were subsequently removed.

8.4.4.16 Criteria for Determining Reliability and Validity

The validity and reliability of a questionnaire are vital to test the construct validity and reliability of the H&S practices and H&S compliance variables. Validity and reliability provide the degree to which latent variables measure what they are supposed to measure. Construct validity comprises numerous sub-dimensions, all of which must be satisfied to achieve construct validity. For construct validity, the current study utilized content validity, convergent validity and discriminant validity. In addition to this, the current study further utilized Cronbach's alpha, corrected item-correlation and construct/composite reliability to measure the reliability of the SME questionnaire instrument.

8.4.4.17 Content Validity

The content validity is the extent to which a constituent variable belongs to its corresponding construct (Pallant, 2007; Leedy & Ormrod, 2010). Since content validity cannot be tested using statistical tools, an in-depth literature survey was necessary to specify the variables and define the latent variables in order to keep the researchers' judgment on the right track. An exploratory study with H&S experts using the Delphi method was conducted.

8.4.4.18 Reliability

Reliability tests were conducted after establishing the content validity and preliminary data analyses empirically. Scale reliability is the correlation between two scores ranging from 0 to 1.00 where the Cronbach's alpha is the most common form of internal consistency reliability coefficient. A lenient cut-off of 0.60 is common in exploratory research. The generally agreed upon lower limit for alpha is 0.70 (Hair et al., 1998) and a cut-off of 0.80 for a good scale (Lingard, Wakefield & Cashin, 2011). The adopted cut-off alpha for this study is 0.70, and measures below 0.70 were eliminated. Composite or construct reliability for CFA was calculated after the re-specification of the measurement model. Composite reliability represents a better choice, and it draws on standardized loadings and measurement error for each item; however, it was reported that only 20 percent of SEM studies reported composite reliability to exceed 0.60 (Shook et al., 2004). This was calculated as follows: Sum of standardized loadings^2 / (Sum of standardized loadings^2 + Sum of indicator measurement error).

8.4.4.19 Convergent Validity

The analysis of convergent validity was carried out after the establishment of the content validity and preliminary data analyses. Convergent validity was tested by determining whether the scores of items in one scale correlate with the scores on the other scales, and converge or load together on a single construct in the measurement model (Hair, Black, Babin, Anderson & Tatham, 2006). Factor loadings of 0.30 and 0.40 are considered significant for sample sizes of 350 and 200, respectively (Hair et al., 2006). No factor loadings were less than 0.30 in the current study. Had there been any, this would have been considered not significant. The current study used a cut-off value of 0.30 to retain the variables as suggested by Hair et al. (2006).

8.4.4.20 Discriminant Validity

The analysis of discriminant validity was carried out after the establishment of the content validity and preliminary data analyses. Discriminant validity is the extent to which items representing a latent variable are unique and capture some phenomena that other measures do not (Hair et al., 2006). Items on one construct should not load or converge too closely with items from other scales. Different latent variables that correlate too highly may indeed measure the same construct instead of different constructs. Items in a construct attaining intercorrelations below 0.90 suggest that there is no multicollinearity, but the constructs have discriminant validity (Hair et al., 1998). Another method of attaining discriminant validity as indicated is the average variance estimate (AVE). The AVE is calculated as follows: Sum of squared standardized loadings / (Sum of squared standardized loadings + Sum of indicator measurement error) (Fornell, 1983). Despite the two methods of analysing discriminant validity, the intercorrelation of the items in a construct-element was used for this study.

8.4.5 Methods of Data Analysis of the Quantitative Survey

Descriptive and multivariate correlation data analyses were conducted after screening the data. Three major steps were carried out in the multivariate correlation analysis: exploratory factor analysis, structural equation modelling which includes confirmatory factor analysis, and structural model testing. Analyses were conducted on selected respondent profiles and firm variables with the elements of the H&S compliance model.

8.4.5.1 Exploratory Factor Analysis

Exploratory factor analysis (EFA) was performed to gather information about the unidimensionality of the factors to yield their factor analysability. The EFA was conducted using SPSS, version 22. EFA being a precursor to SEM was used in the current study to confirm the validity and reliability of the six element-constructs of the proposed construction H&S compliance model. The maximum likelihood, with a minimum eigenvalue of one, together with principal axis factoring with oblimin Kaiser normalization was specified as the analysis method for this study. A bivariate correlation was performed to assess the strength among the research elements of the H&S compliance model.

8.4.5.2 Structural Equation Modelling

Structural equation modelling (SEM) was performed immediately after exploratory factor analysis. The development of the methods of analysis involving SEM with latent variables has provided researchers considerable means to construct, test and modify theories (Anderson & Gerbing, 1988). SEM represents a component of the methodological instrument of the social sciences. It is a comprehensive statistical approach for testing hypotheses about relationships among observed and latent variables (Kline, 2005; Molenaar, Park & Washington, 2009). It assesses whether the sample covariance matrix is consistent with the hypothesised model (Sweeny, 2009). In comparison to other statistical analytical techniques such as factor and regression analysis, SEM is a relatively young field.

The increasing complexity of research questions in the social and behavioural sciences and the appearance of flexible user-friendly computer software programs, especially with the advent of Windows applications (Marcoulides & Hershberger, 1999), has increased the interest in SEM (Kelloway, 1998). It has been argued to be the most important multivariate correlational analysis technique (Byrne, 2010). Recent researchers have shown that SEM has become increasingly popular in the construction H&S literature. For example, Agumba (2013), Musonda (2012), Molenaar et al. (2009), Chanda and Mohamed (2008) and Mohamed (2002) used SEM to develop H&S models. Therefore, it was considered the most suitable method of analysis for the third phase of the current study in testing the conceptualised model. This is because SEM can be used to analyse and test theoretical models (Schumacker & Lomax, 2004).

8.4.6 Structural Equation Modelling Analysis

A two-step process was followed after the EFA to test the structural model. The approach included confirmatory factor analysis and structural model testing (Anderson & Gerbing, 1988). First, it is vital to confirm the measurement model before the structural model can be finalized. The measurement models test relationships which are paths between the measures, also known as manifest or observed variables, and constructs, also termed as latent variables, whilst the structural model clarifies the causal relationships, as well as the degrees of influence (Tabachnick & Fidell, 2007).

8.4.6.1 Structural Equation Modelling Process

Structural model testing can be carried out through five basic steps involved in SEM analysis, namely (1) model specification, (2) model identification, (3) model estimation, (4) model testing and (5) model modification (Schumacker & Lomax, 2004). The study adopted Schumacker and Lomax's (2004) process. However, Hair et al. (1998) proposed a seven-step process for SEM analysis which included: (1) developing a theoretical model, (2) constructing a path diagram of causal relationships, (3) converting the path diagram into a set of structural equations and measurement models, (4) choosing the input matrix type (correlation matrix or covariance matrix) and estimating the proposed model, (5) assessing the identification equations and measurement models, (6) choosing the input matrix type (correlation matrix or covariance matrix) and estimating the proposed model, assessing the identification of model equations, evaluating the results of goodness-of-fit and (7) making the indicated modifications to the model if theoretically justified.

8.4.6.1.1 Step 1: Model Specification

Model specification is the first vital step in SEM. It involves developing a theoretical model (Schumacker & Lomax, 2004). This process must be guided by a combination of theory and empirical results from previous research (Hair et al., 2006), although the role of informed judgement, hunches and dogmatic statements of belief should not be discounted (Kelloway, 1998). This was the first step performed in this study. In particular, attention must be paid to include all relevant variables. If the theoretical model is not consistent with the true model, the theoretical model is said to be mis-specified and lacks validity. This may occur if the researcher failed to include an important variable or an important parameter or, alternatively, if an unimportant parameter or variable was included in error (Schumacker & Lomax, 2004). Having developed the theoretical framework of the model, the next step was to illustrate this in a path diagram, which is a pictorial representation of all relationships in the model. This is a graphical representation of how the various constructs of the model relate to one another. This is essentially the first step in the SEM process (Kline, 2005; Byrne, 2010). While it is not a formal requirement of SEM, construction of a path diagram offers important benefits. Specifically, the hypotheses contained in the model are much more easily comprehended in visual form than in either verbal or mathematical terms. It may also help improve the conceptualization of the model by drawing attention to omitted links or excluded variables, thereby decreasing the possibility of specification error (Diamantopoulos, 1994; Kline, 2005). Path diagrams not only enhance the understanding of structural models but substantially contribute to the creation of the correct input files (Raykov & Marcoulides, 2000).

Path models adhere to common drawing specifications that are utilised in SEM models. The observed variables are enclosed by boxes or rectangles. The relationships between the latent variables and their corresponding indicators are represented by arrows that originate at the latent variable and end at the indicators. Each indicator is also associated with an error term representing the errors in measurement. The error terms associated with the endogenous variables represent error in equations and indicate that the dependent variables in the model are not perfectly explained by the independent variables. A curved double-headed line between two independent variables indicates covariance (Kline, 2005; Byrne, 2010).

8.4.6.1.2 Step 2: Model Identification

The second step of SEM is model identification. This is crucial because identification problems should be resolved prior to the estimation of parameters (Schumacker & Lomax, 2004; Kline, 2005). Identification revolves around the question of whether one has sufficient information to obtain a unique solution for the parameters to be estimated by the model (Byrne, 2010). Identification determines whether it is possible to find unique values for the parameters of the specified model. It concerns the correspondence between the information to be estimated (the free parameters) and the information from which it is to be estimated (the observed variances and covariance).

Models can be under-identified, just-identified or over-identified. A model is considered just-identified if it has only one estimate for each parameter and generates zero degrees of freedom and therefore cannot be rejected (Byrne, 2010). A just-identified model will always provide one unique solution that will be able to perfectly reproduce the correlation matrix (Kline, 2005). However, the solution is uninteresting because it has no generalizability (Hair et al., 2006). An under-identified model is obtained when one or more parameters are not uniquely determined, that is the number of unknowns exceeds the number of equations, therefore there is not much empirical information to allow its unique estimation (Schumacker & Lomax, 2004; Kline, 2005) and therefore its estimation should not be relied upon (Kline, 2005). The most accepted situation is one in which there are more indicators than unknown variables and the model is over-identified and has positive degrees of freedom (Byrne, 2010).

Only models that are identified can be estimated (Kline, 2005). In an over-identified model, there are a number of possible solutions, and the task is to select the one that comes closest to explaining the observed data within some margin of error (Kelloway, 1998). The determination and identification status of a model can be difficult; but it is best to simply count the number of parameters in the model and subtract this from the number of non-redundant elements in the sample correlation matrix (Hoyle, 1995; Raykov & Marcoulides, 2000). The formula for determining a model can be represented as follows:

$$N(N+1)/2 \tag{8.2}$$

where N is the number of observed variables in the model. The resulting difference is referred to as the degrees of freedom. If positive, the model is considered to be identified. The current structural and measurement models were over-identified and were therefore appropriate for testing.

8.4.6.1.3 Step 3: Model Estimation

The third step is model estimation. The purpose of estimation is to generate numerical values for free parameters within the model that produces the implied matrix (Σ) such that the parameter values yield a matrix as close as possible to the sample covariance matrix (S). The estimation process involves the selection of a particular fitting function to minimise the difference between Σ and S. In addition, EQS software can estimate models with regressions among combinations of continuous latent variables and observed variables. It can also estimate factor indicators and other observed dependent variables when they are all continuous. Several fitting functions or estimation procedures are available to be used. EQS can use different estimator choices: maximum likelihood (ML), maximum likelihood with robust standard errors and chi-square (MLR), generalized least squares (GLS) and weighted least squares (WLS) (Muthèn & Muthèn, 2007). This study adopted MLR in the EQS program, because of slight non-normal data identified in the process of data screening and its robustness to rectify non-normality.

8.4.6.1.4 Step 4: Model Testing

Once the parameter estimates are obtained for a SEM model, the fourth step is to determine how well the data fit the model (Schumacker & Lomax, 2004). Assessing whether a specified model fits the data is an important step of SEM (Yuan, 2005), as it determines whether the model being tested should be accepted or rejected. Model fit refers to the extent to which a hypothesised model is consistent with the data (Schumacker & Lomax, 2004). Goodness of fit in SEM is the degree to which the actual/observed input matrix is predicted by the estimation model (Hair et al., 2006). A model is said to fit the observed data to the extent that the covariance matrix it implies is equivalent to the observed covariance matrix (Hoyle, 1995). The process of estimation results in an implied covariance matrix Σ that is as close as possible to sample covariance matrix S; the closer Σ is to S, the better the fit of the model (Schumacker & Lomax, 2004). Model fit represents one of the most controversial areas of SEM (Barrett, 2007). If the model does not fit the data such that the observed covariance matrix is statistically different from the covariance structure of the model, either the model or the data should be rejected (Fornell, 1983).

The issue of model fit assessment has been a real problem (Barrett, 2007). It represents a major challenge facing theory developers and researchers, as different indexes and values are reported (Kline, 2005; Barrett, 2007), unlike many statistical procedures that have a single, powerful fit index, for example, F-test in ANOVA. In SEM there are an increasingly large number of model fit indices (Schumacker & Lomax, 2004). Determining the tests that best suit the model is a matter of discretion. As such, there is a possibility that only those fit measures that fall within the acceptable range and support the proposed model are reported (Kline, 2005). Model fit indices provide no guarantee that a model is useful. Fit indices provide information on a models lack of fit and do not reflect the extent to which the model is plausible (Kline, 2005). Even if a model fits well, data can never confirm a model. They can only fail to disconfirm. Other, equal or better fitting models may exist (Maruyama, 1998).

8.4.6.1.4.1 Discussion of the Selected Fit Indices Assessment of model fit under SEM is an important process (Kline 2005; Yuan, 2005), as fit measures are continually evolving (Kline, 2005). The assessment of model fit under SEM determines whether the proposed model should be accepted or rejected (Hu & Bentler, 1999). It is unnecessary and unrealistic to include every index in the program's output (Hooper, Coughlan & Mullen, 2008). Similarly, the cut-off criteria for fit indexes vary for different studies. It is difficult to designate a specific cut-off value for each fit index, because it does not work equally well with various conditions (Hu & Bentler, 1999). The controversy and discussion on subjective interpretation and appropriateness under specific modelling conditions used is another dimension (Schumacker & Lomax, 2004). It is now common practice to use multiple tests when evaluating and reporting overall model fit (Hu & Bentler, 1999).

Although there is not an agreed and strict list-of-fit indexes to be examined and criteria to be met while evaluating the model fit, Hooper et al. (2008) indicated three types of goodness-of-fit measures used, namely:

- Absolute fit measures or accurate fit measures that measure the overall model fit both measurement and structural models, with no adjustment for the degree of over-fitting that might occur using root mean square error of approximation (RMSEA) and standardised root mean squared residual (SRMR)
- Incremental fit measures that compare the proposed model to a baseline model specified by the researcher using the Tucker-Lewis fit index (TLI) and the comparative fit index (CFI)
- Parsimonious fit measures that adjust the measures of fit to provide a comparison between models with differing numbers of estimated coefficient using chi-square divided by the degree of freedom, that is, χ^2/df

Based on the aforementioned discussions, the following model fit indices were selected for the current study. The most common model fit index is the chi square (χ^2) goodness-of-fit test (Kline, 2005). A significant chi-square indicates the rejection of the null hypothesis, suggesting that the model is not plausible in the population. Goodness-of-fit indexes usually represent indexes ranging from zero to one, with zero indicating a complete lack of fit and one indicating perfect fit. However, with regard to the chi-square test, zero indicates perfect fit and large numbers indicate extreme lack of fit; as such the chi-square test has been referred to as a lack-of-fit test (Mulaik, James, Van Alstine, Bennet, Lind & Stilwell, 1989). However, goodness-of-fit tests tend to be quite sensitive to sample size. This is especially true for the chi-square test (Hair et al., 2006). When the sample size becomes very large, the chi-square test increases in sensitivity and becomes impractical (Kelloway, 1998). Similarly, a poor fit based on a small sample may result in a non-significant chi-square, implying one should accept the model (Kline, 2005). Based on the sensitivity of the chi-square index to sample size (Kelloway, 1998; Schumacker & Lomax, 2004), a number of fit indexes were chosen for this study to supplement the chi-square index. The normed chi-square was used as a supplement index to the chi-square. The normed chi-square reduces the sensitivity of the sample size. Its value is achieved when the value of chi-square (χ^2) is divided by the degrees of freedom, χ^2/df, which results in a lower value called the normed chi-square. The minimal acceptable cut-off value has not been clearly established (Kline, 2005).

Values of less than 2 or 3 have been specified to be standard and applicable (Hsu, Su, Kao, Shu, Lin & Tseng, 2012). However, studies in construction H&S have used cut-off values of less than 5 as acceptable to data fit (Chanda & Mohamed, 2008; Agumba, 2013). A cut-off value less than 5 was adopted in this study.

The root mean square error of approximation (RMSEA) focuses on the discrepancy between the model and population covariance matrices per degree of freedom. It is generally regarded as one of the most informative fit indexes (Kline, 2005). RMSEA has been reported in previous structural equation modelling studies in H&S (Fernandez-Muniz, Montes-Peon & Vazquez-Ordas, 2007; Chanda & Mohamed,

2008; Molenaar et al., 2009; Agumba, 2013). An RMSEA value of 0.05 or less is indicative of good fit. Values between 0.05 and 0.08 are considered reasonable. Values between 0.08 and 0.10 are considered a moderate fit and values greater than 0.10 indicate poor fit (Hsu et al., 2012). A cut-off value of 0.07 is also considered acceptable (Steiger, 2007). A cut-off value of less than 0.05 was adopted for this study. Standardised root mean square residual (SRMR) has a lower bound of zero and an upper bound of one. Generally, values below 0.05 are interpreted as indicating good model fit (Hair et al., 2006; Hsu et al., 2012), whereas values ranging from 0.05 to 0.09 are considered acceptable fit (Hu & Bentler, 1999). Values less than 0.10 are favourable (Kline, 2005). These indexes were therefore adopted for this study.

However, the comparative fit index (CFI) is the revised form of the normed fit index (NFI), which takes into account the sample size (Byrne, 2010). CFI performs well even with small samples (Tabachnick & Fidell, 2007). This was introduced in the 1990s by Bentler and was subsequently included as part of the EQS program (Kline, 2005). The CFI assumes that all latent variables are uncorrelated (null/independence model) and compares the sample covariance matrix with the null model. The statistical values for CFI range between zero and one with values closer to one indicating good fit. A cut-off value for CFI greater than or equal to 0.90 (Hsu et al., 2012) can be accepted as a good fit. However, recent studies have shown that a value greater than 0.90 is needed in order to ensure that mis-specified models are not accepted (Hu & Bentler, 1999). A CFI greater than or equal to 0.95 is presently recognized as indicative of good fit (Hu & Bentler, 1999). The CFI has been reported in previous H&S studies by Agumba (2013), Chanda and Mohammed (2008), Larsson, Pousette and Törner (2008) and Fernandez-Muniz et al. (2007) and therefore was adopted for this study. A cut-off value of greater than 0.80 was adopted for this study.

The Tucker-Lewis index (TLI), also known as the non-normed fit index (Schreiber, Stage, King, Nora & Barlow, 2006), that penalises model complexity was also used for this study. The TLI is one of the indices that are less affected by sample size. A TLI greater than or equal to 0.90 (Hsu et al., 2012) indicates acceptable model fit. Some studies have used the cut-off values of 0.80 since TLI tends to run lower than the goodness-of-fit index (GFI) (Sweeny, 2009). The use of TLI has been supported because it is relatively insensitive to sample size, sensitive to model mis-specifications, relatively insensitive to violations of assumptions of multivariate normality and relatively insensitive to estimation method (Hu & Bentler, 1999). Previous studies in H&S have reported on this index (Chanda & Mohamed, 2008; Molenaar et al., 2009). Therefore, this index was adopted for this current study because of its benefits. A cut-off value of 0.80 was used for this study.

8.4.6.1.5 Step 5: Model Modification

In the final step of SEM, if the fit of the implied theoretical model is not as strong as desired (which is often the case with initial models), then the next step is to modify the model and subsequently evaluate this modified model (Schumacker & Lomax, 2004). Modifications can be made by linking the indicators to the latent variable from fixed to free or vice versa, allowing or constraining correlations among measurement errors or allowing or constraining correlations among latent variables (Hair et al., 2006). This process is known as model modification (Kline, 2005). This is done to improve the model. It implies a better fitting model and/or a more parsimonious model that is substantively more interpretable. To assist in the process, most SEM software provides modification indexes for each fixed parameter. This value indicates the minimum improvement that could be obtained in the chi-square value if that parameter was fixed for estimation (Schumacker & Lomax, 2004; Kline, 2005).

Providing a slightly different approach, the Wald index indicates how much a proposed model's chi-square would increase if a particular parameter were fixed to zero (Raykov & Marcoulides, 2000). Another method is to examine the residual matrix, the difference between the observed covariance matrix (S) and the model implied covariance matrix (Σ). It should be small in magnitude and should not be larger for one variable than another. Large values overall signify general model mis-specification whereas large values for single variables indicate mis-specification for that variable only (Schumacker & Lomax, 2004). Testing and revising of models should not become a procedure completely determined by statistical results, devoid of theoretical underpinnings. Theoretical considerations must guide model modifications (Kline, 2005), because adjusting a model after initial testing increases the chance of making a type I error. Blind use of modification indexes can lead researchers astray from their original goal (Raykov & Marcoulides, 2000). In this current study, the structural model was not modified, but some of the measurement models were modified.

8.4.7 Ethical Considerations

Ethical considerations were relevant to consider the proper conduct of this research. This research has upheld the responsibility to protect the interests of the survey respondents. With regard to the survey respondents, no one was forced to respond to this survey. Respondents were asked to participate of their own free will. They were told of their rights not to participate or to end their participation if they so wished. In addition, they were briefed about the purpose of the study, and how or why they were chosen. As such, they were free from deception or stress that might arise from their participation in this research. The respondents were also guaranteed protection through anonymity and all information that may reveal their identity would be held in strict confidence.

8.5 SUMMARY

This chapter presented the methodology adopted for conducting this research study. It also provided the justifications for the philosophical position and methods of data collection. The research design described in this chapter has linked three important elements of the research methodology, namely the underlying philosophical assumptions, the research methods/approach, and the process followed in the questionnaire administration, as well as an introduction to the data analysis. In testing the theoretical model proposed for this study, the survey method is discussed. The descriptive and multivariate correlation data analyses were also discussed. The methodology for analysing the proposed conceptual model was discussed including the process to be followed when using SEM. The three main steps for conducting the multivariate correlation analysis were discussed, namely exploratory factor analysis (EFA), confirmatory factor analysis (CFA) and structural equation modelling (SEM). Finally, ethical considerations pertaining to the collection of data were discussed, including issues relating to validity and reliability. The next chapter presents the results of the data analysis from the Delphi study.

REFERENCES

Aaker, A., Kumar, V. and George, S. (2009). *Marketing research* (10 edn.). New York: John Wiley & Sons, Inc.

Adenuga, O.A., Soyingbe, A.A. and Ajayi, M.A. (2007). A study on selected safety measures on construction companies in Lagos, Nigeria. RICS (Cobra).

Adler, M. and Ziglio, E. (1996). *Gazing into the oracle: The Delphi method and its application to social policy and public health*. London: Kingsley Publishers.

Agumba, J.N. (2013). A construction health and safety performance improvement model for South African small and medium enterprises. Unpublished DPhil in Engineering Management, University of Johannesburg, South Africa. Available from: www.ujdigispace.uj.ac.za [Accessed 05 October 2015].

Aigbavboa, C.O. (2013). An integrated beneficiary centred housing satisfactory model for publicly funded housing schemes in South Africa. Unpublished DPhil in Engineering Management, University of Johannesburg, South Africa, Available from: www.ujdigispace.uj.ac.za [Accessed 15 June 2014].

Amara, R. (1975). *Some methods of futures research*. Menlo Park, CA: Institute for the Future.

Amaratunga, D., Baldry, D., Sarshar, M. and Newton, R. (2002). Quantitative qualitative research in the environment. *Work Study*, 51(1): 17–31.

Anderson, J. and Gerbing, D. (1988). Structural equation modelling in practice: A review and recommended two-step approach. *Psychological Bulletin*, 103(3): 411–423.

Anderson, J.D. (2006). Imperial COE. Superintendent of Schools. Available from: http://www.icoe.org/ [Accessed 15 April 2014].

Bagozzi, R.P. and Yi, Y. (2012). Specification, evaluation, and interpretation of structural equation models. *Journal of the Academy of Marketing Science*, 40(1): 8–34.

Barrett, P. (2007). Structural equation modeling: Adjudging model fit. *Personality and Individual Differences*, 42(5): 815–824.

Beales, R. (2005). Delphi creates fresh CDS challenge: A new cash-settlement method for simplifying credit derivatives faces a stern test with the latest bankruptcy. *The Financial Times*, 43.

Bell, S. (1996). *Learning with information systems: Learning cycles in information system development*. New York: Routledge.

Boote, D.D. and Beile, P. (2005). Scholars before researchers: On the centrality of the dissertation literature review in research preparation. *Educational Researcher*, 34(5): 3–15.

Brace, I. (2008). *Questionnaire design: How to plan, structure and write survey material for effective market research*. Philadelphia: Kogan Page Publishers.

Brieschke, P.A. (1992). Reparative praxis: Rethinking the catastrophe that is social science. *Theory into Practice*, 31(2): 173–180.

Brill, J., Bishop, M. and Walker, A. (2006). The competencies and characteristics required of an effective project manager: A web-based Delphi study. *Educational Technology Research and Development*, 54(2): 115–140.

Bryman, A. (2001). *Social research methods*. Oxford: University Press.

Buckley, C. (1995). Delphi: A methodology for preferences more than predictions. *Library Management*, 16(7): 16–19.

Burns, A.C. and Bush, R.F. (2002). *Marketing research: Online research applications*. Upper Saddle River, NJ: Prentice Hall.

Burrell, G. and Morgan, G. (1994). *Sociological paradigms and organisational analysis* (6th edn.). London: Heinemann, 1–37.

Byrne, M.B. (2010). *Structural equation modeling with AMOS: Basic concepts, applications and programming* (2nd edn.). New York: Routledge/Taylor & Francis Group.

Cavalli-Sforza, V. and Ortolano, L. (1984). Delphi forecasts of land use: Transportation interactions. *Journal of Transportation Engineering*, 110(3): 324–339.

Chanda, T. and Mohamed, S. (2008). Structural equation model of construction safety culture. *Engineering, Construction and Architectural Management*, 15(2): 114–131.

Comrey, A.L. and Lee, H.B. (1992). *A first course in factor analysis*. Hillsdale, NJ: Lawrence Erlbaum Associates.

Cooper, D. and Schindler, P. (2006). *Marketing research*. New York: McGraw-Hill.

Cooper, D.R. and Emory, C.W. (1995). *Business research methods*. Chicago: Irwin.

Cresswell, J.W. (2003). *Research design: Qualitative, quantitative and mixed methods approaches* (2nd edn.). Thousand Oaks, CA: Sage.

Creswell, J.W., Clark, V.L., Gutmann, M.L. and Hanson, W.E. (2007). Advanced mixed methods research designs. In Tashakkori, A. and Teddlie, C. (Eds.), *Handbook of mixed methods in social and behavioural research*. Thousand Oaks, CA: Sage, 209–240.

Crisp, J., Pelletier, D., Duffield, C., Adams, A. and Nagy, S. (1997). The Delphi method. *Nursing Research*, 46: 116–118.

Critcher, C. and Gladstone, B. (1998). Utilizing the Delphi technique in policy discussion: A case study of a privatized utility in Britain. *Public Administration*, 76(3): 431–449.

Crotty, M. (1998). *The foundation of social research: Meaning and perspective in the research process*. London: Sage.

Cuhls, K. (2003). Delphi methods. http://www.unido.org/ [Accessed 15 March 2015].

Dalkey, N.C. and Helmer, O. (1963). An experimental application of the Delphi method to the use of experts. *Journal of the Institute of Management Sciences*, 9: 458–467.

Delbecq, A.L., Van de Ven, A.H. and Gustafson, D.H. (1975). *Group techniques for program planning: A guide to nominal group and Delphi processes*. Glenview, IL: Scott, Foresman & Company.

Diamantopoulous, A. (1994). Modelling with LISREL: A guide for uninitiated. *Journal of Marketing Management*, 10(1–3): 105–136.

Dion, P.A. (2008). Interpreting structural equation modelling results: A reply to Martin and Cullen. *Journal of Business Ethics*, 83: 365–368.

Dunn, W.N. (1994). *Public policy analysis: An introduction*. Englewood Cliffs, NJ: Prentice Hall.

Ellram, L. (1996). The use of the case study method in logistics research. *Journal of Business Logistics*, 17(8): 93–138.

Els, D.A. and Delarey, R.P. (2006). Developing a holistic wellness model. *South African Journal of Human Resources Management*, 4(2): 46–56.

Fernandez-Muniz, B., Montes-Peon, M.J. and Vazquez-Ordas, J.C. (2007). Safety culture: Analysis of the causal relationships between its key dimensions. *Journal of Safety Research*, 38(6): 627–641.

Fornell, C. (1983). Issues in the application of covariance structure analysis: A comment. *Journal of Construction Research*, 9(4): 443–448.

Forzano, G. (2008). *Research methods for the behavioural sciences*. Belmont, CA: Cengage Learning.

Garland, R. (1991). The mid-point on rating scale: Is it desirable? *Marketing Bulletin*, 2: 66–70.

Garrod, B. (2008). The Delphi technique. Institute of Rural Sciences. University of Wales: Aberystwyth, UK.

agreement and that round three successfully refined the discussion to the point that clear points of consensus or a lack thereof could be determined; therefore, a fourth round was not necessary.

The median, mean, standard deviation (SD), percentages and interquartile deviation (IQD) scores of each question were calculated. In situations where the score was two points from the median score, the experts were requested to give detail explanations on the responses. If a consensus had not been formed at the third round, the data from the third round would have been analysed again by calculations for median, mean, standard deviation and percentages scores and sent to the experts for consideration in responding to a fourth round. A fourth round was not needed, and the participants were informed of this when the third round questionnaires were sent out. The goal of the research technique was to cycle the questions towards a consensus amongst the experts. During each round of questionnaires, the experts were given the results of the median of the previous round. It was anticipated that by the third round, responses would converge to indicate a consensus from the experts. A consensus is achieved with 100 percent of the participants in agreement, but two-thirds in agreement is also considered a common consent (Stitt-Gohdes & Crews, 2004).

The goal for this study was that each question or statement should have a consensus, but common consent would be acceptable. Common consent was obtained if 60 percent of the experts agreed on each statement that was achieved in the study. All statements were examined individually for consensus. The quantitative results were statistically analysed after each round of questionnaires to determine whether a consensus had been reached for each question or statement using the provided scale for each question or statement. If consensus was reached prior to the final round, that question or statement was no longer required (asked/required) in the next rounds. After the third round of the Delphi survey, consensus was reached regarding most of the attributes that determine health and safety compliance in small to medium-sized construction companies in Ghana.

Based on the findings of the analyses of responses from the Delphi rounds, a list of attributes that determine health and safety compliance was prepared, which informs the conceptual framework for the broader study. Issues relating to health and safety compliance in Ghana were highlighted, which responded accordingly to the set objectives of the Delphi study. The results of the Delphi study are presented regarding the specific objectives of the Delphi in the next section.

9.3 FINDINGS FROM THE DELPHI STUDY

9.3.1 To Identify the Main and Sub-Attributes That Determine Health and Safety Compliance in Small to Medium-Sized Construction Companies in Ghana

A set of main attributes and sub-attributes that are relevant to health and safety compliance were extracted through a comprehensive review of literature. Although, the reviewed literature was based on studies from the developed countries, they were collectively used to examine the attributes that could determine health and safety compliance in the Ghanaian construction industry. The influence of the attributes on health and safety compliance was obtained as a product of the impact on small to medium-sized construction companies in Ghana. The main attributes were based on the level of influence, as categorized on the questionnaire used for the Delphi study. This was established by assessing the extent to which the listed attributes would determine the health and safety compliance among small to medium-sized construction companies. The impact of the sub-attributes in determining health and safety compliance was likewise assessed if they were present or lacking. The rating was based on an ordinal scale of 1 to 10 with 1 being low influence or no impact and 10 being high influence or very high impact. The levels of influence and impact were obtained as a product of the consensus achieved as detailed in the proceeding chapter.

Five attributes from the six identified main attributes that determine health and safety compliance were considered by the experts to have a high influence, with the exception of one attribute (government support). Only one attribute, safe working condition, reached consensus with the IQD cut-off (IQD ≤1) score set to reach consensus (Table 9.1). By applying the median to determine whether an attribute reached consensus or not, all six attributes were considered by the experts to have reached consensus. Five of the attributes had good consensus, with the exception of government support, which had weak consensus.

Eight attributes were identified from the first sub-attribute (safe environment). The experts considered seven attributes out of eight to have high influence on the attributes that determine H&S compliance. None of the eight attributes had an IQD cut-off (IQD ≤1) score set to achieve consensus. Seven out of the eight attributes had high impact (7–8) on H&S compliance. One attribute (safe transportation of building materials) had medium impact (5–6) on H&S compliance (Table 9.2). Seven attributes had good consensus, with the exception of one, safe transportation of building materials, which had weak consensus. The outcome of these results is depicted in Table 9.2.

TABLE 9.1
Health and Safety Compliance Main Attributes

Health and Safety Compliance Main Attributes	Median	Mean	SD	IQD ≤1
Safe environment	8	7.43	1.4	1.50
Safe act of workers	8	8	0.93	2.00
Safe working condition	8	8.29	0.7	1.00
Reaction of workers to safe condition	8	7.29	1.48	2.25
Government support	6	5.86	2.17	2.75
Contractors' organisation culture	7	7.14	2.1	1.50

Notes: SD, standard deviation; IQD, interquartile deviation.

TABLE 9.2
Safe Environment

Safe Environment	Median	Mean	SD	IQD ≤1
Safe and healthy work environment	8	8	1.07	1.25
Safe storage of equipment	8	7.71	1.03	1.25
Safe storage of building materials	8	7.29	1.75	3.25
Safe storage of formwork and false work	7	6.57	1.68	2.07
Safe transportation of building materials	6	6.29	1.39	1.5
Safe transportation of formwork and false work	7	6.14	1.55	1.25
Safe transportation of equipment	7	6.83	1.57	1.08
Provision of warning systems	8	7.57	1.5	2.25

Notes: SD, standard deviation; IQD, interquartile deviation.

Of the sixteen identified attributes under the second sub-attributes (safe acts of workers), only three attributes (ensure equipment/tools are in good condition before usage, ensure the use of personal protective equipment [PPE] and ensure proper positioning of tasks) were considered by the experts to have reached consensus with an IQD cut-off (IQD ≤1) score set to achieve consensus (Table 9.3). Fourteen attributes were considered to have reached consensus when the median was used to achieve consensus. Two attributes (avoid annoyance and horseplay at the workplace) failed to reach consensus (Table 9.3). Four attributes had strong consensus, while ten had good consensus and the remaining two attributes had weak consensus in deterring H&S compliance.

Of the eighteen identified attributes under the third sub-attribute (safe working condition), only six attributes were considered by the experts to have reached consensus with an IQD cut-off (IQD ≤1) score set to achieve consensus (Table 9.4). However, sixteen of the attributes reached consensus under the median score. Salary and payment of Social Security and National Insurance Trust (SSNIT) did not reach consensus under the median score to determine H&S compliance. Three of the attributes had very high impact (9–10). Eight other attributes had high impact (7–8.99). The remaining four attributes had medium impact (5–6.99) on health and safety compliance (Table 9.4).

Of the eight attributes identified under the sub-attributes (reaction of workers to safe condition), only three attributes were considered by the experts to have reached consensus with an IQD cut-off (IQD ≤1) score set to achieve consensus (Table 9.5). All seven attributes reached consensus under the median score, with the exception of one attribute (adhere to regular use of provided change room). The seven attributes had high impact (7–8.99). One attribute had medium impact (5–6.99). Seven attributes had good consensus and one had weak consensus (Table 9.5).

TABLE 9.3
Safe Act of Workers

Safe Act of Workers	Median	Mean	SD	IQD ≤1
Inspect workplace before commencing any activity	9	7.86	1.73	1.36
Tidy up workplace at the end of any activity	7	6.5	1.71	2
Use appropriate tools/equipment	8	7.57	1.5	2.25
Do not work under the influence of alcohol and other drugs	8	8.14	1.46	2.25
Do not smoke in flammable materials store	9	8.43	1.68	2.5
Ensure equipment/tools are in good condition before usage	9	8.14	1.36	1
Use correct proper lifting, handling or moving of objects	8	7.86	1.36	1.36
Ensure proper stacking of objects/ materials in safe locations	8	7.57	1.4	1.07
Avoid annoyance and horseplay at the workplace	6	6.57	1.84	1.5
Ensure the use of personal protective equipment (PPE)	9	9	0.93	0.25
Do not remove safety guards from the workplace or equipment	8	7.87	1.25	1.25
Do not throw or accidentally drop objects from high levels	7	7.43	1.51	1.5
Ensure proper positioning of tasks	7	7.29	1.38	1
Do not service equipment that is in operation	6	6.14	2.61	2.5
Concentrate on the task at hand	7	7	2	2
Work in good physical conditions	8	8.29	1.5	2

Notes: SD, standard deviation; IQD, interquartile deviation.

Three attributes of the sub-attribute (government support) as identified were considered by the experts to have reached consensus with an IQD cut-off (IQD ≤1) score set to achieve consensus (Table 9.6). All the five sub-attributes were considered by the experts to have reached consensus under the median and had high impact (7–8.99). All the five attributes had good consensus (Table 9.6).

Of the eleven attributes under the sub-attribute (contractor's organisational culture), only one attribute (communication of H&S information to workers) was considered by the experts to have reached consensus with an IQD cut-off (IQD ≤1) score set to achieve consensus (Table 9.7). Four of the attributes were considered by the experts to have very high impact (9–10) under the median score. The remaining seven attributes were considered by the experts to have high impact (7–8.99) on health and safety compliance (Table 9.7).

TABLE 9.4
Safe Working Condition

Safe Working Condition	Median	Mean	SD	IQD ≤1
Provision of training	9	9.26	0.45	0.46
Good inspection programme	9	9	0.82	1
Provision of incentives to workers	7	6.8	0.75	0.8
Provide safety regulations of equipment	9	8.5	1.12	1.25
Good company safety policies	8	8.71	0.88	1.25
Good salary	6	6.43	1.59	1.5
Payment of Social Security and National Insurance Trust (SSNIT)	5	5.43	1.18	1.5
Provision of sufficient lighting system for enclosed areas	7	7.57	1.18	1
Safe movement around workplace	8	7.86	0.69	0.5
Provision of guidance on the recommended illumination level for various types of task	8	7.43	1.13	1.5
Workers should be given proper ventilation	7	6.57	1.9	1.5
Provision of adequate facilities, toilet, drinking water, washing and canteen	8	7.43	2.07	3.5
Provision of facilities that are clean, safe and accessible to all workers	8	7.43	1.81	3
Availability of facilities within a reasonable distance from the work area	6	6	1	0.5
Provision of change room for workers	6	6.14	1.57	2.5
Facilities must be available for both day and night workers	6	6.17	1.72	2.75
Provision of safe means to facilitates all the time	7	7	1.41	1
Provision of break periods for workers to access the facilities	8	7.14	1.95	2.5

Notes: SD, standard deviation; IQD, interquartile deviation.

TABLE 9.5
Reaction of Workers to Safe Condition

Reaction of Workers to Safe Condition	Median	Mean	SD	IQD ≤1
Attend safety education programme	8	7.43	2.32	1.68
Attend safety training programme	8	7.57	1.99	1.57
Adhere to warning signs and notices	8	7.43	1.59	1.25
Follow safety regulations	8	7.74	1.92	0.93
Adhere to company safety policies	8	7.43	1.92	0.93
Adhere to guidance on recommended illumination level for various tasks	7	7	1.51	1.25
Put to proper use of the available facilities (toilet, drinking water, washing and canteen)	7	6.71	1.39	0.46
Adhere to regular use of provided change room	6	5.86	1.64	1.61

Notes: SD, standard deviation; IQD, interquartile deviation.

TABLE 9.6
Government Support

Government Support	Median	Mean	SD	IQD ≤1
Formulate H&S policy for construction activities	8	7.71	0.7	1
Implementation of H&S policy by government representatives	7	7.57	1.18	1
Monitoring of H&S policy implementation by the government representatives	8	7.71	1.16	1
Provision of H&S policy update by government representatives	8	7.86	1.12	2
Provision of H&S training by government representatives	7	7.57	1.59	3

Notes: SD, standard deviation; IQD, interquartile deviation.

TABLE 9.7
Contractor's Organisational Culture

Contractor's Organisational Culture	Median	Mean	SD	IQD ≤1
Provision of personal protective equipment (PPE)	9	8	1.51	2.25
Provision of signs/notices on sites	7	7.71	1.58	2
Training of workers on health and safety (H&S)	9	8.43	1.05	1.25
Involve workers in H&S program	9	8	1.2	2
H&S staffing	7	7	1.69	2.25
H&S inspection	8	7.71	1.48	1.25
Company H&S policy	8	7.67	1.11	1.5
Management commitment in H&S	8	7.57	1.84	2.25
Appropriateness of site for erection of residential building	8	7.71	1.58	3
Communication of H&S information to workers	9	8.57	1.05	0.82
Update on H&S information to workers	8	8	1.31	2

Notes: SD, standard deviation; IQD, interquartile deviation.

9.3.2 To Determine the Factors That Enable Small to Medium-Sized Construction Companies to Comply with Health and Safety

Of the twelve identified factors that were considered to enable small to medium-sized enterprise (SME) contractors to comply with health and safety, only three factors (changes in company structure, changes in ownership at various stages of growth and short track records of H&S regulations) were identified by the experts to have reached consensus with an IQD cut-off (IQD ≤1) score on health and safety compliance (Table 9.8). Four of the factors had high impact (7–8.99), while the remaining eight factors had medium impact (5–6.99) on health and safety compliance (Table 9.8).

9.3.3 To Identify the Factors That Affect Small to Medium-Sized Construction Companies in Not Complying with Health and Safety Regulations

Of the ten identified factors that were considered to affect small to medium-sized construction companies in not complying with H&S regulations, only four factors (limited company resources, unavailable H&S policy, limited knowledge of occupational health and safety [OHS] and limited access to body responsible for the implementation of H&S policy) were considered by the experts to have reached consensus with an IQD cut-off (IQD ≤1) score on health and safety compliance (Table 9.9). Nine of the factors had high impact (7–8.99) on health and safety compliance. Only one factor had medium impact (5–6.99) on health and safety compliance (Table 9.9).

TABLE 9.8
Factors That Enable Small to Medium-Sized Construction Companies to Comply with H&S Regulations

Factors of Compliance	Median	Mean	SD	IQD ≤1
Changes in company structure	6	5.5	1.26	0.75
Changes in ownership at various stages of growth	5	5.17	0.69	0.58
Lack of H&S experts	6	5.67	2.87	4.5
Lack of finance in the management of H&S regulation	6	5.86	2.42	2.75
Lack of personnel to monitor changing legal requirements	6	5.86	2.47	3.5
Short track records of H&S regulations	5	5.29	1.67	1
Company limited cash flow	5	5.71	2.31	2.5
Lack of enforcement from the legislative bodies overseeing the implementation of OSH Act	7	6.43	2.5	1.93
Lack of knowledge to understand H&S regulations	7	6.29	2.6	3
Lack of knowledge to interpret H&S rules	7	6	2.62	3.25
Lack of knowledge to identify hazards or risks	7	6.29	2.49	2.25
Lack of interest in compliance with environmental health regulations	6	5.57	2.61	2.75

Notes: SD, standard deviation; IQD, interquartile deviation.

TABLE 9.9
Factors That Affect Small to Medium-Sized Construction Companies in Not Complying with H&S Regulations

Factors of Non-Compliance	Median	Mean	SD	IQD ≤1
Limited company resources	7.00	7.43	1.05	0.82
Unavailable H&S policy	8.00	7.71	1.28	0.71
Limited knowledge of occupational health and safety (OHS)	8.00	7.71	1.28	0.71
Inability to employ H&S personnel	7.00	7.00	1.07	1.25
Inability to train employees on H&S regulations	8.00	7.43	1.59	3.00
Lack of knowledge on H&S policy implementation	7.00	6.86	1.64	1.61
Lack of coordination of the implementation of H&S policy within the organisation	7.00	6.57	1.92	2.25
Limited access to body responsible for the implementation of H&S policy	7.00	7.00	0.93	0.25
Lack of cooperation from client	6.00	6.00	1.93	1.50
Management bottleneck	7.00	6.43	1.84	1.50

Notes: SD, standard deviation; IQD, interquartile deviation.

9.3.4 To Identify the Effects of Health and Safety Non-Compliance on Small to Medium-Sized Construction Companies

Of the seven factors identified by the experts as the effects of H&S non-compliance on small to medium-sized construction companies, only four factors (loss of funds due to accident compensation, loss of funds in the training of new employees, loss of funds in the employment of H&S personnel and payment of government representatives on H&S education) were considered to have reached consensus with an IQD cut-off (IQD ≤1) score on health and safety compliance (Table 9.10). Three of the factors had high impact (7–8.99) on health and safety compliance. The remaining four factors had medium impact (5–6.99) on health and safety compliance (Table 9.10).

9.3.5 To Evaluate the Management Issues That Affect the Government in the Implementation of Health and Safety Policies

Of the nine identified management issues that affect the government in the implementation of H&S policies, only one management issue (short track of records of H&S regulations) was considered to have reached consensus with an IQD cut-off (IQD ≤1) score on health and safety compliance (Table 9.11). All nine management issues had high impact (7–8.99) on health and safety compliance (Table 9.11). These representations are provided in Table 9.11.

TABLE 9.10
Effects of H&S Non-Compliance on SME Contractors

Effects of H&S Non-Compliance on SME Contractors	Median	Mean	SD	IQD ≤1
Reduction in profit margin	5.00	5.43	2.72	1.75
Loss of funds due to accident compensation	7.00	7.00	1.51	1.00
Loss of funds in the training of new employees	5.00	4.29	1.58	0.54
Loss of funds in the employment of H&S personnel	5.00	4.29	1.28	0.54
Payment of government representatives on H&S education	5.00	4.29	1.28	0.54
Loss of company reputation	7.00	7.71	0.88	1.25
Loss of clients	7.00	6.86	1.55	1.11

Notes: SD, standard deviation; IQD, interquartile deviation.

TABLE 9.11
Management Issues That Affect Government in the Implementation of H&S Policies

Management Issues That Affect Government in the Implementation of H&S Policies	Median	Mean	SD	IQD ≤1
Lack of finance in the management of H&S	7.00	6.57	2.19	1.82
Lack of H&S experts	8.00	7.00	1.85	1.25
Limited cash flow	8.00	7.33	1.37	1.00
Lack of capacity	8.00	7.29	0.95	1.50
Lack of personnel to monitor changing legal requirements	7.00	6.29	1.80	2.00
Short track of records of H&S regulations	7.00	7.14	1.21	0.50
Lack of knowledge to understand H&S regulations	7.00	6.57	1.90	1.50
Lack of knowledge to interpret H&S rules	6.00	6.29	1.98	2.00
Lack of knowledge to identify hazards or risks	7.00	6.71	1.98	2.00
Lack of interest to comply with environmental health regulations	7.00	5.57	2.51	3.00

Notes: SD, standard deviation; IQD, interquartile deviation.

9.4 DISCUSSION OF THE DELPHI STUDY RESULTS

The first objective of the Delphi study was to identify the main and sub-attributes that determine H&S compliance of small to medium-sized construction companies in Ghana. Findings emanating from the survey revealed that the attributes that determine health and safety compliance among small to medium-sized construction companies in Ghana are similar to other cultural contexts. Consensus was only reached for safe working conditions under the main attributes that determine H&S compliance, with an IQD score of 1.00, mean 8.29 and median 8.00. which indicate strong consensus. Even though five other attributes had a median (7–8.99) indicating good consensus, their IQD fell between 1.50 and 2.25. Only one attribute (government support) showed weak consensus among the six attributes that determine consensus, with a median ≤6.99, mean ≤5.99, and IQD ≥2.1≤3. The SD for the five attributes that had a high influence on H&S compliance, falling between 0.7 and 2.1. The only attribute without high influence on H&S compliance had the highest SD of 2.17.

The assessment of the sub-attributes of six major determinants of H&S compliance showed that the sub-attributes that determine H&S compliance in the Ghanaian construction industry are identical to other cultural contexts. Of the sixty-six sub-attributes, consensus was reached for only fifteen, which had been found to be the strong determinants of H&S compliance in small to medium-sized construction companies in Ghana. Among these sub-attributes which were determined by the experts to have reached consensus are ensure proper positioning of tasks, ensure equipment/tools are in good condition before usage (good inspection programme), provision of incentives to workers, provision of sufficient lighting system for enclosed areas, safe movement around workplace, and availability of facilities within a reasonable distance from the work area. Other sub-attributes which were considered by the experts to have strong influence on H&S compliance are facilities must be available for day and night workers, follow safety regulations, adhere to company safety policies, put to proper use the available facilities (toilet, drinking, water, washing and canteen), formulate H&S policy for construction activities, implementation of H&S policy by the government representatives, monitoring of H&S policy implementation by the government representatives and communication of H&S information to workers.

The assessment of the sub-attributes findings replicated the results of a majority of studies on health and safety compliance in relation to adequate training and education and, in general, a safe work environment (Adenuga, Soyingbe, & Ajayi 2007; Othman, 2012; Windapo & Oladapo, 2012; Idubor & Oisamoje, 2013) upon which the current study's sub-attributes were also based. Provision of personal protective equipment (PPE), training of workers on H&S, involve workers in H&S programmes and communication of H&S information to workers were rated as variables with a very high impact, as these affect the quality of services on the safety of employees. Provision of training and good inspection programmes also have a very high impact in relation to the safe working condition of the employees. This finding agreed with previous findings from the studies conducted by Idubor and Oisamoje (2013) and Windapo and Oladapo (2012). Further findings as related to inspection of the workplace before commencing any activity, prohibition of smoking in flammable materials storage and good equipment/tools usage imply that the sub-variables play a vital role in employees' health and safety compliance.

In conclusion the results suggest that the attributes that bring about H&S compliance in small to medium-sized construction companies in Ghana are similar to the determinants in other cultural contexts. Furthermore, H&S compliance is ensured if more attention is given to these attributes in the development of an integrated H&S compliance model for small to medium-sized construction companies in Ghana. Particular attention should be given to the fifteen sub-attributes determined by the experts, which have all been described to have significant influence and high impact in determining H&S compliance.

The second objective of the Delphi study was to determine the factors that enable small to medium-sized construction companies to comply with health and safety practices and regulations. The reason for this is that most of the requirements for a fully fledged contractor, in terms of finance and policies, are lagging among small to medium-sized construction companies in Ghana. The assessment of the nine factors of major determinants of H&S compliance showed that three factors were considered by the experts to have achieved strong consensus, with an IQD score between 0.58 and 1. But, both the means (5.5, 5.17 and 5.29) and the medians (6.00, 5.00 and 5.00) were below the range that classify the identified factors to have attained strong consensus. These factors have also medium impact (5–6.99). These results contradict the choice of the experts for the major determinants of H&S compliance based on the IQD score.

The third objective of the Delphi study was to identify the factors that affect small to medium-sized construction companies in not complying with H&S regulations. Small to medium-sized construction companies' inability to comply with H&S regulations might be due to several factors, some of which may be beyond their control and other out of their reach owing to limited resources. The assessment of the ten factors of major determinants of H&S compliance showed that four factors were considered by the experts to have achieved strong consensus with IQD less or equal to one, median (9–10) and mean (8–10). All four factors had a high impact (7–8.99) on the H&S compliance, with varying standard deviations ranging from 0.93 to 1.28. Most of the factors have high influence on H&S compliance among SME contractors in Ghana.

The fourth objective of the Delphi study was to identify the effects of small to medium-sized construction companies in not complying with H&S regulations. The major effects of small to medium-sized construction companies' inability to comply with H&S regulations have been attributed to their weak financial resources. The assessment of the seven factors of major determinants of H&S compliance showed that four factors were considered by the experts to have achieved consensus with IQD less or equal to one. But, the median and the mean scores for all four of the factors fell below the range for median (9–10) and mean (8–10) to reach strong consensus. Of the seven factors, only factors had high impact (7–8.99) on H&S compliance, with SD ranging from 0.88 to 155.

The fifth objective of the Delphi study was to evaluate the management issues that affect the government in the implementation of H&S policies in small to medium-sized construction companies. Financial constraint was found to be the major contributor to the government's inability to recruit competent personnel, and organise training and workshops for small to medium-sized construction companies' employees to enable them to initiate H&S policies. The assessment of the ten management issues as considered by the experts to be the major determinants of H&S compliance showed that only two management issues (limited cash flow and short track records of H&S regulations) were considered by the experts to have achieved consensus with an IQD less or equal to one. Nine management issues as considered by the experts had high influence (7–8.99) on H&S compliance. Their SD ranged from 0.95 to 2.51 and their mean ranged from 6.29 to 7.29. These indicate how significant management issues are towards the implementation of H&S policies in small to medium-sized construction companies.

9.5 SUMMARY

This chapter presented a summary of results and discussions of the results from all the Delphi rounds. Computation for each question element was made for the influence and impact of the attributes in predicting H&S compliance and how this will contribute to eliminating the occurrence of accidents in small to medium-sized construction companies in Ghana. The influence or impact of the absence or presence of a particular H&S practice element on the overall H&S compliance of other elements was presented. Others are ineffective implementation of H&S regulations hindering the successful performance of the construction industry in Ghana. The chapter concluded with a summative discussion of the findings based on the objectives of the Delphi study. The findings from the expert participants revealed a consistence discussion on H&S compliance in Ghana with consensus being reached in most cases and in others with a discrete conclusion. The result of the Delphi study assisted in the determination of key factors and constructs that are of critical significance in determining H&S compliance in the construction industry and were achieved through the results obtained from the Delphi study. The outcome led to the development of the holistically integrated conceptual model for H&S compliance among small to medium-sized construction companies in Ghana. The next chapter of the book provides the evaluation of these factors and their interrelationships.

REFERENCES

Adenuga, O.A., Soyingbe, A.A. and Ajayi, M. A. (2007). A study on selected safety measures on construction companies in Lagos, Nigeria. RICS (Cobra).

Idubor, E.E. and Oisamoje, M.D. (2013). An exploration of health and safety management issues in Nigeria's effort to industrialise. *European Scientific Journal* 9(12): 154–169.

Othman, A.A.E. (2012). A study of the causes and effect of contractors' noncompliance with the health and safety regulations in the South African construction industry. *Architectural Engineering and Design Management*, 8: 180–191.

Stitt-Gohdes, W.L. and Crews, T.B. (2004). The Delphi technique: A research strategy for career and technical education. *Journal of Career and Technical Education*, 20(2): 55–67.

Windapo, A. and Oladapo, A. (2012). Determinant of construction firms' compliance with health and safety regulations in South Africa. In: Smith, S.D. (Ed.), *Proceedings of 28th Annual Association of Researchers in Construction Management (ARCOM) Conference*, 3–5 September, Edinburgh, UK, 433–444.

Section VIII

Conceptual Integrated Health and Safety Compliance Model for Small to Medium-Sized Construction Companies

10 Conceptual Integrated Health and Safety Compliance Model for Small to Medium-Sized Construction Companies

10.1 INTRODUCTION

This chapter of the book presents the discussion of findings from the review of literature relating to the variable selection for the conceptual model. The conceptual model theory forms the basis of the discussion in this chapter. Based on an in-depth review of the previous models as presented in Chapter 2 of this book, the hypothesised integrated holistic health and safety compliance model is presented in this chapter (except for government support and contractor's organisational culture, which have already been discussed in Chapter 3 of this book) as variable constructs identified as gaps in health and safety compliance research. Detailed discussions of the holistic, integrated model and the variables of the model, identification of the model and the justification for the selected variables are all presented in this chapter.

10.2 SELECTION OF VARIABLES FOR HEALTH AND SAFETY COMPLIANCE

Both objective and subjective attributes have been combined in the health and safety compliance study models for the assessment of health and safety (H&S) compliance in small to medium-sized construction companies. Behavioural safety compliance can be improved through a strong effort by both employees and employers by complying with safety requirements (Mat Zin & Ismail, 2012). The three constructs proposed by the accident root causes tracing model (ARCTM) in addition to the two constructs from Heinrich's (1930) theory are supported and adopted for the current study. But both models have one construct in common which was also adopted for the current study. The H&S compliance bundle was also considered in the current study, as is typical for the construction industry. The H&S compliance bundle thus contains safe environment with seven variables; safe acts of workers with seventeen variables; safe condition of workers with seventeen variables; and the reaction of workers to safe condition with eight variables.

Almost all H&S compliance studies have the same constructs as conceptualised for the current study. However, the current study brings into focus that government support and the contractor's organisational culture are also critical in determining the compliance of small to medium-sized construction companies to H&S issues. These identified gaps are thus found to be peculiar to developing countries, and Ghana in particular, as the support from the government is relevant for the growth of the category of contractors and likewise the companies' organisational culture. The constructs that influence the level of H&S compliance in small to medium-sized construction companies will be explained in detail in the next sections of this chapter. The additional two constructs are explained in Chapter 3.

10.2.1 Safe Environment

Safe work creates no obstacles to being competitive and successful (Construction Regulations, 2003). In fact, no country or industry has been able to jump to a high level of productivity without making sure that the work environment is safe. It is the responsibility of the employer to provide a safe work environment for all employees, free from any hazards and complying with all state and federal laws. Health and safety in the workplace is about preventing work-related injury and disease, and designing an environment that promotes well-being for everyone at work (Heinrich, 1930; Safe Work Australia, 2013). Knowledge is the key ingredient in providing a safe work environment; if everyone knows the correct procedures, then accidents and injuries can be minimised. For instance, small to medium-sized construction companies can achieve a safe working environment through the provision of safe equipment and safe storage and transportation of dangerous substances. Certain environmental factors such as recklessness and undesirable traits of characters may be passed on through inheritance; whilst inheritance and environment may cause fault within the construction industry. But with a carefully planned environment, small to medium-sized construction companies can achieve a safe work environment. Incidences can be prevented by providing a safe work environment to achieve health and safety compliance in the construction industry. The current study will look at a safe work environment that has been hypothesised for the development of a holistic health and safety compliance model and is summarised in Table 10.1.

10.2.2 Safe Act of Workers

Ignorant behaviour, and the attitude of employers and employees contribute to a rise of issue on behavioural safety and non-compliance to requirements of the Occupational Safety and Health Act (OSHA) 1994 published by the United Kingdom (Mat Zin & Ismail, 2012). The practicing of standards of safety and health at work to eliminate workplace accidents has been stipulated in OSHA 1994. OSHA is identified as an approach providing a legislative framework to enforce human behaviour towards safety compliance (Mat Zin & Ismail, 2012). Safety behaviour describes the behaviour that supports practices and activities. Safety training and safety compliance that should be carried by employees according to occupational, safety and health requirements is required to achieve a safe act by construction workers.

TABLE 10.1
Conceptual Model Latent Features

Latent Variable Construct	Measurement Variables
Safe environment (SE)	Safe and healthy work environment
	Safe storage of equipment
	Safe storage of building materials
	Safe storage of formwork and false work
	Safe transportation of building materials
	Safe transportation of formwork and false work
	Safe transportation of equipment
	Provision of the warning system
Safe act of workers (SAW)	Inspect workplace before commencing any activity
	Tidy up workplace at the end of any activity
	Use appropriate tools and equipment
	Do not work under the influence of alcohol and other drugs
	Do not smoke in flammable materials stores
	Ensure equipment or tools are in good condition before usage
	Ensure proper lifting, handling or moving of objects
	Ensure proper stacking of objects or materials in safe locations
	Avoid annoyance and horseplay at the workplace
	Ensure the use of personal protective equipment (PPE)
	Do not remove safety guards against the workplace or equipment
	Do not throw or accidentally drop objects from high levels
	Ensure proper positioning of tasks
	Do not work under the effects of alcohol and other drugs.
	Do not service equipment that is in operation
	Concentrate on the task at hand
	Work in good physical conditions
Safe working condition (SWC)	Provision of training
	Good inspection program
	Provision of incentives to workers
	Provision of safety regulation of equipment
	Good company safety policies
	Good salaries
	Payment of Social Security and National Insurance Trust (SSNIT)
	Provision of sufficient lighting system for enclosed areas
	Safe movement around the workplace
	Provision of guidance on the recommended illumination level for various types of task
	Workers should have adequate ventilation
	Provision of adequate facilities (toilet, drinking water, washing and canteen)
	Availability of facilities within a reasonable distance from the work area
	Provision of change room for workers
	Facilities must be available to both day and night workers
	Provision of safe means to facilities all the time
	Provision of break periods for workers to access facilities

(Continued)

TABLE 10.1 (CONTINUED)
Conceptual Model Latent Features

Latent Variable Construct	Measurement Variables
Reaction of workers to safe condition (RWSC)	Attend safety education programmes
	Attend safety training programmes
	Adhere to warning signs and notices
	Follow safety regulations
	Adhere to company safety policies
	Adhere to guidance on recommended illumination level for various tasks
	Put to proper use available facilities (toilet, drinking water, washing and canteen)
	Adhere to regular use of provided change room
Government support (GS)	Formulate H&S policy for construction
	Implementation of H&S policy by government representatives
	Monitoring of H&S policy implementation by government representatives
	Provision of H&S policy update by government representatives
	Provide health and safety training by government representatives
Contractor's organisational culture (COC)	Provision of personal protective equipment (PPE)
	Provision of signs and notices on sites
	Training of workers on H&S
	Involve workers in H&S programmes
	Health and safety staffing
	Health and safety inspection
	Company health and safety policy
	Management commitment in H&S
	Assessment of hazard identification and risk
	Consultation on H&S information to workers
	Update on H&S information to workers
Health and safety compliance (HSC)	Accidents on sites will be minimised
	Compensations paid on accident victims will be reduced
	Reduce the cost of training on H&S
	Limited number of H&S education by government representatives
	Limited number of H&S monitoring by government representatives
	Improvement in H&S performance
	Increase in productivity

Accident causation theories address the human (worker) as the main problem of accident causation on site. An accident happens owing to the human characteristic, the combination of extreme environment and an overload of human capability, and conditions (Abdelhamid & Everett, 2000). Working with safety devices such as personal protective equipment (PPE), using equipment that is in good condition and following the correct work procedure at any time will lead to safety. Employees should have in-depth knowledge on the activities they undertake and also obey workplace procedures during execution of work. Smallwood (2010) identified workers' attitude as one of the factors leading to an unsafe act of a worker. Workers' safety behaviour will contribute to safety practices, therefore there is a need to orient workers to ensure the act of safe work is practiced in executing their task. The worker conducts safe acts under the condition that he or she has undergone safety training and has been provided with health and safety equipment to protect him or her from any harm. Hence, employees' safe acts is a major contributing factor to health and safety compliance in small to medium-sized construction companies. Employees should always follow the occupational, safety and health requirements to prevent the occurrence of an accident at the workplace.

10.2.3 Safe Working Condition

The resultant effects of unsafe acts or unsafe conditions are accidents. The carelessness or fault of a person is the negative features of a person's personality. Although, regardless of how these unwanted characteristics have been acquired, they can be corrected. Errors and technical failures as a result of unsafe acts or mechanical or physical conditions can also be corrected to prevent accidents from occurring by performing safe acts and under safe conditions (Abdul Hamid, Yusuf & Singh, 2003; Hosseinian & Torghabeh, 2012:54). Accident prevention is an integral programme, a series of coordinated activities, directed to the control of unsafe personal performance and unsafe mechanical conditions, based on certain knowledge, attitudes and abilities of the workers. Accidents can be prevented from occurring if the chain of sequence in the domino theory is disturbed, for example, the unsafe acts or unsafe conditions (Abdul Hamid et al., 2003; Hosseinian & Torghabeh, 2012). Workers are the main reasons for accidents and management has the responsibility of preventing accidents (having the power and authority). It is mandatory to provide employees with safe working conditions to enable them to abide by health and safety regulations and perform well at their respective workplaces. The safe conditions that are considered for the present study are summarised in Table 10.1.

10.2.4 Reaction of Workers to Safe Condition (RWSC)

Accident root causes tracing model (ARCTM) derived most of its important rules from the efforts of Heinrich (1930), Petersen, Bird and Ferrell (in Heinrich, Petersen, & Roos, 1980), Petersen (2000), Hosseinian and Torghabeh (2012), Jha (2011), and Fang, Choudhry and Hinze (2006). ARCTM indicates that the unsafe condition which contributes to the occurrence of an accident due to the employees' inability to identify the existence of the unsafe condition before the activity is carried out can be prevented if these actions are performed under safe conditions. The reaction of the employee to safe conditions depends on the fact that the employee should identify the safe conditions before any activity is carried out (Abdulhamid & Everett, 2000; Fang et al., 2006). The employee should identify the safe conditions and conduct his or her activities under the H&S regulations. According to Hosseinian and Torghabeh (2012:59), 'in order to be at a safe side in conducting any activity, the worker should identify a safe condition for his activity. When the worker realises the existence of the unsafe condition and any related hazard, he or she must quickly ignore the work to prevent any accident occurring'. The activity can only take place when the condition at the said place of the activity is rectified. The health and safety features that are considered for the current study are summarised in Table 10.1.

10.3 MODEL SPECIFICATION AND JUSTIFICATION

This study aims to build a conceptual health and safety compliance model. The theoretical conceptual framework for the current research is built on the work of Heinrich (1930) and the ARCTM, which was also built on the previous accident models, as discussed in Chapter 2 of this book. Heinrich (1930) theorised that ancestry and social environment, the fault of a person, and unsafe acts and conditions lead to accidents. The reason for the cause of the accident is people, and management handles the prevention of accidents. It has been revealed that the majority of accidents are due to human error and that accidents can be prevented if management provides a conducive environment for employees to work. Adams in the year 1976 shared a similar view with Heinrich (1930). Adams' emphasis was on management's organisational structure and reflects the relationship involved with the causes and effects of all incidents and accidents that have direct management involvement. The role of management in accident prevention was also emphasised in a broader sense taking into consideration the root of unsafe acts or conditions (Heinrich, Petersen & Roos, 1980). The ARCTM pointed to the fact that an unsafe condition, the reaction of the worker to an unsafe condition and unsafe acts of workers can lead to an accident. Petersen (2000) also conceptualised that accidents are due to unsafe acts and unsafe conditions. The non-compliance level of health and safety in small to medium-sized construction companies is related to the environment, unsafe acts, unsafe conditions, the reaction of the worker to an unsafe condition and unsafe acts of the worker (Table 10.2).

Both Heinrich (1930) and Petersen (2000), as well as the ARCTM, emphasised unsafe acts and unsafe conditions as the main causes of accidents in the construction industry. The two basic components of the model are safe acts and safe conditions. Based on the fundamental underpinning of two models, and the incorporated theoretical perspectives, which have been adopted in other similar studies, they are therefore useful for conceptualizing the present study as a variety of health and safety studies and health and safety compliance have been conceptualized within the broader theoretical framework. Therefore, the conceptual framework for this thesis is primarily based on the approach used by Heinrich (1930) and the ARCTM. Based on the fundamental factors and constructs associated with all the previous models as revealed in Table 10.1, the present model for the study looks at the safe work environment, safe acts, safe conditions, reaction of workers to safe conditions and safe acts of worker. These factors have been measured in the majority of the previous studies, but consideration has not been given to government support and contractor's organisational culture; which have been classified as the exogenous variables and their role in predicting overall health and safety compliance in small to medium-sized construction companies, which is the endogenous variable. These will in turn, predict small to medium-sized construction companies' health and safety compliance.

The study aims to forecast on the relative predictive power of these different variables for health and safety compliance in order to determine whether health and

TABLE 10.2
Factors of Non-Compliance with Health and Safety

Elements of Accident Causation Model	Heinrich, 1959	Petersen, 1971	ARCTM	Weaver, 1971	Ferrell, 1997	Bird, 1974
Environment	X					
Personal factors	X					
Unsafe acts/conditions	X	X	X			
Lack of control						X
Basic factor						X
Immediate factor						X
Sub-causes						
Reaction of workers to unsafe condition		X				
Unsafe acts of worker			X			
Operational errors				X		
Overload					X	
Incorrect response					X	
Improper activity					X	
Accident	X	X	X	X	X	X
Injury	X	X	X	X	X	X

safety compliance depends on the supposed features of the variables, taking into account the needs of the construction industry and their compliance with the policies and codes in Ghana, as indicated in the other frameworks. It is apparent that some of the aforementioned variables should be measured by objective means, some by subjective means and some will include both forms of measurement.

The conceptual model theorises that health and safety compliance is established by the relationship that exists between the exogenous variables, which include the basic elements by which the subjective and objective measurements are linked. These variables, identified from the review of the literature, are considered the major determinants of H&S compliance. The determinants identified have been adopted to fit into the health and safety compliance for small to medium-sized companies in the Ghanaian construction industry.

10.4 STRUCTURAL COMPONENT OF THE MODEL

The integrated health and safety (H&S) compliance model for small to medium-sized companies in the Ghanaian construction industry, in the case of developing countries, is derived from a safe environment (SE), safe acts of workers (SAW), safe work condition (SWC), reaction of workers to safe condition (RWSC), government support (GS) and contractor's organisational culture (COC) in the process of achieving safety in the construction industry. The health and safety compliance (HSC) model to be tested in the postulated hypothesis is not based on prior study or any examination, is composed of SE, SAW, SWC, RWSC, GS and COC, and it is a multidimensional structure. The postulated model is presented in Figure 10.1 (Model 1.0).

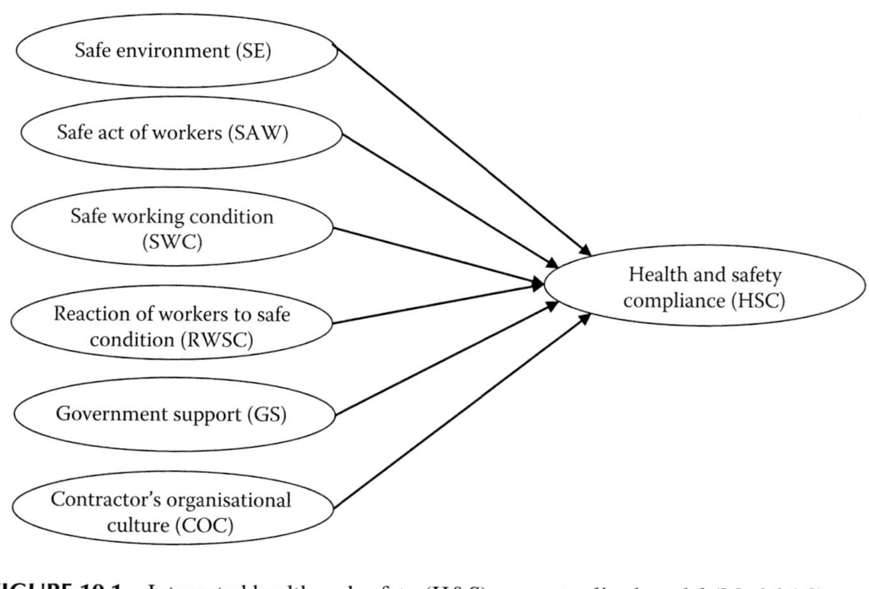

FIGURE 10.1 Integrated health and safety (H&S) conceptualised model (Model 1.0).

The theoretical underpinning relating to this is derived from the works of Heinrich (1930), Petersen's model of 2000 and the ARCTM. Most of the important rules of the ARCTM were derived from the efforts of Heinrich (1930), Petersen's model of 2000, Bird's model of 1974, Heinrich et al. (1980), Petersen (2000), Hosseinian and Torghabeh (2012), Jha (2011), and Fang et al. (2006), as discussed in Chapter 3. The conceptualised model is the notion that compliance of health and safety in small to medium-sized construction companies is related to the evaluation of the variables, such as SE, SAW, SWC, RWSC, GS and COC. It is difficult to discuss the principal variable without reference to variables of government support and the contractor's organisational culture as well as the inclusion of the other exogenous variables. The evaluation will depend on the compliance assessment of several indicator variables under each of the exogenous variables. In this book, the objective evaluation of health and safety compliance is assessed by measuring the actual condition of the construction industry that is an exogenous variable in the model.

10.5 MEASUREMENT COMPONENT OF THE MODEL

The measurement component of the hypothesised model comprises of the following health and safety compliance factors: SE, 8 measurement variables; SAW, 17 measurement variables; SWC, 17 measurement variables; RWSC, 8 measurement variables; GS, 5 measurement variables; COC, 11 measurement variables; and HSC, 7 manifest measurement variables. The success for the consideration of health and safety compliance for the benefit of the construction industry has been theorised in the present model.

10.6 SUMMARY

The theorised conceptual model for small to medium-sized construction companies was discussed in this chapter. The discussion was that the health and safety compliance (HSC) model for small to medium-sized construction companies is a multidimensional structure composed of the seven latent variables of safe work environment, safe acts, safe condition, reaction of workers to safe condition, safe action of worker, government support and contractor's organisational culture. These factors were derived from the literature review. The explanation of the selected variables for the construction of the integrated H&S compliance model has been highlighted in the theoretical framework in this chapter. The next chapter of the book presents the discussion of the survey results.

REFERENCES

Abdelhamid, T.S. and Everett, J.G. (2000). Identifying root causes of construction accidents. *Journal of Construction Engineering and Management*, 126(1): 52–60.

Abdul Hamid, A.R., Yusuf, W.Z.W. and Singh, B. (2003). Hazards at construction sites. Proceedings of the 5th Asia-Pacific Structural Engineering and Construction Conference (APSEC 2003), 26–28 August, Johor Bahru, Malaysia.

Construction Regulations. (2003). Safe Working Practice, Occupational Health and Safety Group. Available from: www.safepractice.co.za [Accessed: 8 April 2014].

Fang, D., Choudhry, R.M. and Hinze, J.W. (2006). Proceedings of CIB W99 International Conference on Global Unity for Safety & Health in Construction, 28–30 June, Beijing, China.

Heinrich, H.W. (1930). *Industrial accident prevention*. New York: McGraw-Hill.

Heinrich, H.W., Petersen, D. and Roos, N. (1980). *Industrial accident prevention*. New York. McGraw-Hill.

Hosseinian, S.S. and Torghabeh, Z.J. (2012). Major theories of construction accident causation models: A literature review. *International Journal of Advances in Engineering & Technology*, 4(2): 53–66.

Jha, K.N. (2011). *Construction project management: Theory and practice*. Delhi: Pearson Education in South Asia.

Mat Zin, S. and Ismail, F. (2012). Employers' behavioural safety compliance factors toward occupational, safety and health improvement in the construction industry. *Procedia: Social and Behavioral Sciences* 36: 742–751. Available from: www.Sciencedirect.com. [Assessed 25 September 2014].

Petersen, D. (2000). The behavioural approach to safety management. *Professional Safety*, 37–39.

Safe Work Australia. (2013). Available from: http://www.jobaccess.gov.au [Accessed 04 October 2014].

Smallwood, J.J. (2010). Excavation health and safety (H&S): A South African perspective. In: Egbu, C. (Ed.). *Proceedings of the 26th Annual Association of Researchers in Construction Management (ARCOM) Conference, 6–8 September 2010*, Leeds, UK. 233–241.

Section IX

Results from the Questionnaire Survey: Findings Part II

11 Survey Results

11.1 INTRODUCTION

The theoretical conceptual model for the study is presented in Figure 10.1 (Model 1.0) (see Chapter 10). The hypothesised integrated health and safety (H&S) compliance model theory is based on factors extracted from the reviewed literature and the views of experts obtained during the Delphi study. The views of the experts have been described in detail in Chapter 9. This chapter presents descriptive statistics, inferential statistics and hypotheses testing results based on the questionnaire analysis. Statistical analyses techniques used to test the validity and reliability of the measuring instrument have been discussed in this section. Another statistical technique used is the empirical testing of the proposed conceptual model presented in Chapter 10. The results obtained from the quantitative survey are provided and discussed.

The data analysis was conducted in two stages:

1. Descriptive data analysis
2. Multivariate correlational data analysis including exploratory factor analysis (EFA), confirmatory factor analysis (CFA) and structural equation modelling (SEM)

A sample size of 558 cases was returned at the end of the survey. Greater model fit bias has been attributed to sample size, which most times affects the SEM fit. Therefore, the sample size obtained for the current study is considered as large (Kline, 2005). A sample size less than 100 cases would be difficult to analyse when the analytical tool to be used is SEM. Sometimes, the appropriate sample size may depend on observed variables (MacCallum, Browne & Sugawara, 1996; Tong, 2007). The minimum sample of 200 for SEM analysis is considered as adequate for analysis (Bentler & Chou, 1987). The variable ratio of ideal SEM is suggested to be at least 5:1 (Tong, 2007). This implies, SEM with 10 observed variables should have more than 50 respondents. For the purpose of this study 558 responses were collected. There are 66 hypothesised observed variables and the ratio to sample size for the current study is 8.45:1. Therefore, the variable ratio to sample size meets the requirement recommendation (Tong, 2007). The sample data (558) was first taken through random sampling before carrying out the EFA and CFA. From the initial check, 269 samples were realised for the EFA analysis and 289 samples for the CFA analysis.

11.2 DESCRIPTIVE STATISTICS

This section provides demographic information on the individual respondents. The analysed results for the descriptive data were: the respondents' background information including their individual information and the company information. Descriptive statistics such as percentages, means and standard deviation were used in the analysis.

11.2.1 Demographic Profile of the Respondents and Firms

A total of 269 samples were realised for the EFA after the random sampling. The responses represent 82.2 percent ($N = 221$) males and 17.8 percent ($N = 48$) females, as shown in Table 11.1.

Table 11.2 shows that a majority of the respondents (32.7%; $N = 88$) were between the ages of 26 and 30 years, followed by the age group of 31 to 35 years (20.8%), and the age ranges 20 to 26 years and 36 to 40 years constituted 14.1 percent each of the sample.

The highest education level of the majority of the sample respondents was a national diploma or certificate (34.9%; $N = 94$) and the fewest number of sample respondents held a post-graduate degree (10.0%; $N = 27$) as shown in Table 11.3.

Table 11.4 shows that a large number of the respondents (38.3%) have worked in their respective firms between 2 to 5 years, followed by the group with 6 to 10 years at their place of work (28.6%) and the fewest number of respondents (1.5%) have been at their company 31 years or more.

The majority of the respondents (27.1%; $N = 73$) indicated their firms have been in existence in the range of 15 to 20 years, followed by the year range of 5 to 9 years

TABLE 11.1
Gender

Gender	Frequency	Percentage
Male	221	82.2
Female	48	17.8

TABLE 11.2
Age Group

Age Group	Frequency	Percentage
Less than 20 years	7	2.6
20–25 years	38	14.1
26–30 years	88	32.7
31–35 years	56	20.8
36–40 years	38	14.1
41–45 years	23	8.6
46 years and above	18	6.7
Missing	1	0.4

TABLE 11.3
Highest Qualification

Qualification	Frequency	Percentage
Senior school certificate	42	15.6
National diploma or certificate	94	34.9
Bachelor's degree	90	33.5
Post-graduate	27	10.0
Missing	16	5.9

TABLE 11.4
Tenure in the Firm (Small to Medium-Sized Construction Company)

Tenure	Frequency	Percentage
2–5 years	103	38.3
6–10 years	77	28.6
11–15 years	32	11.9
16–20 years	30	11.2
21–25 years	11	4.1
26–30 years	10	3.7
31 years and above	4	1.5
Missing	2	0.7

($N = 64$). The firms with the year range of 21 to 30 years constituted the least (11.9%; $N = 32$) of the sample, as shown in Table 11.5.

Table 11.6 shows that the majority of the respondents were employed by private firms (62.1%; $N = 167$), followed by those employed in public firms (24.5%, $N = 66$). Sole proprietorships were represented by the least number of respondents (13.0%; $N = 35$) of the sample.

The majority of the ongoing projects were under public liability (50.9%; $N = 137$), followed by private firms (39.8%; $N = 107$), and sole proprietorships constituted the least (5.6%; $N = 15$) of the sample, as shown in Table 11.7.

TABLE 11.5
Existence of Firm (Small to Medium-Sized Construction Companies)

Existence	Frequency	Percentage
5–9 years	64	23.8
10–14 years	62	23.0
15–20 years	73	27.1
21–30 years	32	11.9
31 years and above	35	1.3.0
Missing	3	1.1

TABLE 11.6
Firm (Small to Medium-Sized Construction Companies) Ownership

Ownership	Frequency	Percentage
Private	167	62.1
Public liability	66	24.5
Sole proprietorship	35	13.0
Missing	1	0.4

TABLE 11.7
Ongoing Projects

Ongoing	Frequency	Percentage
Private	107	39.8
Public liability	137	50.9
Sole proprietorship	15	5.6
Missing	10	3.7

As indicated in Table 11.8, the majority of the respondents (66.9%; N = 180) were carrying out building construction works, followed by civil engineering works (30.1%; N = 81), and other works had the fewest respondents (1.5%; N = 4) of the sample.

The majority of the respondents (43.9%; N = 118) indicated the national level as the geographical spread of their firm, followed by regional (26.8%; N = 72). Few of the firms were international (7.8%; N = 21), as shown in Table 11.9.

Table 11.10 shows that a large number of the respondents (31.2%; N = 84) worked with the D2/K2 class of contractors, followed by the D3/K3 class (30.9%; N = 83), and the D4/K4 class constituted the least (9.3%; N = 25) of the sample.

The majority of the respondents (30.1%; N = 81) had contracts valued in the range of 2 to 5 million cedis (GH¢), followed by contracts valued at 6 to 10 million cedis (GH¢) (29.0%; N = 78) and the contract value in the range of 16 million-plus cedis (GH¢) constituted (16.0%) the least of the sample, as shown in Table 11.11.

TABLE 11.8
Type of Projects

Type	Frequency	Percentage
Building construction	180	66.9
Civil engineering	81	30.1
Other	4	1.5
Missing	4	1.5

TABLE 11.9
Geographical Spread

Geographical Spread	Frequency	Percentage
Metropolitan	57	21.2
Regional	72	26.8
National	118	43.9
International	21	7.8
Missing	1	0.4

TABLE 11.10
Classification of Firm (Small to Medium-Sized Construction Companies)

Classification	Frequency	Percentage
D1/K1	67	24.9
D2/K2	84	31.2
D3/K3	83	30.9
D4/K4	25	9.3
Missing	10	3.7

TABLE 11.11
Project Value (GH¢)

Value (in Millions)	Frequency	Percentage
2–5	81	30.1
6–10	78	29.0
11–15	66	24.5
16 and above	43	16.0

Table 11.12 shows that a large number of the projects (52.8%; $N = 142$) were for the public domain, followed by property developers (34.6%; $N = 93$). Other projects constituted the least (2.6%; $N = 7$) of the sample.

Table 11.13 indicates the factors influencing small to medium-sized enterprise (SME) contractors' non-compliance with H&S features in terms of percentage responses on a scale of 1 (strongly disagree) to 5 (strongly agree), and a mean score (MS) ranging between 1.00 and 5.00. All MSs are above the midpoint score of 3.00, which indicates that the respondents agreed with the government support features of health and safety compliance. It is notable that all 14 respondents ranked factors influencing SME contractors' non-compliance with H&S features have a MS $> 3.50 \leq 5.00$, which

TABLE 11.12
Client

Client	Frequency	Percentage
Property developers	93	34.6
Parastatal organisation	25	9.3
Public	142	52.8
Other	7	2.6
Missing	2	0.7

indicates that the respondents perceive the factors influencing SME contractors' non-compliance with H&S features to be between 'neutral' and 'agree'. All the rankings of factors influencing SME contractors' non-compliance with H&S features indicate MS > 3.50 ≤ 3.80. The relatively low MS achieved suggests that all the factors influencing SME contractors' non-compliance with H&S features are not particularly significant in driving health and safety compliance among SME contractors.

Table 11.14 indicates the factors influencing H&S compliance of SME contractor features in terms of percentage responses on a scale of 1 (strongly disagree) to 5 (strongly agree), and a MS ranging between 1.00 and 5.00. All MSs are above the midpoint score of 3.00, which indicates that the respondents agreed with the factors influencing H&S compliance of SME features. It is notable that the first eight ranked factors influencing H&S compliance of SME features have an MS > 4.00 ≤ 5.00, which indicates that the respondents perceive the factors influencing H&S compliance of SME features to be between 'agree' and 'strongly agree'. However, the ranking of factors influencing H&S compliance of SMEs features from the ninth to the sixteenth variable indicates MS > 3.50 ≤ 4.00, which indicates that the respondents perceive the factors influencing H&S compliance of SME features to be between 'neutral' and 'agree'. The relatively high MS (4.18 to 4.02) achieved from eight variables suggest that these variables are very significant in driving health and safety compliance among SME contractors.

Table 11.15 indicates the safe environmental features in terms of percentage responses on a scale of 1 (strongly disagree) to 5 (strongly agree), and a MS ranging between 1.00 and 5.00. All MSs are above the midpoint score of 3.00, which indicates that the respondents agreed with the safe environmental features of health and safety compliance. It is notable that all the safe environmental features have an MS > 4.00 ≤ 5.00, which indicates that the respondents perceive the safe environmental features to be between 'agree' and 'strongly agree'. The relatively high MS (4.28 to 4.03) achieved suggests that these variables are very significant in driving health and safety compliance among SME contractors.

Table 11.16 indicates the safe act of workers features in terms of percentage responses on a scale of 1 (strongly disagree) to 5 (strongly agree), and an MS ranging between 1.00 and 5.00. All MSs are above the midpoint score of 3.00, which indicates that the respondents agreed with the safe act of workers features of health and safety compliance. It is notable that fifteen ranked safe act of workers features have

TABLE 11.13
Factors Influencing SME Contractors' Non-Compliance with H&S

Variable	Strongly Disagree ... Strongly Agree					MS	SD	Rank
	1	2	3	4	5			
Lack of enforcement from the legislative bodies overseeing the implementation of OSH Act	8.0	7.7	10.1	52.3	22.0	3.89	1.130	1
Lack of interest in compliance with environmental health regulations	7.3	9.0	7.6	47.6	28.5	3.86	1.160	2
Lack of knowledge to understand H&S regulations	6.3	12.2	7.3	53.5	20.6	3.82	1.118	3
Lack of H&S experts	8.0	10.4	10.4	52.1	19.1	3.80	1.142	4
Lack of knowledge to identify hazards or risks	6.3	13.9	6.3	48.3	25.3	3.80	1.168	4
Lack of finance in the management of H&S regulation	7.7	9.1	10.1	52.8	20.3	3.79	1.126	5
Short track records of H&S regulations	5.6	9.4	14.0	52.1	18.9	3.78	1.058	6
Lack of specialist skills workers in H&S	7.4	11.2	9.5	46.7	25.3	3.78	1.175	6
Lack of knowledge to interpret H&S rules	6.3	9.7	9.0	51.4	23.6	3.77	1.107	7
Lack of personnel to monitor changing legal requirements	5.9	11.8	13.5	48.3	20.5	3.77	1.109	7
Changes in structure	2.8	8.7	20.3	49.3	18.9	3.76	0.960	8
Limited cash flow	6.3	11.8	11.5	49.8	20.6	3.75	1.119	9
Changes in ownership at various stages of growth	4.5	11.1	19.1	48.6	16.7	3.54	1.033	10

TABLE 11.14
Factors Influencing H&S Compliance of SME Contractors

Variable	Strongly Disagree ... Strongly Agree							
	1	2	3	4	5	MS	SD	Rank
Interest in compliance with environmental health regulations	1.7	3.8	7.7	53.5	33.3	4.18	0.841	1
Available H&S experts	1.4	3.9	11.2	56.1	27.4	4.14	0.817	2
Enough capacity						4.11		3
Reduction in accident rate	2.5	8.1	7.7	54.9	26.8	4.10	0.944	4
Available knowledge to interpret H&S rules	1.1	6.3	8.4	59.6	24.6	4.08	0.824	5
Available skills to identify hazards or risk	2.1	2.4	8.0	55.2	32.2	4.06	0.821	6
Available knowledge to understand H&S regulations	1.8	4.2	12.3	57.2	24.6	4.04	0.835	7
Available personnel to monitor changing legal requirements	2.5	4.6	15.4	52.3	25.3	4.02	0.899	8
Enough capacity	2.1	4.9	13.7	54.7	24.6	3.97	0.877	9
Good track records of H&S regulations	2.5	5.3	11.6	56.8	29.8	3.96	0.899	10
Good track of H&S information	1.4	6.6	7.7	51.7	32.5	3.94	0.886	11
Incur extra cost on training of new employees	2.5	9.1	11.9	52.6	23.9	3.87	0.964	12
Incur extra cost of accident compensation	6.6	7.3	9.8	51.7	24.5	3.85	1.095	13
Incur extra cost on the employment of H&S personnel	2.5	9.3	11.4	51.8	25.0	3.81	0.974	14
Incur extra cost for the payment of government representatives on H&S education	4.2	15.8	14.8	47.8	17.6	3.76	1.081	15
Reduction in their profit margin	6.6	12.9	18.5	46.2	15.7	3.56	1.107	16

TABLE 11.15
Safe Environment

	Strongly Disagree ... Strongly Agree							
Variables	**1**	**2**	**3**	**4**	**5**	**MS**	**SD**	**Rank**
Provision of warning system	1.4	2.8	4.2	49.1	42.5	4.28	0.792	1
Safe and healthy work environment	2.1	1.4	4.2	53.6	38.8	4.26	0.780	2
Safe storage of equipment	1.4	3.8	2.8	56.8	35.2	4.21	0.786	3
Safe transportation of equipment	1.0	1.7	7.0	57.5	32.8	4.19	0.725	4
Safe storage of materials	1.7	3.5	3.8	56.1	34.9	4.19	0.805	4
Safe transportation of materials	1.0	4.9	3.8	57.6	32.6	4.16	0.793	5
Safe transportation of formworks and false work	3.1	3.5	7.3	59.2	26.8	4.03	0.874	6
Safe storage of formworks and false work	2.5	4.2	10.5	53.3	29.5	4.03	0.890	6

TABLE 11.16
Safe Act of Workers

Variable	Strongly Disagree ... Strongly Agree 1	2	3	4	5	MS	SD	Rank
Ensure the use of personal protective equipment (PPE)	1.0	1.0	3.5	45.8	48.6	4.40	0.706	1
Do not work under the influence of alcohol and other drugs	2.8	1.4	4.2	38.2	53.5	4.38	0.859	2
Work in good physical conditions	0.3	2.1	3.5	48.4	45.6	4.37	0.687	3
Use appropriate tools/equipment	0.7	1.7	3.5	47.9	46.2	4.37	0.702	3
Ensure equipment/tools are in good condition before usage	0.1	1.7	4.5	49.3	44.4	4.36	0.654	4
Do not smoke in flammable materials store	2.8	3.1	3.1	39.9	51.0	4.33	0.899	5
Inspect workplace before commencing any activity	1.4	2.8	4.5	50.0	41.3	4.27	0.789	6
Ensure proper stacking of objects/materials in safe locations	0.7	2.1	5.2	54.9	37.2	4.26	0.711	7
Concentrate on the task at hand	1.4	3.5	5.2	47.2	42.7	4.26	0.824	7
Use proper means of lifting, handling or moving of objects	1.0	1.0	5.9	56.6	35.4	4.24	0.701	8
Do not throw or accidentally drop objects from high levels	2.4	2.1	6.3	47.6	41.7	4.24	0.852	8
Do not remove safety guards from the workplace or equipment	0.3	3.8	8.0	48.3	39.6	4.23	0.781	9
Ensure proper use of work guidelines	2.8	5.0	3.9	47.7	40.6	4.18	0.933	10
Avoid annoyance and horseplay at the workplace	1.4	2.4	9.8	54.4	32.1	4.13	0.791	11
Ensure proper positioning of tasks	2.1	5.2	5.9	56.3	30.4	4.08	0.871	12
Do not service equipment that is in operation	6.6	7.3	8.3	45.5	32.3	3.90	1.134	13

an MS > 3.90 ≤ 5.00, which indicates that the respondents perceive the safe act of workers features to be between 'agree' and 'strongly agree'. However, the ranking of safe act of workers features of the seventeenth variable indicates MS > 4.00 ≤ 4.40. The relatively high MS (4.40 to 4.08) achieved suggests that these variables are very significant in driving health and safety compliance among SME contractors, with the exception of one variable with MS = 3.90.

Table 11.17 indicates the safe working condition features in terms of percentage responses on a scale of 1 (strongly disagree) to 5 (strongly agree), and an MS ranging between 1.00 and 5.00. All MSs are above the midpoint score of 3.00, which indicates that the respondents agreed with the safe working condition features of health and safety compliance. It is notable that all the eighteen ranked safe working condition features have a MS > 4.00 ≤ 5.00, which indicates that the respondents perceive the safe working condition features to be between 'agree' and 'strongly agree'. The relatively high MS (4.32 to 4.01) achieved suggests that these variables are very significant in driving health and safety compliance among SME contractors.

Table 11.18 indicates the reaction of workers to safe condition features in terms of percentage responses on a scale of 1 (strongly disagree) to 5 (strongly agree), and an MS ranging between 1.00 and 5.00. All the MS are above the midpoint score of 3.00, which indicates that the respondents agreed with the reaction of workers safe condition features of health and safety compliance. It is notable that all the eight ranked reaction of workers to safe working condition features have an MS > 4.10 ≤ 5.00, which indicates that the respondents perceive the reaction of workers to safe working condition features to be between 'agree' and 'strongly agree'. The relatively high MS (4.32 to 4.09) achieved suggest that these variables are very significant in driving health and safety compliance among SME contractors.

Table 11.19 indicates the government support features in terms of percentage responses on a scale of 1 (strongly disagree) to 5 (strongly agree), and an MS ranging between 1.00 and 5.00. All the MSs are above the midpoint score of 3.00, which indicates that the respondents agreed with the government support features of health and safety compliance. It is notable that all the five ranked government support features have an MS > 3.80 ≤ 5.00, which indicates that the respondents perceive the government support features to be between 'neutral' and 'agree'. The relatively low MS (3.90 to 3.80) suggests that these variables are not very significant in driving health and safety compliance among SMEs contractors.

Table 11.20 indicates the contractor's organisational culture features in terms of percentage responses on a scale of 1 (strongly disagree) to 5 (strongly agree), and an MS ranging between 1.00 and 5.00. All the MS are above the midpoint score of 3.00, which indicates that the respondents agreed with the contractor's organisational culture features of health and safety compliance. It is notable that all the eleven ranked contractors' organisational culture features have an MS > 4.10 ≤ 5.00, which indicates that the respondents perceive the contractor's organisational culture features to be between 'agree' and 'strongly agree'. The relatively high MS (4.31 to 4.16) achieved suggests that all the contractor's organisational culture features are very significant in driving health and safety compliance among SME contractors.

Table 11.21 indicates the health and safety compliance features in terms of percentage responses on a scale of 1 (strongly disagree) to 5 (strongly agree), and a MS

TABLE 11.17
Safe Working Condition

Variable	Strongly Disagree ... Strongly Agree					MS	SD	Rank
	1	2	3	4	5			
Safe movement around workplace	0.3	1.0	6.6	50.5	41.5	4.32	0.675	1
Provide safety regulations of equipment	0.7	3.1	4.2	52.3	39.7	4.27	0.745	2
Good company safety policies	0.3	3.8	6.6	51.6	37.6	4.22	0.761	3
Provision of break periods for workers to access the facilities	0.7	3.2	5.3	56.0	34.8	4.21	0.742	4
Good inspection programme	0.4	3.2	6.3	56.5	33.7	4.20	0.721	5
Provision of facilities that are clean, safe and accessible to all workers	1.1	2.1	5.7	59.0	32.2	4.19	0.724	6
Provision of sufficient lighting system for enclosed areas	0.7	3.8	5.9	54.7	34.8	4.19	0.767	6
Availability of facilities within a reasonable distance from the work area	0.7	2.1	7.4	58.5	31.3	4.18	0.711	7
Workers should be given adequate ventilation	1.4	2.1	4.6	60.9	31.0	4.18	0.732	7
Provision of training	1.0	4.5	5.9	53.7	34.8	4.17	0.811	8
Provision of adequate facilities (toilet, drinking water, washing and canteen)	1.1	3.5	6.4	56.2	32.9	4.16	0.778	9
Good salary	0.7	5.2	10.8	44.6	38.7	4.15	0.864	10
Facilities must be available for all workers, day and night	1.1	2.1	7.7	59.9	29.2	4.14	0.729	11
Provision of safe means of facilities all the time	0.7	3.5	6.3	60.9	28.5	4.13	0.733	12
Payment of Social Security and National Insurance Trust (SSNIT)	0.7	4.5	11.1	49.1	34.5	4.12	0.829	13
Provision of change room for workers	0.7	5.7	6.7	57.2	29.7	4.10	0.804	14
Provision of guidance on the recommended illumination level for various types of tasks	0.7	2.8	11.7	60.8	24.0	4.05	0.730	15
Provision of incentive to workers	2.1	7.0	8.8	52.3	29.8	4.01	0.927	16

TABLE 11.18
Reaction of Workers to Safe Condition

	Strongly Disagree ... Strongly Agree							
Variable	**1**	**2**	**3**	**4**	**5**	**MS**	**SD**	**Rank**
Follow safety regulations	0.4	2.1	4.9	50.7	41.9	4.32	0.697	1
Adhere to warning signs and notices	0.7	2.1	4.6	55.8	36.8	4.26	0.704	2
Adhere to company safety policies	0.4	0.7	7.8	57.6	33.6	4.23	0.649	3
Attend safety training programme	1.1	2.8	6.0	53.0	37.2	4.22	0.768	4
Adhere to guidance on recommended illumination level for various tasks	0.4	2.5	7.7	60.7	28.8	4.15	0.688	5
Put to proper use of the available facilities (toilet, drinking water, washing and canteen)	0.4	4.9	5.6	57.2	31.9	4.15	0.763	5
Attend safety education programme	1.1	3.9	7.7	55.8	31.6	4.13	0.792	6
Adhere to regular use of provided change room	2.1	3.5	7.4	56.8	30.2	4.09		7

TABLE 11.19
Government Support

Variable	Strongly Disagree ... Strongly Agree					MS	SD	Rank
	1	2	3	4	5			
Formulate H&S policy of construction	4.2	5.2	8.4	55.6	26.6	3.95	0.968	1
Monitoring of H&S policy implementation by government representatives	3.5	7.0	11.5	53.1	24.8	3.89	0.974	2
Implementation of H&S policy by government representatives	4.9	7.3	10.8	52.4	24.5	3.84	1.033	3
Provision of H&S policy update by government representatives	5.2	7.0	12.2	51.0	24.5	3.83	1.045	4
Provision of H&S training by government representatives	5.6	8.0	10.1	53.1	23.1	3.80	1.059	5

TABLE 11.20
Contractor's Organisational Culture

Variable	Strongly Disagree ... Strongly Agree					MS	SD	Rank
	1	2	3	4	5			
Provision of signs/notices on sites	0.3	2.1	4.5	51.9	41.1	4.31	0.689	1
Provision of personal protective equipment (PPE)	0.7	3.8	1.7	55.6	38.1	4.31	0.739	1
Update on H&S information to workers	1.4	2.8	4.9	50.2	40.8	4.26	0.792	2
Communication of H&S information to workers	1.4	1.7	6.3	54.4	36.2	4.22	0.757	3
Involve workers in H&S programme	0.7	2.8	7.7	53.0	35.9	4.21	0.755	4
Assessment of hazard identification and risk	1.0	2.8	5.9	55.1	35.2	4.21	0.759	4
Company health and safety policy	0.4	2.1	7.4	57.2	33.0	4.20	0.693	5
Management commitment in H&S	0.4	2.5	7.4	57.9	31.9	4.19	0.700	6
Health and safety inspection	0.3	3.5	6.6	56.3	33.2	4.19	0.733	6
Health and safety staffing	0.3	3.8	7.7	56.1	32.1	4.16	0.748	7

TABLE 11.21
Health and Safety Compliance

Variable	Strongly Disagree ... Strongly Agree							
	1	2	3	4	5	MS	SD	Rank
Improve in performance	0.3		5.2	47.2	47.2	4.41	0.624	1
Reduce accident on sites	1.7	1.4	5.2	47.2	44.4	4.31	0.783	2
Increase in productivity	0.7	2.1	7.4	50.7	39.1	4.25	0.742	3
Compensations paid on accident victims will be reduced	5.6	5.9	11.8	45.1	31.6	3.91	1.080	4
A limited number of H&S education by government representatives	1.7	10.4	15.3	52.4	20.1	3.79	0.991	5
Limited number of H&S monitoring by government representatives	4.2	10.1	10.4	58.7	16.7	3.74	0.941	6
Reduce the cost of training on H&S	5.9	12.9	12.5	48.1	20.6	3.64	1.122	7

ranging between 1.00 and 5.00. All MSs are above the midpoint score of 3.00, which indicates that the respondents agreed with the health and safety compliance features. It is notable that the first three ranked health and safety compliance variables have an MS $> 4.20 \leq 5.00$, which indicates that the respondents perceive the health and safety compliance variables to be between 'agree' and 'strongly agree'. However, the ranking of health and safety compliance variables from the fourth to the seventh variable indicates MS $> 3.60 \leq 3.90$, which indicates that the respondents perceive the health and safety compliance variables to be between 'neutral' and 'agree'. The relatively high MS (4.40 to 4.20) achieved from the three health and safety compliance variables suggests that these variables are very significant in achieving health and safety compliance among SME contractors.

11.3 INFERENTIAL STATISTICS

11.3.1 Conceptual Health and Safety Compliance Model for Small to Medium-Sized Construction Companies

Fourteen factors were realised from the EFA based on the 269 cases, as shown in Table 11.22; further explanation on how they were achieved is provided in the following sections. These are as follows: factor one (F1) with five indicator variables, factor two (F2) with five indicator variables, factor three (F3) with three indicator variables, factor four (F4) with six indicator variables, factor five (F5) with six indicator variables, factor six (F6) with five indicator variables, factor seven (F7) with four indicator variables, factor eight (F8) with four indicator variables, factor nine (F9) with four indicator variables, factor ten (F10) with seven indicator variables, factor eleven (F11) with four indicator variables, factor twelve (F12) with four indicator variables, factor thirteen (F13) with four indicator variables and factor fourteen (F14) with three indicator variables.

11.3.2 Results of Exploratory Factor Analysis Data

The theoretical conceptual model elements were analysed using EFA. SPSS software, version 22, was used to evaluate the reliability, discriminant validity and convergent validity of the instrument. Principal axis factoring with oblimin rotation (PAF oblimin) was the method of factor extraction used to determine the unidimensionality of the elements. Bartlett's test of sphericity and the Kaiser-Meyer-Olkin (KMO) measure of sampling adequacy were used to assess the factor analysability of the data. The KMO should range from 0 to 1 and a minimum value of 0.60 was suggested as good for factor analysis (Tabachnick & Fidell, 2007). A value greater than 0.50 as a minimum cut-off value and a desirable cut-off value of 0.80 or higher is recommended. A cut-off value of KMO greater than or equal to 0.70 is likewise recommended. For the purpose of this study, data with KMOs of ≥ 0.70 ($p < 0.05$) was considered factor analysable. In addition, the eigenvalue was computed to establish the factors within the items proposed. A minimum eigenvalue of 1 was considered significant and used to explain the variance captured by a factor. Eigenvalues of less than 1 were considered insignificant and therefore excluded (Hair, Anderson, Tatham & Black, 1998).

TABLE 11.22
Conceptual Model Indicator Variables

Indicator Variable	Measurement Variable
F1SAW 10	Ensure the use of personal protective equipment
FISAW 16	Work in good physical conditions
FISAW 7	Ensure proper lifting, handling or moving of objects
FISAW 8	Ensure proper stacking of objects or materials in safe locations
FISAW 9	Avoid annoyance and horseplay at the workplace
F2GS 1	Formulate H&S policy for construction
F2GS 2	Implementation of H&S policy by government representatives
F2GS 3	Monitoring of H&S policy implementation by government representatives
F2GS 4	Provision of H&S policy update by government representatives
F2GS 5	Provide health and safety training by government representatives
F3SAW 11	Do not remove safety guards from the workplace or equipment
F3SAW 12	Do not throw or accidentally drop objects from high levels
F3SE 2	Safe storage of equipment
F4SAW 13	Ensure equipment or tools are in good condition before usage
F4SAW 14	Do not service equipment which is in operation
F4SE 3	Safe storage of building materials
F4SE 4	Safe storage of formwork and false work
F4SE 5	Safe transportation of building materials
F4SE 6	Safe transportation of formwork and false work
F5COC 10	Consultation on H&S information to workers
F5COC 11	Update on H&S information to workers
F5COC 6	Health and safety inspection
F5COC 7	Company health and safety policy
F5COC 8	Management commitment in H&S
F5COC 9	Assessment of hazard identification and risk
F6SAW 1	Inspect workplace before commencing any activity
F6SAW 3	Use appropriate tools and equipment
F6SAW 4	Do not work under the influence of alcohol and other drugs
F6SAW 5	Do not smoke in flammable materials stores
F6SAW 6	Ensure equipment/tools are in good condition before usage
F7SWC 14	Availability of facilities within a reasonable distance from the work area
F7SWC 6	Good salaries
F7SWC 7	Payment of Social Security and National Insurance Trust
F7SWC 8	Provision of sufficient lighting system for enclosed areas
F8RWSC 3	Adhere to warning signs and notices
F8RWSC 4	Follow safety regulations
F8RWSC 5	Adhere to company safety policies
F8RWSC 6	Adhere to guidance on recommended illumination level for various tasks
F9RWSC 1	Attend safety education programme
F9RWSC 2	Attend safety training programme
F9SWC 1	Provision of training
F9SWC 2	Good inspection programme

(Continued)

TABLE 11.22 (CONTINUED)
Conceptual Model Indicator Variables

Indicator Variable	Measurement Variable
F10RWSC 7	Put to proper use available facilities (toilet, drinking water, washing and canteen)
F10RWSC 8	Adhere to regular use of provided change room
F10SAW 15	Concentrate on task at hand
F10SWC 13	Provision of sufficient lighting system for enclosed areas
F10SWC 15	Provision of change room for workers
F10SWC 16	Facilities must be available to both day and night workers
F10SWC 17	Provision of safe means to facilities all the time
F11COC 1	Provision of personal protective equipment
F11COC 2	Provision of signs and notices on sites
F11SWC 4	Provision of safety regulation of equipment
F11SWC 5	Good company safety policies
F12COC 3	Training of workers on health and safety
F12COC 4	Involve workers in H&S programmes
F12COC 5	Health and safety staffing
F12SWC 9	Safe movement around workplace
F13SWC 10	Provision of guidance on the recommended illumination level for various types of tasks
F13SWC 11	Workers should have adequate ventilation
F13SWC 12	Provision of adequate facilities (toilet, drinking water, washing and canteen)
F13SWC 18	Provision of break periods for workers to access facilities
F14SE 1	Safe and healthy work environment
F14SE 7	Safe transportation of equipment
F14SE 8	Provision of warning system

11.3.2.1 Exploratory Factor Analysis: Dimensionality of Health and Safety Compliance Elements

In the following section, the measures of reliability, convergent validity and discriminant validity for each of the factors realised through EFA (Table 11.22) are discussed.

11.3.2.1.1 Factor One (F1)

EFA was conducted to assess the unidimensionality and reliability of H&S compliance. Principal axis factoring with oblimin rotation (PAF oblimin) was specified as the extraction and rotation method. There were four items measuring factor one (F1). The result of F1 is reported in Table 11.23. The corrected item-total correlation was greater than the suggested cut-off value of 0.30, suggesting that the items were good measures of the element and the Cronbach's alpha was greater than 0.807 at 0.808, indicating acceptable internal reliability. The KMO of 0.886 with Bartlett's test of sphericity of $p < 0.000$ was also obtained, indicating consistency with the recommended KMO cut-off value of 0.70 and Bartlett's test of sphericity of $p < 0.05$

TABLE 11.23
Factor One (F1)

Item	Question	Factor Loading	Corrected Item-Total Correlation	Cronbach's Alpha If Item Deleted
SAW 10	Ensure the use of personal protective equipment (PPE)	0.5592	0.573	0.778
SAW 16	Work in good physical conditions	0.6391	0.487	0.800
SAW 7	Use proper means of lifting, handling or moving of objects	0.7122	0.690	0.739
SAW 8	Ensure proper stacking of objects/materials	0.7105	0.666	0.749
SAW 9	Avoid annoyance and horseplay at the workplace	0.6898	0.567	0.779

(Hair et al., 2010). These results suggested that factor analysis could be conducted with the data. All the five items (SAW 10, SAW 16, SAW 7, SAW 8 and SAW 9) are expected to measure H&S compliance loaded together on this factor. The factor loadings for all items were greater than 0.5593 reported in Table 11.23, which were greater than the recommended value of 0.40 (Hair et al., 1998; Field, 2005).

11.3.2.1.2 Factor Two (F2)

EFA was conducted to assess the unidimensionality and reliability of H&S compliance. Principal axis factoring with oblimin rotation (PAF oblimin) was specified as the extraction and rotation method. There were five items measuring factor two (F2). The result of F2 is reported in Table 11.24. The corrected item-total correlation was greater than the suggested cut-off value of 0.30, suggesting that the items were good measures of the element and the Cronbach's alpha was greater than 0.905 at 0.902, indicating acceptable internal reliability. The KMO of 0.886 with Bartlett's test of sphericity of $p < 0.000$ was also obtained, indicating consistency with the recommended KMO cut-off value of 0.70 and Bartlett's test of sphericity of $p < 0.05$ (Hair, Black, Babin, Anderson & Tatham, 2006). These results suggested that factor analysis could be conducted with the data. All five items (GS 1, GS 2, GS 3, GS 4 and GS 5) are expected to measure H&S compliance loaded together on this factor. The factor loadings for all items were greater than 0.4182 reported in Table 11.24, which were greater than the recommended value of 0.40 (Hair et al., 1998; Field, 2005).

11.3.2.1.3 Factor Three (F3)

EFA was conducted to assess the unidimensionality and reliability of H&S compliance. Principal axis factoring with oblimin rotation (PAF oblimin) was specified as

TABLE 11.24
Factor Two (F2)

Item	Question	Factor Loading	Corrected Item-Total Correlation	Cronbach's Alpha if Item Deleted
GS 1	Formulate H&S policy of construction	0.4182	0.680	0.896
GS 2	Implementation of H&S policy by government representatives	0.5056	0.818	0.875
GS 3	Monitoring of H&S policy implementation by the government representatives	0.5155	0.841	0.872
GS 4	Provision of H&S policy update by government representatives	0.4761	0.834	0.873
GS 5	Provision of H&S training by government representatives	0.4674	0.811	0.876

the extraction and rotation method. There were three items measuring factor three (F3). The result of F3 is reported in Table 11.25. The corrected item-total correlation was greater than the suggested cut-off value of 0.30, suggesting that the items were good measures of the element and the Cronbach's alpha was greater than 0.598 at 0.595, indicating acceptable internal reliability. The KMO of 0.886 with Bartlett's test of sphericity of $p < 0.000$ was also obtained, indicating consistency with the recommended KMO cut-off value of 0.70 and Bartlett's test of sphericity of $p < 0.05$ (Hair et al., 2006). These results suggested that factor analysis could be conducted with the data. All the three items (SAW 11, SAW 12 and SE 2) are expected to measure H&S compliance loaded together on this factor. The factor loadings for all

TABLE 11.25
Factor Three (F3)

Item	Question	Factor Loading	Corrected Item-Total Correlation	Cronbach's Alpha if Item Deleted
SAW 11	Do not remove safety guards from the workplace or equipment	0.5807	0.541	0.294
SAW 12	Do not throw or accidentally drop objects from high levels	0.5712	0.465	0.406
SE 2	Safe storage of equipment	0.5793	0.241	0.714

items were greater than 0.5712 reported in Table 11.25, which were greater than the recommended value of 0.40 (Hair et al., 1998; Field, 2005).

11.3.2.1.4 Factor Four (F4)

EFA was conducted to assess the unidimensionality and reliability of H&S compliance. Principal axis factoring with oblimin rotation (PAF oblimin) was specified as the extraction and rotation method. There were six items measuring factor four (F4). The result of F4 is reported in Table 11.26. The corrected item-total correlation was greater than the suggested cut-off value of 0.30, suggesting that the items were good measures of the element and the Cronbach's alpha was greater than 0.852 at 0.864, indicating acceptable internal reliability. The KMO of 0.886 with Bartlett's test of sphericity of $p < 0.000$ was also obtained, indicating consistency with the recommended KMO cut-off value of 0.70 and Bartlett's test of sphericity of $p < 0.05$ (Hair et al., 2006). These results suggested that factor analysis could be conducted with the data. All six items SAW 13, SAW 14, SE 3, SE 4, SE 5 and SE 6 are expected to measure H&S compliance loaded together on this factor. The factor loadings for all items were greater than 0.3502 as reported in Table 11.26, which were greater than the recommended value of 0.40 (Hair et al., 1998; Field, 2005).

11.3.2.1.5 Factor Five (F5)

EFA was conducted to assess the unidimensionality and reliability of H&S compliance. Principal axis factoring with oblimin rotation (PAF oblimin) was specified as the extraction and rotation method. There were six items measuring factor five (F5). The result of F5 is reported in Table 11.27. The corrected item-total correlation was greater than the suggested cut-off value of 0.30, suggesting that the items were good measures of the element and the Cronbach's alpha was greater than

TABLE 11.26
Factor Four (F4)

Item	Question	Factor Loading	Corrected Item-Total Correlation	Cronbach's Alpha if Item Deleted
SAW 13	Ensure proper positioning of tasks	0.5126	0.664	0.822
SAW 14	Do not service equipment that is in operation	0.3502	0.504	0.868
SE 3	Safe storage of materials	0.4662	0.563	0.842
SE 4	Safe storage of formworks and false work	0.5881	0.785	0.799
SE 5	Safe transportation of materials	0.6085	0.694	0.820
SE 6	Safe transportation of formworks and false work	0.5677	0.720	0.812

TABLE 11.27
Factor Five (F5)

Item	Question	Factor Loading	Corrected Item-Total Correlation	Cronbach's Alpha if Item Deleted
COC 10	Communication of H&S information to workers	0.6811	0.681	0.850
COC 11	Update on H&S information to workers	0.6968	0.636	0.857
COC 6	Health and safety inspection	0.6253	0.663	0.852
COC 7	Company health and safety policy	0.6089	0.680	0.850
COC 8	Management commitment in H&S	0.6382	0.709	0.844
COC 9	Assessment of hazard identification and risk	0.5993	0.676	0.851

0.872 at 0.873, indicating acceptable internal reliability. The KMO of 0.886 with Bartlett's test of sphericity of $p < 0.000$ was also obtained, indicating consistency with the recommended KMO cut-off value of 0.70 and Bartlett's test of sphericity of $p < 0.05$ (Hair et al., 2006). These results suggested that factor analysis could be conducted with the data. All six items (COC 10, COC 11, COC 6, COC 7, COC 8 and COC 9) are expected to measure H&S compliance loaded together on this factor. The factor loadings for all items were greater than 0.5993 reported in Table 11.27, which were greater than the recommended value of 0.40 (Hair et al., 1998; Field, 2005).

11.3.2.1.6 Factor Six (F6)

EFA was conducted to assess the unidimensionality and reliability of H&S compliance. Principal axis factoring with oblimin rotation (PAF oblimin) was specified as the extraction and rotation method. There were five items were measuring factor six (F6). The result of F6 is reported in Table 11.28. The corrected item-total correlation was greater than the suggested cut-off value of 0.30, suggesting that the items were good measures of the element and the Cronbach's alpha was greater than 0.800 at 0.804, indicating acceptable internal reliability (Nunnally & Bernstein, 1994). The KMO of 0.886 with Bartlett's test of sphericity of $p < 0.000$ was also obtained, indicating consistency with the recommended KMO cut-off value of 0.70 and Bartlett's test of sphericity of $p < 0.05$ suggested by Hair et al. (2006). These results suggested that factor analysis could be conducted with the data. All the five items (SAW 1, SAW 3, SAW 4, SAW 5 and SAW 6) are expected to measure H&S compliance loaded together on this factor. The factor loadings for all items were greater than 0.5641 reported in Table 11.28, which were greater than the recommended value of 0.40 as suggested by Field (2005) and Hair et al. (1998).

TABLE 11.28
Factor Six (F6)

Item	Question	Factor Loading	Corrected Item-Total Correlation	Cronbach's Alpha if Item Deleted
SAW 1	Inspect workplace before commencing any activity	0.6466	0.529	0.784
SAW 3	Use appropriate tools/ equipment	0.6426	0.579	0.764
SAW 4	Do not work under the influence of alcohol and other drugs	0.6167	0.602	0.756
SAW 5	Do not smoke in flammable materials store	0.6142	0.662	0.736
SAW 6	Ensure equipment/tools are in good condition before usage	0.5641	0.563	0.769

11.3.2.1.7 Factor Seven (F7)

EFA was conducted to assess the unidimensionality and reliability of H&S compliance. Principal axis factoring with oblimin rotation (PAF oblimin) was specified as the extraction and rotation method. There were four items measuring factor seven (F7). The result of F7 is reported in Table 11.29. The corrected item-total correlation was greater than the suggested cut-off value of 0.30, suggesting that the items were good measures of the element and the Cronbach's alpha was greater than 0.770 at

TABLE 11.29
Factor Seven (F7)

Item	Question	Factor Loading	Corrected Item-Total Correlation	Cronbach's Alpha if Item Deleted
SWC 14	Availability of facilities within a reasonable distance from the work area	0.5889	0.454	0.770
SWC 6	Good salary	0.5553	0.646	0.672
SWC 7	Payment of Social Security and National Insurance Trust (SSNIT)	0.5932	0.715	0.629
SWC 8	Provision of sufficient lighting system for enclosed areas	0.6278	0.493	0.753

0.764, indicating acceptable internal reliability. The KMO of 0.886 with Bartlett's test of sphericity of $p < 0.000$ was also obtained, indicating consistency with the recommended KMO cut-off value of 0.70 and Bartlett's test of sphericity of $p < 0.05$ (Hair et al., 2006). These results suggested that factor analysis could be conducted with the data. All four items (SWC 14, SWC 6, SWC 7 and SWC 8) are expected to measure H&S compliance loaded together on this factor. The factor loadings for all items were greater than 0.5553 reported in Table 11.29, which were greater than the recommended value of 0.40 (Hair et al., 1998; Field, 2005).

11.3.2.1.8 Factor Eight (F8)

EFA was conducted to assess the unidimensionality and reliability of H&S compliance. Principal axis factoring with oblimin rotation (PAF oblimin) was specified as the extraction and rotation method. There were four items measuring factor eight (F8). The result of F8 is reported in Table 11.30. The corrected item-total correlation was greater than the suggested cut-off value of 0.30, suggesting that the items were good measures of the element and the Cronbach's alpha was greater than 0.835 at 0.835, indicating acceptable internal reliability. The KMO of 0.886 with Bartlett's test of sphericity of $p < 0.000$ was also obtained, indicating consistency with the recommended KMO cut off value of 0.70 and Bartlett's test of sphericity of $p < 0.05$ (Hair et al., 2006). These results suggested that factor analysis could be conducted with the data. All the four items (RWSC 3, RWSC 4, RWSC 5 and RWSC 6) are expected to measure H&S compliance loaded together on this factor. The factor loadings for all items were greater than 0.5468 as reported in Table 11.30, which were greater than the recommended value of 0.40 (Hair et al., 1998; Field, 2005).

11.3.2.1.9 Factor Nine (F9)

EFA was conducted to assess the unidimensionality and reliability of H&S compliance. Principal axis factoring with oblimin rotation (PAF oblimin) was specified as the extraction and rotation method. There were four items measuring factor nine (F9).

TABLE 11.30
Factor Eight (F8)

Item	Question	Factor Loading	Corrected Item-Total Correlation	Cronbach's Alpha if Item Deleted
RWSC 3	Adhere to warning signs and notices	0.6957	0.721	0.766
RWSC 4	Follow safety regulations	0.6566	0.716	0.770
RWSC 5	Adhere to company safety policies	0.6244	0.645	0.801
RWSC 6	Adhere to guidance on recommended illumination level for various tasks	0.5468	0.587	0.825

TABLE 11.31
Factor Nine (F9)

Item	Question	Factor Loading	Corrected Item-Total Correlation	Cronbach's Alpha if Item Deleted
RWSC 1	Attend safety education programme	0.7123	0.738	0.790
RWSC 2	Attend safety training programme	0.7196	0.731	0.793
SWC 1	Provision of training	0.6545	0.645	0.830
SWC 2	Good inspection programme	0.6290	0.659	0.825

The result of F9 is reported in Table 11.31. The corrected item-total correlation was greater than the suggested cut-off value of 0.30, suggesting that the items were good measures of the element and the Cronbach's alpha was greater than 0.851 at 0.852, indicating acceptable internal reliability (Nunnally & Bernstein, 1994). The KMO of 0.886 with Bartlett's test of sphericity of $p < 0.000$ was also obtained, indicating consistency with the recommended KMO cut-off value of 0.70 and Bartlett's test of sphericity of $p < 0.05$ suggested by Hair et al. (2006). These results suggested that factor analysis could be conducted with the data. All four items (RWSC 1, RWSC 2, SWC 1 and SWC 2) are expected to measure H&S compliance loaded together on this factor. The factor loadings for all items were greater than 0.6290 as reported in Table 11.31, which were greater than the recommended value of 0.40 as suggested by Field (2005) and Hair et al. (1998).

11.3.2.1.10 Factor Ten (F10)

EFA was conducted to assess the unidimensionality and reliability of H&S compliance. Principal axis factoring with oblimin rotation (PAF oblimin) was specified as the extraction and rotation method. There were seven items measuring factor ten (F10). The result of F10 is reported in Table 11.32. The corrected item-total correlation was greater than the suggested cut-off value of 0.30, suggesting that the items were good measures of the element and the Cronbach's alpha was greater than 0.896 at 0.896, indicating acceptable internal reliability (Nunnally & Bernstein, 1994). The KMO of 0.886 with Bartlett's test of sphericity of $p < 0.000$ was also obtained, indicating consistency with the recommended KMO cut-off value of 0.70 and Bartlett's test of sphericity of $p < 0.05$ suggested by Hair et al. (2006). These results suggested that factor analysis could be conducted with the data. All seven items (RWSC 7, RWSC 8, SAW 15, SWC 13, SWC 15, SWC 16 and SWC 17) are expected to measure H&S compliance loaded together on this factor. The factor loadings for all items were greater than 0.5907 as reported in Table 11.32, which were greater than the recommended value of 0.40 as suggested by Field (2005) and Hair et al. (1998).

TABLE 11.32
Factor Ten (F10)

Item	Question	Factor Loading	Corrected Item-Total Correlation	Cronbach's Alpha if Item Deleted
RWSC 7	Put to proper use of the available facilities (toilet, drinking water, washing and canteen)	0.6067	0.721	0.877
RWSC 8	Adhere to regular use of provided change room	0.5907	0.669	0.884
SAW 15	Concentrate on the task at hand	0.6625	0.626	0.888
SWC 13	Provision of facilities that are clean, safe and accessible to all workers	0.6861	0.678	0.883
SWC 15	Provision of change room for workers	0.6304	0.756	0.873
SWC 16	Facilities must be available for all workers, day and night	0.6608	0.696	0.881
SWC 17	Provision of safe means of facilities all the time	0.6549	0.738	0.875

11.3.2.1.11 Factor Eleven (F11)

EFA was conducted to assess the unidimensionality and reliability of H&S compliance. Principal axis factoring with oblimin rotation (PAF oblimin) was specified as the extraction and rotation method. There were four items measuring factor eleven (F11). The result of F11 is reported in Table 11.33. The corrected item-total correlation was greater than the suggested cut-off value of 0.30, suggesting that the items were good measures of the element and the Cronbach's alpha was greater than 0.748 at 0.751, indicating acceptable internal reliability (Nunnally & Bernstein, 1994). The KMO of 0.886 with Bartlett's test of sphericity of $p < 0.000$ was also obtained, indicating consistency with the recommended KMO cut-off value of 0.70 and Bartlett's test of sphericity of $p < 0.05$ suggested by Hair et al. (2006). These results suggested that factor analysis could be conducted with the data. All four items (COC 1, COC 2, SWC 4 and SWC 5) are expected to measure H&S compliance loaded together on this factor. The factor loadings for all items were greater than 0.5591 as reported in Table 11.33, which were greater than the recommended value of 0.40 as suggested by Field (2005) and Hair et al. (1998).

11.3.2.1.12 Factor Twelve (F12)

EFA was conducted to assess the unidimensionality and reliability of H&S compliance. Principal axis factoring with oblimin rotation (PAF oblimin) was specified as

TABLE 11.33
Factor Eleven (F11)

Item	Question	Factor Loading	Corrected Item-Total Correlation	Cronbach's Alpha if Item Deleted
COC 1	Provision of personal protective equipment (PPE)	0.7008	0.565	0.679
COC 2	Provision of signs/notices on sites	0.7099	0.569	0.678
SWC 4	Provide safety regulations of equipment	0.6748	0.577	0.672
SWC 5	Good company safety policies	0.5591	0.471	0.731

the extraction and rotation method. There were four items measuring factor twelve (F12). The result of the factor is reported in Table 11.34. The corrected item-total correlation was greater than the suggested cut-off value of 0.30, suggesting that the items were good measures of the element and the Cronbach's alpha was greater than 0.713 at 0.709, indicating acceptable internal reliability (Nunnally & Bernstein, 1994). The KMO of 0.886 with Bartlett's test of sphericity of $p < 0.000$ was also obtained, indicating consistency with the recommended KMO cut-off value of 0.70 and Bartlett's test of sphericity of $p < 0.05$ suggested by Hair et al. (2006). These results suggested that factor analysis could be conducted with the data. All four items (COC 3, COC 4, COC 5 and SWC 9) are expected to measure H&S compliance loaded together on this factor. The factor loadings for all items were greater than 0.5829 reported in Table 11.34, which were greater than the recommended value of 0.40 as suggested by Field (2005) and Hair et al. (1998).

TABLE 11.34
Factor Twelve (F12)

Item	Question	Factor Loading	Corrected Item-Total Correlation	Cronbach's Alpha if Item Deleted
COC 3	Training of workers on health and safety (H&S)	0.6322	0.639	0.563
COC 4	Involve workers in H&S programme	0.6594	0.575	0.602
COC 5	Health and safety staffing	0.6173	0.504	0.648
SWC 9	Safe movement around workplace	0.5829	0.299	0.757

11.3.2.1.13 Factor Thirteen (F13)

EFA was conducted to assess the unidimensionality and reliability of H&S compliance. Principal axis factoring with oblimin rotation (PAF oblimin) was specified as the extraction and rotation method. There were four items measuring factor thirteen (F13). The result of the factor is reported in Table 11.35. The corrected item-total correlation was greater than the suggested cut-off value of 0.30, suggesting that the items were good measures of the element and the Cronbach's alpha was greater than 0.819 at 0.818, indicating acceptable internal reliability (Nunnally & Bernstein, 1994). The KMO of 0.886 with Bartlett's test of sphericity of $p < 0.000$ was also obtained, indicating consistency with the recommended KMO cut-off value of 0.70 and Bartlett's test of sphericity of $p < 0.05$ suggested by Hair et al. (2006). These results suggested that factor analysis could be conducted with the data. All four items (SWC 10, SWC 11, SWC 12 and SWC 18) are expected to measure H&S compliance loaded together on this factor. The factor loadings for all items were greater than 0.5865 as reported in Table 11.35, which were greater than the recommended value of 0.40 as suggested by Field (2005) and Hair et al. (1998).

11.3.2.1.14 Factor Fourteen (F14)

EFA was conducted to assess the unidimensionality and reliability of H&S compliance. Principal axis factoring with oblimin rotation (PAF oblimin) was specified as the extraction and rotation method. There were three items measuring factor fourteen (F14). The result of F14 is reported in Table 11.36. The corrected item-total correlation was greater than the suggested cut-off value of 0.30, suggesting that the items were good measures of the element and the Cronbach's alpha was greater than 0.795 at 0.796, indicating acceptable internal reliability (Nunnally & Bernstein, 1994). The

TABLE 11.35
Factor Thirteen (F13)

Item	Question	Factor Loading	Corrected Item-Total Correlation	Cronbach's Alpha if Item Deleted
SWC 10	Provision of guidance on the recommended illumination level for various types of task	0.6260	0.714	0.737
SWC 11	Workers should be given adequate ventilation	0.5950	0.670	0.758
SWC 12	Provision of adequate facilities toilet, drinking water, washing, and canteen	0.6954	0.655	0.766
SWC 18	Provision of break periods for workers to access the facilities	0.5865	0.527	0.822

TABLE 11.36
Factor Fourteen (F14)

Item	Question	Factor Loadings	Corrected Item-Total Correlation	Cronbach's Alpha if Item Deleted
SE 1	Safe and healthy work environment	0.6034	0.607	0.755
SE 7	Safe transportation of equipment	0.5817	0.652	0.709
SE 8	Provision of warning system	0.5335	0.658	0.698

KMO of 0.886 with Bartlett's test of sphericity of $p < 0.000$ was also obtained, indicating consistency with the recommended KMO cut-off value of 0.70 and Bartlett's test of sphericity of $p < 0.05$ suggested by Hair et al. (2006). These results suggested that factor analysis could be conducted with the data. All three items (SE 1, SE 7 and SE 8) are expected to measure H&S compliance loaded together on this factor. The factor loadings for all items were greater than 0.5335 as reported in Table 11.36, which were greater than the recommended value of 0.40 as suggested by Field (2005) and Hair et al. (1998).

11.3.2.2 Comparative Analysis of Frequencies for Exploratory Factor Analysis and Confirmatory Factor Analysis

Table 11.37 shows the valid percentages for the frequencies of EFA and CFA of the survey samples. Items were selected at random from the frequencies conducted on both EFA with 269 cases and CFA with 289 cases for a comparative analysis. Table 11.37 shows that the valid percentages for the selected items from both the EFA and

TABLE 11.37
Comparative Analysis of Frequencies for EFA and CFA Samples

Item		EFA Valid Percentage (%)	CFA Valid Percentage (%)
Gender	Male	82.2	82.6
	Female	17.8	17.4
Project type	Private firm	41.3	38.2
	Public liability	52.9	56.2
	Sole proprietor	5.8	5.7
Classification of firm	D1K1	25.9	23.6
	D2K2	32.4	37.0
	D3K3	32.0	31.2
	D4K4	9.7	8.3

CFA are almost similar. The similarity in the results obtained from the frequencies of both EFA and CFA gives a strong indication for the CFA to be conducted on the factors with the 269 cases. Kline (2005) indicated that when the construct has few items or indicator variables, less than three, the model will not be over-identified for testing, using SEM. The minimum number of measurement variables obtained in each latent construct after EFA was three and this meets the recommended number of items per element of three to enable robust structural equation modelling testing (Kline, 2005).

Table 11.38 gives details of the selected model after a series of CFA on the fourteen factors realised from the EFA. Finally, six factors were reported from the CFA based on the 289 cases (Table 11.38). Different types of models were realised during the CFA at one stage or the other through the following processes:

1. Merging of factors
2. Removing of items which have low R^2 or cross-loadings with other items
3. Removing cases to achieve stronger models fit statistics

However, at the conclusion of the stated processes, one model was reported in the current study. During the CFA processes, model 4 had its factors merged. Items were then removed from model 5. Hence, model 5 had cases removed and model 6 was then refined by removing items. Model 8 was refined by removing items to arrive at a model with six factors. Finally, safe act of workers (SAW) and safe working condition (SWC) from the six-factor model were combined to form one factor as shown (Table 11.38) to give a five-factor model. However, the five-factor model was deemed fit for the current study, two of the original conceptual H&S latent constructs names from Figure 10.1 (Model 1.0) (Chapter 10) were retained and another two renamed. The exception was the latent construct of safe act of workers (SAW) and safe working condition (SWC), which were combined to arrive at one factor, safe act and working condition. The retained latent constructs are government support (GS), safe environment (SE), contractor's organisational culture (COC) and reaction of workers to safe condition (RWSC).

11.4 STRUCTURAL EQUATION MODELLING

The reason why SEM is preferred to other statistical approaches, such as analysis of variance (ANOVA) and regression, is because it displays better conditions to demonstrate causality. Three conditions to demonstrate causality are association, isolation and directionality (Hoyle, 1995). SEM is not distinctive in the first aspect. For isolating and putative causes, SEM is more flexible and comprehensive than any univariate or multivariate modelling approach. SEM provides a means of controlling not only for extraneous or confounding variables, but for measurement error as well. SEM can be shown with many statistical procedures because it comes from theory (research design) and sample logic. In the case of a model as a whole, SEM produces a good fit, and the result greatly supports the individual causal relationships within the model. The current research has clearly shown that measuring H&S compliance of small to medium-sized construction companies is a complex construct. Therefore,

TABLE 11.38
Conceptual Model Indicator Variables

Latent Construct	Indicator Variable	Measurement Variable	Label
Safe act and working condition (SAWC)	F1SAW 7	Ensure proper lifting, handling or moving of objects	SAWC 1
	F1SAW 8	Ensure proper stacking of objects or materials in safe locations	SAWC 2
	F1SAW 9	Avoid annoyance and horseplay at the workplace	SAWC 3
	F7SWC 14	Availability of facilities within a reasonable distance from the work area	SAWC 4
	F10SAW 15	Concentrate on task at hand	SAWC 5
	F10SWC 13	Provision of sufficient lighting system for enclosed areas	SAWC 6
	F10SWC 16	Facilities must be available to both day and night workers	SAWC 7
	F13SWC 12	Provision of adequate facilities (toilet, drinking water, washing and canteen)	SAWC 8
Government support (GS)	F2GS 1	Formulate H&S policy for construction	GS 1
	F2GS 2	Implementation of H&S policy by government representatives	GS 2
	F2GS 3	Monitoring of H&S policy implementation by government representatives	GS 3
	F2GS 4	Provision of H&S policy update by government representatives	GS 4
	F2GS 5	Provide health and safety training by government representatives	GS 5
Contractor's safety policy (CSP)	F3SE 2	Safe storage of equipment	CSP 1
	F14SE 1	Safe and healthy work environment	CSP 2
	F4SE 3	Do not service equipment which is in operation	CSP 3
	F4SE 5	Safe storage of formwork and false work	CSP 4
Contractor's organisational culture (COC)	F5COC 10	Consultation on H&S information to workers	COC 1
	F5COC 11	Update on H&S information to workers	COC 2
	F5COC 6	Health and safety inspection	COC 3
	F5COC 7	Company health and safety policy	COC 4
	F5COC 8	Management commitment in H&S	COC 5
	F12COC 4	Involve workers in H&S programmes	COC 6
	F11COC 1	Provision of personal protective equipment	COC 7
	F11COC 2	Provision of signs and notices on sites	COC 8
Adherence to safety regulations (ASR)	F8RWSC 3	Adhere to warning signs and notices	ASR 1
	F8RWSC 4	Follow safety regulations	ASR 2
	F8RWSC 5	Adhere to company safety policies	ASR 3
Health and safety compliance (HSC)	F6HSC 2	Compensation paid to accident victims will be reduced	HSC 1
	F6HSC 3	Reduce the cost of training on health and safety (H&S)	HSC 2
	F6HSC 4	A limited number of H&S education by government representatives	HSC 3
	F6HSC 5	Limited number of H&S monitoring by government representatives	HSC 4

in order to examine the factors that determine H&S compliance among small to medium-sized construction companies in Ghana, EQS, version 6.2, software was used to investigate the measurement model adequacy and structural model goodness of fit.

11.4.1 Structural Equation Modelling Analytic Strategy

This study aims to test a model of H&S compliance in a sample of small to medium-sized construction companies in the major cities of Ghana. Three steps were conducted in the analysis – EFA, CFA and SEM – using EQS, version 6.2, for analysis. First, a series of CFA was used to test for the measurement equivalency for each of the five latent constructs in Table 11.38 and the manifest or composite variables of H&S compliance in the hypothesised model of H&S compliance. The CFA results defined the relations between the observed and unobserved variables. The CFA provided the link between scores on a measuring instrument and the underlying constructs they are designed to measure. The CFA was carried out to reaffirm the factor structure of the observed and unobserved variables, hence, the construct validity. Second, the fit of the entire measurement model underlying the hypothesised structural model was tested. The structural model defined the relationship amongst the different exogenous variables and specified the manner by which each exogenous variable directly or indirectly influences the changes in the values of other exogenous constructs in the model. Thus, the endogenous variables were defined (H&S compliance in small to medium-sized construction companies). All analyses were performed using EQS, including the testing of the hypothesised SEM. In SEM, a covariance matrix generated from a particular sample is compared with the covariance matrix generated from the hypothesised model, and fit statistics are used to determine the acceptability of the solution obtained. A combination of fit statistics for model evaluation is deemed adequate for an SEM study (Table 11.39) as adopted for the current study.

11.4.2 The Measurement Model: Confirmatory Factor Analysis

After the constructs demonstrated sufficient evidence of unidimensionality and reliability using EFA, a CFA was then performed. EQS, version 6.2, with maximum likelihood and the use of a robust estimator was used to analyse the construct validity of the measurement models. The types of goodness-of-fit indexes and their acceptable cut-off values selected for this study are shown in Table 11.39.

11.4.2.1 Statistics on Structural Equation Modelling Assumptions: Outliers and Missing Data

Data sets investigation revealed that some data sets had missing values. A detailed examination of the pattern of missing data revealed that the missing data was missing at random (MAR) and not missing completely at random (MCAR). The condition that data was MCAR is a situation where the presence or absence of the observation is independent of other observed variables and the variable itself

TABLE 11.39
Cut-Off Criteria for Fit Indexes

Fit Index	Acronym	Acceptable Thresholds for Continuous Data	Reference
Chi-square test	χ^2 test	Low χ^2 relative df with an insignificant *p*-value ($p > 0.05$)	Hooper et al. (2008), Hsu et al. (2012)
Normed chi-square	χ^2/df ratio	Ratio of (χ^2) to df ≤2 or 3 good fit Ratio of (χ^2) to df ≤5 acceptable	Hsu et al. (2012), Kline (2005)
Root mean square error of approximation	RMSEA	Values less than 0.05 with confidence interval (CI) 0.00–0.05 'good fit' Values greater than 0.06 to 0.08 with confidence interval 0.00–0.05 'acceptable fit' Values greater than 0.08 to 1.00 with confidence interval 0.00–0.05 'mediocre fit' Values greater than 1.00 with confidence interval 0.00–0.05 'poor fit'	Schreiber et al. (2006), Hsu et al. (2012)
Comparative fit index	CFI	Equal or greater than 0.95 'good fit' Equal or greater than 0.90 'acceptable fit'	Schreiber et al. (2006), Hu and Bentler (1999)
Tucker-Lewis index (or non-normed fit index)	TLI (or NNFI)	Equal or greater than 0.95 'good fit' Equal or greater than 0.90 'acceptable fit'	Schreiber et al. (2006), Hu and Bentler (1999)
Standardised root mean square residual	SRMR	Equal or less than 0.05 'good fit' Equal or less than 0.08 'acceptable fit'	Schreiber et al. (2006), Hu and Bentler (1999)
Normed fit index	NFI	Greater than 0.90 'good fit'	Bentler and Bonnet (1980)

(McDonald & Ho, 2002). The condition of MCAR is a very strict assumption that may be difficult to justify in practice (McDonald & Ho, 2002). Therefore, the assumption of the condition of MAR was adopted. Hence, the robust maximum likelihood estimation solution in EQS was used to address the problem. This method produces better results compared to other methods (Kline, 2005). Consequently, cases with missing variables were skipped and not included in the analysis. Further examination of the data set revealed that there were a few outliers in the data. The EQS result output included case numbers with the largest contribution to Yuan, Lambert and Fouladi's normalized multivariate kurtosis. Examination of these case numbers showed the case numbers that include outliers and it was upon these inspections that the conclusion was based. It was concluded that there were a few outliers in the data. The robust maximum likelihood (RML) was adopted for estimation which was adequate in addressing the problems of outliers.

11.4.2.2 Statistics on Structural Equation Modelling Assumptions: Data Distribution Characteristics

The model analysis began after the distribution characteristics of the data was established through the maximum likelihood estimation. This method assumes multivariate normality. The EQS result output included univariate statistics such as mean, skewness, kurtosis and the respective standard deviations. Similarly, the multivariate kurtosis formed part of the result output. Analysis of the univariate statistics and Yuan, Lambert and Fouladi based on normalized multivariate kurtosis suggested non-normality in the sample data set. The result shows that all the Yuan, Lambert and Fouladi estimates of normalized multivariate kurtosis were greater than the upper limit value of 3.0. The outcome of the results led to descriptions of the data to be highly kurtotic. However, the adoption of the robust maximum likelihood estimation method of the postulated model was due to the non-normality of the data. The results in the preceding sections are reported using the robust statistics (Satorra-Bentler scaled statistics) for the chi-square (Satorra & Bentler, 1988). The first item of each factor is fixed to establish the factors' scale in the models. Yuan, Lambert and Fouladi's coefficient and other univariate statistics are presented Table 11.40.

11.4.2.3 Statistics on Structural Equation Modelling Assumptions: Identifiability of the Model

According to Boomsma (2000:466), 'the researcher has the responsibility of examining a model to ascertain whether it is theoretically identified or not'. Therefore, a SEM analysis was conducted to identify the structural model. In addition to this, it must fulfil the conditions of model identification.

A model can be analysed when it has been identified. Therefore, it is necessary to carry out such an identification process. A model is said to be identified if it is theoretically possible to derive a unique estimate for each parameter (Kline, 2005). The sample size does not matter if a model is not identified, as it will be difficult to analyse it. A model can then be said to be identified if it has at least as many observations as free model parameters (namely the degree of freedom ≥0) and that every unobserved variable must be assigned a scale. However, a model could be just-identified, over-identified or under-identified (Byrne, 2010). An over-identified model is one in which the number of parameters to be estimated is less than the number of data variances and covariances of the observed variables. It therefore, results in a positive degree of freedom (df). The significance of an over-identified model is that it allows for a model to be rejected and thereby, rendering it of scientific value. A just-identified model cannot be rejected, and it is not possible to obtain a solution for an under-identified model. The EQS result outputs showed that the lowest value for the degree of freedom was 2.0 and the highest value was 4.0 for the current study. This result shows an over-identified model because the scores showed a positive value of degree of freedom.

TABLE 11.40
Univariate Statistics and Yuan, Lambert and Fouladi's Normalized Multivariate Estimates

Latent Construct	Indicator Variable	Mean ($\bar{X}$)	Skewness (G1)	Skewness (G2)	SD (σx)	Yuan, Lambert and Fouladi's Coefficient
Safe act and working condition (SAWC)	SAWC 1	4.2985	−1.1351	4.3625	0.6364	262.0696
	SAWC 2	4.3097	−1.1535	4.2998	0.6398	
	SAWC 3	4.2060	−1.0242	2.6524	0.7036	
	SAWC 4	4.2386	−0.7813	2.4076	0.6348	
	SAWC 5	4.3195	−1.6851	4.1830	0.7911	
	SAWC 6	4.2395	−1.1685	3.9765	0.6651	
	SAWC 7	4.1780	−1.0414	3.2818	0.6773	
	SAWC 8	4.2281	−1.2026	3.3075	0.7052	
Government support (GS)	GS 1	4.0113	−1.5323	2.9505	0.9090	
	GS 2	3.8722	−1.2451	1.3996	1.0087	
	GS 3	3.9173	−1.1858	1.3608	0.9677	
	GS 4	3.8647	−1.1483	1.0441	1.0226	
	GS 5	3.8421	−1.2281	1.1976	1.0413	
Contractor's safety policy (CSP)	CSP 1	4.2734	−1.5639	5.3230	0.6965	
	CSP 2	4.3086	−1.4598	5.0613	0.6840	
	CSP 3	4.2528	−1.5004	4.5522	0.7197	
	CSP 4	4.2463	−1.0054	2.2091	0.6697	
Contractor's organisational culture (COC)	COC 1	4.2734	−0.9825	2.4551	0.6690	
	COC 2	4.3184	−1.3655	3.6343	0.7096	
	COC 3	4.2368	−0.7696	1.3233	0.6618	
	COC 4	4.2717	−0.5259	0.9174	0.6106	
	COC 5	4.2264	−0.9210	2.6996	0.6525	
	COC 6	4.2622	−1.1938	3.1958	0.7035	
	COC 7	4.3271	−1.1943	3.8775	0.6400	
	COC 8	4.3558	−0.9892	2.8814	0.6289	
Adherence to safety regulations (ASR)	ASR 1	4.3019	−1.0573	3.1936	0.6451	
	ASR 2	4.3674	−1.0288	2.7432	0.6389	
	ASR 3	4.2471	−0.3881	0.4448	0.6083	
Health and safety compliance (HSC)	HSC 1	3.9132	−1.1734	0.9323	1.0801	16.8968
	HSC 2	3.6446	−0.8331	−0.0955	1.1215	
	HSC 3	3.7882	−0.8247	0.3516	0.9409	
	HSC 4	3.7361	−1.1095	0.8697	0.9912	

11.4.3 Confirmatory Factor Analysis of the Latent Construct

The first step in assessing measurement invariance is to conduct separate CFAs of the latent constructs. Therefore, a CFA was carried out on the exogenous variables (safe act of workers, safe working condition, government support, contractor's safety policy, contractor's organisation culture and adherence to safety regulations) and the endogenous variable (H&S compliance of small to medium-sized construction companies) to determine whether the measures used were sufficient indicators to assess the coefficients and to reaffirm the factor structure of each construct.

The fit of the items to the latent variables was further conducted with EQS to explore the measurement model. If the fit of each of these models is good and the item loading is acceptable, it can be assumed that the indicators underlying the factor are tapping into the construct at hand in each of the latent constructs. This is in line with practice established by McDonald and Ho (2002) and supported by the SEM experts' recommendations. The evaluation of models as attested by the experts should be derived from a large number of criteria, rather than a single 'magic index' (Kline, 2005; Byrne, 2010). Various goodness-of-fit indexes were considered in the study to determine the goodness of fit for the CFA models.

11.4.3.1 Fit Statistics on Measurement Models (Confirmatory Factor Analysis)

11.4.3.1.1 Measurement Model for Safe Act and Working Condition Features (SAWC) Construct

The unidimensional model for safe act and working condition (SAWC) features are presented (Table 11.38). From the 289 cases analysed for this construct, eight indicator variables (F1SAW 7, F1SAW 8, F1SAW 9, F7SWC 14, F10SAW15, F10SWC13, F10SWC16 and F13SWC12) made up of two factors were realised and renamed safe act and working condition (SAWC) constructs as one factor and numbered systematically (Table 11.38). All eight indicator variables obtained were used for the CFA. In order for a variable to be included in a CFA, thus enabling the model to be described as well-fitting, the distribution of residuals covariance matrix should be symmetrical and centred around zero (Byrne, 2010). The five-indicator model provides good measures of residual matrix and evidence of convergent validity.

The CFA results further revealed that the safe act and working condition features had eight dependent variables, nine independent variables and sixteen free parameters. The number of fixed non-zero parameters was nine. The eight dependent indicator variables for the safe act and working condition are ensure proper lifting, handling or moving of objects; ensure proper stacking of objects or materials in safe locations; avoid annoyance and horseplay at the workplace; availability of facilities within a reasonable distance from the work area; concentrate on task at hand; provision of sufficient lighting system for enclosed areas; facilities must be available to both day and night workers; and provision of adequate facilities (toilet, drinking water, washing and canteen). These indicator variables are presented in Table 11.41. The safe act and working condition features measurement model shown in Figure 11.1 was analysed before it could be included in the full latent variable model. In order to establish how well the model fit the sample data and the strength of the hypothesised relationship between the variables, results on residual covariance matrix (unstandardised and standardised), distribution of standardised residuals, fit statistics and statistical significance at a probability level of 5 percent were examined. In addition, the Cronbach's alpha and the rho coefficient of internal consistency were examined to determine the score reliability. Results of these statistics are presented in the next section for the safe act of workers features.

TABLE 11.41
Postulated Safe Act and Working Condition Features Model

Latent Construct	Indicator Variable	Label
Safe act and working condition (SAWC)	Ensure proper lifting, handling or moving of objects	SAWC 1
	Ensure proper stacking of objects or materials in safe locations	SAWC 2
	Avoid annoyance and horseplay at the workplace	SAWC 3
	Availability of facilities within a reasonable distance from the work area	SAWC 4
	Concentrate on task at hand	SAWC 5
	Provision of sufficient lighting system for enclosed areas	SAWC 6
	Facilities must be available to both day and night workers	SAWC 7
	Provision of adequate facilities (toilet, drinking water, washing and canteen)	SAWC 8

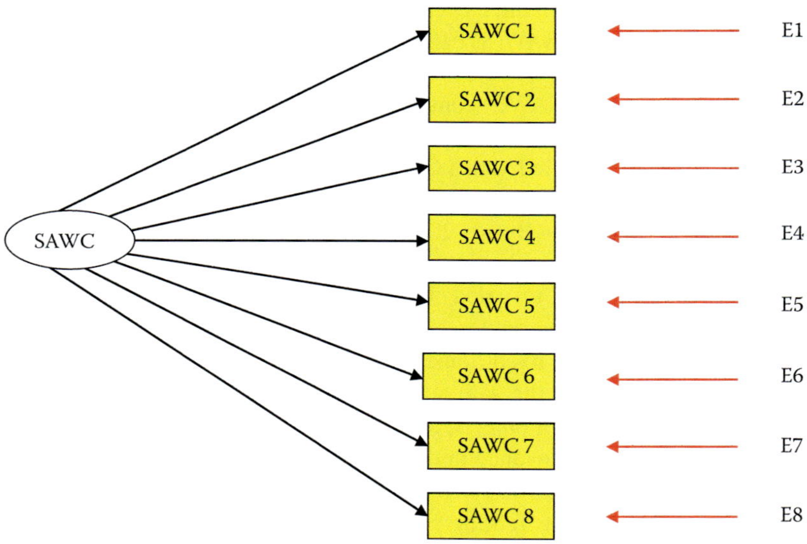

FIGURE 11.1 Measurement model of safe act and working condition.

11.4.3.1.1.1 Diagnostic Fit Analysis: Analysis of Residual Covariance Estimate The unstandardised and standardised absolute residual matrix values of the safe act and working condition features are presented in Tables 11.42 and 11.43. The result reveals that the absolute residual values and the average off-diagonal absolute residual values were close to zero. The unstandardised average off-diagonal residual was 0.0223, while the standardised average off-diagonal residual was found to be 0.0222. A residual value greater than 2.58 is described as large (Byrne, 2010). The results obtained for the safe act and working condition features measurement model suggested a fairly

TABLE 11.42
Residual Covariance Matrix for Safe Act and Working Condition Model (Unstandardised)

	SAWC 1	SAWC 2	SAWC 3	SAWC 4	SAWC 5	SAWC 6	SAWC 7	SAWC 8
SAWC 1	0.000							
SAWC 2	0.053	0.000						
SAWC 3	0.006	0.021	–0.001					
SAWC 4	–0.020	–0.021	–0.008	–0.001				
SAWC 5	0.030	0.007	0.006	–0.003	0.000			
SAWC 6	0.023	–0.020	–0.018	0.016	–0.017	–0.001		
SAWC 7	–0.019	–0.034	0.003	0.021	0.011	0.014	0.001	
SAWC 8	–0.032	–0.032	–0.010	–0.004	–0.007	0.055	0.009	–0.001

Note: Average absolute residual = 0.0208. Average off-diagonal absolute residual = 0.0223. % falling between –0.1 and +0.1 = 99.99%.

TABLE 11.43
Residual Covariance Matrix for Safe Act and Working Condition Model (Standardised)

	SAWC 1	SAWC 2	SAWC 3	SAWC 4	SAWC 5	SAWC 6	SAWC 7	SAWC 8
SAWC 1	0.000							
SAWC 2	0.053	0.000						
SAWC 3	0.007	0.021	0.000					
SAWC 4	–0.019	–0.021	–0.007	0.000				
SAWC 5	0.031	0.007	0.006	–0.003	0.000			
SAWC 6	–0.023	–0.019	–0.016	0.017	–0.015	0.000		
SAWC 7	–0.018	–0.033	0.003	0.021	0.009	0.014	0.000	
SAWC 8	–0.030	–0.030	–0.008	–0.003	–0.008	0.055	0.009	0.000

Note: Average absolute residual = 0.0206. Average off-diagonal absolute residual = 0.0222. % falling between –0.1 and +0.1 = 99.99%.

acceptable fit to the sample data because the absolute residuals were all less than 2.58. In order for a model to be described as well-fitting, the distribution of standardised residuals should be symmetrical and centred around zero (Byrne, 2010).

Further review of the frequency distribution reveals most residual values (99.99%) fall between –0.1 and 0.1, which is in the acceptable range. Of the remaining residuals, 0.01% fell outside the –0.1 to 0.1 range.

From this information, the results suggested a measurement model that was well-fitting despite a minimal discrepancy in fit between the hypothesised model and the sample data. Therefore, since this diagnostic fit analysis indicated a good fit, further tests of goodness of fit were possible to conclusively make a decision on the fit and appropriateness of the measurement model.

11.4.3.1.1.2 Goodness-of-Fit Statistics: Robust Maximum Likelihood (RML) The analysis strategy of goodness of fit for the safe act of workers followed a three-statistics strategy of fit indexes (Hu & Bentler, 1999). The sample data on safe act of workers measurement model yield the $S - B\chi^2$ of 3249.5 with 1861 degrees of freedom (df) with a probability of $p = 0.0000$. This chi-square value indicated that the departure of the sample data from the postulated measurement model was significant and, hence, indicative of good fit. The chi-square test is very sensitive to sample size and is used more as a descriptive index of fit rather than as a statistical test (Kline, 2005). The normed chi-square value is usually adopted by most researchers. The normed chi-square is the procedure of dividing the chi-square by the degrees of freedom. The normed values of up to 3.0 or even 5.0 are recommended (Kline, 2005).

Values for NFI range between 0 and 1 with recommended values greater than 0.90 indicating a good fit (Bentler & Bonnet, 1980). A cut-off criterion of NFI ≥.95 is also accepted (Hu & Bentler, 1999). This index is sensitive to sample size, under-estimating fit for samples less than 200 (Mulaik, James, Van Alstine, Bennet, Lind & Stilwell, 1989; Bentler, 1990), and is thus not recommended to be solely relied on (Kline, 2005). The non-normed fit index (NNFI; also known as the Tucker-Lewis index) is an index that prefers simpler models. The value of the NNFI can indicate poor fit despite other statistics pointing towards good fit if small samples are used (Bentler, 1990; Kline, 2005; Tabachnick & Fidell, 2007). Owing to the non-normed nature, NNFI values can go above 1.0 and can thus be difficult to interpret. The chi-square and degrees of freedom were found to be 1.75. This ratio was lower than the limit of 3.00 or 5.0 advocated for by some authors (Kline, 2005).

The CFI value was found to be 0.794, which was lower than the cut-off limit of 0.95, meaning that the model has an acceptable fit. The NFI value was 0.629, which is within the given range, but the given cut-off value of NFI ≥.95 is shown in Table 11.44. Therefore, the model is acceptable. The NNFI value obtained is 0.777, which is also below the cut-off value of 0.80. These fit indexes for the safe act and working condition model suggested that the postulated model adequately describes the sample data and could therefore be included in the full latent variable model analysis (Table 11.44).

11.4.3.1.1.3 Statistical Significance of Parameter Estimates Table 11.45 shows the correlation values, standard errors and the test of statistics. All the correlation values were less than 1.00, and all the *Z*-statistics were greater than 1.96 and show appropriate signs. The estimates were therefore deemed reasonable, as well as statistically significant. The parameter with the highest standardised coefficient was the indicator with variable SAW 4 (ensure proper lifting, handling or moving of objects) and its parameter coefficient was 0.747.

TABLE 11.44
Robust Fit Indexes for Safe Act and Working Condition Features Construct

Fit Index	Cut-Off Value	Estimate	Comment
$S - B\chi^2$	3249.5		
df	0≥	1861	Good fit
CFI	0.90≥ acceptable; 0.95≥ good fit	0.794	Acceptable
RMSEA 95%	Less than 0.05 with confidence interval (CI) 0.00–0.05 'good fit'	0.051	Good fit
NFI	Greater than 0.90 'good fit'	0.629	Acceptable
NNFI	Greater than 0.80 'good fit'	0.777	Acceptable
RMSEA 95% CI		0.048–0.054	Acceptable range

TABLE 11.45
Factor Loadings and *Z*-Statistics of Safe Act and Working Condition Measurement

Indicator Variable	Unstandardised Coefficient (λ)	Standardised Coefficient (λ)	*Z*-Statistics	R^2	Significant at 5% Level?
SAWC 1	0.766	0.642	9.990	0.587	Yes
SAWC 2	0.745	0.668	10.183	0.554	Yes
SAWC 3	0.718	0.696	10.363	0.516	Yes
SAWC 4	0.665	0.747	10.603	0.442	Yes
SAWC 5	0.733	0.681	10.240	0.537	Yes
SAWC 6	0.797	0.604	9.540	0.537	Yes
SAWC 7	0.720	0.694	10.290	0.635	Yes
SAWC 8	0.722	0.692	10.259	0.519	Yes

Note: Robust statistical significance at 5% level.

Most of the parameter estimates had high correlation values close to 1.00. The high correlation values suggest a high degree of linear association between the indicator variables and the unobserved variable (safe act and working condition features). In addition, the R^2 values were also close to the desired value of 1.00 indicating that the factors explained more of the variance in the indicator variables. The results, therefore, suggest that the indicator variables significantly predict the unobserved construct, because all the measured variables are significantly associated with the safe act of worker features.

11.4.3.1.1.4 Internal Reliability and Validity of Scores The internal consistency and reliability of scores for the safe act of workers features construct was determined from the rho and the Cronbach's alpha coefficient. The reliability coefficient

TABLE 11.46
Reliability and Construct Validity of Safe Act and Working Condition Features Model

Factor	Indicator Variable	Factor Loading	Cronbach's Alpha	Rho Coefficient
Safe act and working condition features	SAWC 1	0.6189	0.937	0.964
	SAWC 2	0.6013		
	SAWC 3	0.5799		
	SAWC 4	0.5370		
	SAWC 5	0.5917		
	SAWC 6	0.6437		
	SAWC 7	0.5816		
	SAWC 8	0.5830		

Note: Parameter estimates are based on standardised solutions.

should fall between 0 and 1.00 (Kline, 2005). Values close to 1.00 are desired. The rho coefficient of internal consistency was found to be 0.964. This value was above the minimum required value of 0.70. Likewise, the Cronbach's alpha was above the minimum acceptable value of 0.70. The Cronbach's alpha was found to be 0.937 (Table 11.46). Both of these values revealed a high level of internal consistency and therefore reliability, suggesting that the indicator variables represent the same latent construct (safe act of workers).

Further, construct validity was determined from the magnitude and reasonableness of the parameter coefficients (factor loading). The parameter coefficients represent the magnitude of correlation or covariance between an item and a construct. Higher parameter coefficients show that the indicator variables have a stronger relationship with a construct and thus converge at a common point. Parameter coefficients of greater than 0.5 indicate a close relationship between the construct and an indicator variable. A parameter coefficient of 0.5 is interpreted as 25 percent of the total variance in the indicator variable being explained by the latent variable (factor). Hence, a parameter coefficient should be 0.5 or higher, and ideally 0.7 or greater, to explain about 50 percent of the variance in an indicator variable (Hair et al., 1998). The standardised parameter coefficient presented in Table 11.45 revealed that all coefficients were significantly higher with the lowest being 0.642 for safe act and working condition features. The magnitude of the parameter estimate was above the 50 percent minimum. This indicates a strong relationship between the indicator variables and the factors of the safe act and working condition features construct. Therefore, the safe act features satisfied both internal reliability and the construct criteria. The rho value was above the minimum value of 0.70 and the magnitude, signs and statistical significance of the parameter estimates were appropriate (Table 11.46).

11.4.3.1.1.5 Summary of Safe Act and Working Condition Features Measurement Model The CFA revealed that the residual covariance estimates fell within the acceptable range. Likewise, the robust fit indexes met the cut-off index criteria and all the parameter estimates were statistically significant and feasible. Considering these criteria, the measurement model for the safe act and working condition features was found to adequately fit the sample data. Therefore, there was no need to improve the measurement model before it could be included in the full latent variable model. Hence, the safe act and working condition features construct was adequately measured by the indicator variables and could be used in the analysis of the full latent variable model.

11.4.3.1.2 Measurement Model for Government Support Features (GS) Construct

The unidimensional model for government support (GS) features are presented in Table 11.38. From the 289 cases analysed for this construct, five indicator variables (F2GS 1, F2GS 2, F2GS 3, F2GS 4 and F2GS 5) made up of the same factor were realised and the name GS was maintained (Table 11.38). All five indicator variables obtained were used for the CFA (Byrne, 2010). In order for a variable to be included in a CFA, thus enabling the model to be described as well-fitting, the distribution of residuals covariance matrix should be symmetrical and centred around zero (Byrne, 2010). The five-indicator model provides good measures of residual matrix and evidence of convergent validity.

The CFA results further revealed that the government support features had five dependent variables, six independent variables and ten free parameters. The number of fixed non-zero parameters was six. The five dependent indicator variables for the government support are formulate H&S policy for construction, implementation of H&S policy by government representatives, monitoring of H&S policy implementation by government representatives, provision of H&S policy update by government representatives and provide health and safety training by government representatives. These indicator variables are presented in Table 11.47. The government support features measurement model shown in Figure 11.2 was analysed before it could be included in the full latent variable model.

TABLE 11.47
Postulated Government Support Features Model

Latent Construct	Indicator Variable	Label
Government support (GS)	Formulate H&S policy for construction	GS 1
	Implementation of H&S policy by government representatives	GS 2
	Monitoring of H&S policy implementation by government representatives	GS 3
	Provision of H&S policy update by government representatives	GS 4
	Provide health and safety training by government representatives	GS 5

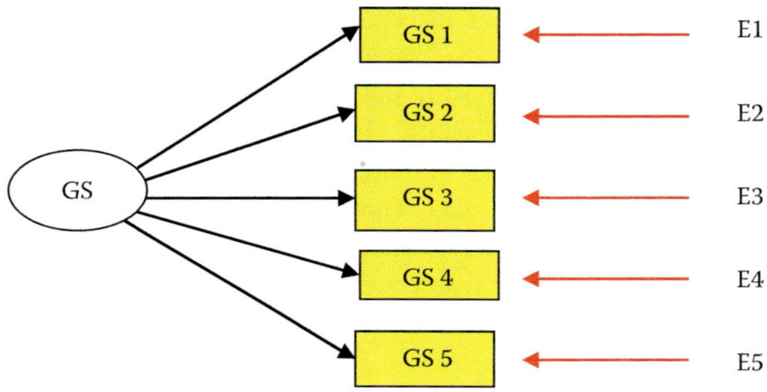

FIGURE 11.2 Measurement model of government support.

In order to establish how well the model fit the sample data and the strength of the hypothesised relationship between the variables, results on the residual covariance matrix (unstandardised and standardised), distribution of standardised residuals, fit statistics and statistical significance at a probability level of 5 percent were examined. In addition, the Cronbach's alpha and the rho coefficient of internal consistency were examined to determine the score reliability. Results of these statistics are presented in the next section for the government support features.

11.4.3.1.2.1 Diagnostic Fit Analysis: Analysis of Residual Covariance Estimate The unstandardised and standardised absolute residual matrix values of the government support features are presented in Tables 11.48 and 11.49. The result reveals that all the absolute residual values and the average off-diagonal absolute residual values were close to zero. The unstandardised average off-diagonal residual was 0.0223, while the standardised average off diagonal residual was found to be 0.0222. A residual value greater than 2.58 is described as large (Byrne, 2010). The results obtained for the government support features measurement model suggested a fairly acceptable fit to the sample data because the absolute residuals were all less than 2.58. In order for a model to be described as well-fitting, the distribution of standardised residuals should be symmetrical and centred around zero (Byrne, 2010).

TABLE 11.48
Residual Covariance Matrix for Government Support Model (Unstandardised)

	GS 1	GS 2	GS 3	GS 4	GS 5
GS 1	–0.001				
GS 2	0.049	–0.001			
GS 3	–0.017	0.001	–0.001		
GS 4	–0.031	0.037	0.022	0.001	
GS 5	–0.026	0.001	–0.013	0.037	0.001

Note: Average absolute residual = 0.0208. Average off-diagonal absolute residual = 0.0223. % falling between –0.1 and +0.1 = 99.99%.

TABLE 11.49
Residual Covariance Matrix for Government Support Model (Standardised)

	GS 1	GS 2	GS 3	GS 4	GS 5
GS 1	0.000				
GS 2	0.049	0.000			
GS 3	–0.016	0.001	0.000		
GS 4	–0.030	0.036	0.023	0.000	
GS 5	–0.025	0.002	–0.012	0.037	0.000

Note: Average absolute residual = 0.0206. Average off-diagonal absolute residual = 0.0222. % falling between –0.1 and +0.1 = 99.99%.

Further review of the frequency distribution reveals most residual values (99.99%) fall between –0.1 and 0.1, which is in the acceptable range. Of the remaining residuals, 0.01% fell outside the –0.1 to 0.1 range.

From this information, the results suggested a measurement model that was well-fitting despite a minimal discrepancy in fit between the hypothesised model and the sample data. Therefore, since this diagnostic fit analysis indicated a good fit, further tests of goodness of fit were possible to conclusively make a decision on the fit and appropriateness of the measurement model.

11.4.3.1.2.2 Goodness-of-Fit Statistics: Robust Maximum Likelihood (RML) The analysis strategy of goodness-of-fit for the safe act of workers followed a three-statistics strategy of fit indexes (Hu & Bentler, 1999). The sample data on safe act of workers measurement model yielded the $S - B\chi^2$ of 3249.5 with 1861 degrees of freedom with a probability of $p = 0.0000$. This chi-square value indicated that the departure of the sample data from the postulated measurement model was significant and, hence, indicative of good fit. The chi-square test is very sensitive to sample size and is used more as a descriptive index of fit rather than as a statistical test (Kline, 2005). The normed chi-square value is usually adopted by most researchers. The normed chi-square is the procedure of dividing the chi-square by the degrees of freedom. The normed values of up to 3.0 or even 5.0 is recommended (Kline, 2005).

Values for NFI range between 0 and 1, whilst recommended values greater than 0.90, indicating a good fit (Bentler & Bonnet, 1980). A cut-off criterion of NFI ≥.95 is also acceptable (Hu & Bentler, 1999). This index is sensitive to sample size, underestimating fit for samples less than 200 (Mulaik et al., 1989; Bentler, 1998), and is thus not recommended to be solely relied on (Kline, 2005). The NNFI is an index that prefers simpler models. The value of the NNFI can indicate poor fit despite other statistics pointing towards good fit if small samples are used (Bentler, 1990; Kline, 2005; Tabachnick & Fidell, 2007). Owing to the non-normed nature, NNFI values can go above 1.0 and can thus be difficult to interpret (Byrne, 1998). The chi-square and degrees of freedom was found to be 1.75. This ratio was lower than the limit of 3.00 or 5.0 (Kline, 2005).

TABLE 11.50
Robust Fit Indexes for Government Support Features Construct

Fit Index	Cut-Off Value	Estimate	Comment
$S - B\chi^2$		3249.5	
df	0≥	1861	Good fit
CFI	0.90≥ acceptable; 0.95≥ good fit	0.794	Acceptable
RMSEA 95%	Less than 0.05 with confidence interval (CI) 0.00–0.05 'good fit'	0.051	Good fit
NFI	Greater than 0.90 'good fit'	0.629	Acceptable
NNFI	Greater than 0.80 'good fit'	0.777	Acceptable
RMSEA 95% CI		0.048: 0.054	Acceptable range

The CFI value was found to be 0.794, which was lower than the cut-off limit of 0.95, so the model is described to have an acceptable fit. The NFI value was 0.629, which is within the given range, but the given cut-off value of NFI ≥.95 is shown in Table 11.50. Therefore, the model is acceptable. The NNFI value obtained is 0.777, which is also below the cut-off value of 0.80. These fit indexes for the government support model suggested that the postulated model adequately describe the sample data and could, therefore, be included in the full latent variable model analysis (Table 11.50).

11.4.3.1.2.3 Statistical Significance of Parameter Estimates Table 11.51 shows the correlation values, standard errors and the test of statistics. All correlation values were less than 1.00, and all *Z*-statistics were greater than 1.96 and show appropriate signs. The estimates were therefore deemed reasonable, as well as statistically significant. The parameter with the highest standardised coefficient was the indicator with variable GS 5 (safe and healthy work environment) and its parameter coefficient was 0.631.

TABLE 11.51
Factor Loadings and *Z*-Statistics of Government Support Measurement

Indicator Variable	Unstandardised Coefficient (λ)	Standardised Coefficient (λ)	*Z*-Statistics	R^2	Significant at 5% Level?
GS 1	0.923	0.384	7.733	0.839	Yes
GS 2	0.867	0.499	9.565	0.853	Yes
GS 3	0.855	0.519	9.767	0.751	Yes
GS 4	0.843	0.539	7.922	0.731	Yes
GS 5	0.776	0.631	9.422	0.710	Yes

Note: Robust statistical significance at 5% level.

Most of the parameter estimates had high correlation values close to 1.00. The high correlation values suggest a high degree of linear association between the indicator variables and the unobserved variable (government support features). In addition, the R^2 values were also close to the desired value of 1.00 indicating that the factors explained more of the variance in the indicator variables. The results therefore, suggest that the indicator variables significantly predict the unobserved construct, because all measured variables are significantly associated with the government support features.

11.4.3.1.2.4 Internal Reliability and Validity of Scores The internal consistency and reliability of scores for the government support features construct was determined from the rho and the Cronbach's alpha coefficient. The reliability coefficient should fall between 0 and 1.00 (Hu & Bentler, 1998; Kline, 2005). Values close to 1.00 are desired. The rho coefficient of internal consistency was found to be 0.964. This value was above the minimum required value of 0.70. Likewise, the Cronbach's alpha was above the minimum acceptable value of 0.70. The Cronbach's alpha was found to be 0.937 (Table 11.52). Both of these values revealed a high level of internal consistency and therefore reliability, suggesting that the indicator variables represent the same latent construct (government support).

Further, construct validity was determined from the magnitude and reasonableness of the parameter coefficients (factor loading). The parameter coefficients represent the magnitude of correlation or covariance between an item and a construct. Higher parameter coefficients show that the indicator variables have a stronger relationship with a construct and thus converge at a common point. Parameter coefficients of greater than 0.5 indicate a close relationship between the construct and an indicator variable. A parameter coefficient of 0.5 is interpreted as 25 percent of the total variance in the indicator variable being explained by the latent variable (factor). Hence, a parameter coefficient should be 0.5 or higher, and ideally 0.7 or greater, to explain about 50 percent of the variance in an indicator variable (Hair et al., 1998).

The standardised parameter coefficient presented in Table 11.51 revealed that all coefficients were significantly higher with the lowest being 0.384 for government support features. The magnitude of the parameter estimate was below the 50 percent

TABLE 11.52
Reliability and Construct Validity of Government Support Features Model

Factor	Indicator Variable	Factor Loading	Cronbach's Alpha	Rho Coefficient
Government support features	GS 1	0.5335	0.937	0.964
	GS 2	0.6390		
	GS 3	0.6440		
	GS 4	0.6047		
	GS 5	0.5964		

Note: Parameter estimates are based on standardised solutions.

minimum. This indicates a weak relationship between the indicator variables and the factors of the government support features construct. Therefore, the government support features satisfied both internal reliability and the construct criteria. The rho value was above the minimum value of 0.70, and the magnitude, signs and statistical significance of the parameter estimates were appropriate (Table 11.52).

11.4.3.1.2.5 Summary of Government Support Feature Measurement Model The CFA revealed that the residual covariance estimates fell within the acceptable range. Likewise, the robust fit indexes met the cut-off index criteria and all the parameter estimates were statistically significant and feasible. Considering these criteria, the measurement model for the government support feature was found to adequately fit the sample data. Therefore, there was no need to improve the measurement model before it could be included in the full latent variable model. Hence, the government support feature construct was adequately measured by the indicator variables and could be used in the analysis of the full latent variable model.

11.4.3.1.3 Measurement Model for Contractor's Safety Policy Features Construct

The unidimensional model for contractor's safety policy (CSP) features are presented in Table 11.38. From the 289 cases analysed for this construct, four indicator variables (F3SE 2, F14SE 1, F4SE 3 and F4SE 5) made up of three factors were realised and renamed CSP as one factor and numbered systematically (Table 11.38). All four indicator variables obtained were used for the CFA. In order for a variable to be included in the CFA, thus enabling the model to be described as well-fitting, the distribution of residuals covariance matrix should be symmetrical and centred around zero (Byrne, 2010). The four-indicator model provides good measures of residual matrix and evidence of convergent validity.

The CFA results further revealed that the contractor's safety policy features had four dependent variables, five independent variables and eight free parameters. The number of fixed non-zero parameters was five. These are the four dependent indicator variables for the contractor's safety policy: safe storage of equipment, safe and healthy work environment, do not service equipment which is in operation, and safe storage of formwork and false work. These indicator variables are presented in Table 11.53. The contractor's safety policy features measurement model shown in Figure 11.3 was analysed before it could be included in the full latent variable model.

TABLE 11.53
Postulated Contractor's Safety Policy Features Model

Latent Construct	Indicator Variable	Label
Contractor's safety policy (CSP)	Safe storage of equipment	CSP 1
	Safe and healthy work environment	CSP 2
	Do not service equipment which is in operation	CSP 3
	Safe storage of formwork and false work	CSP 4

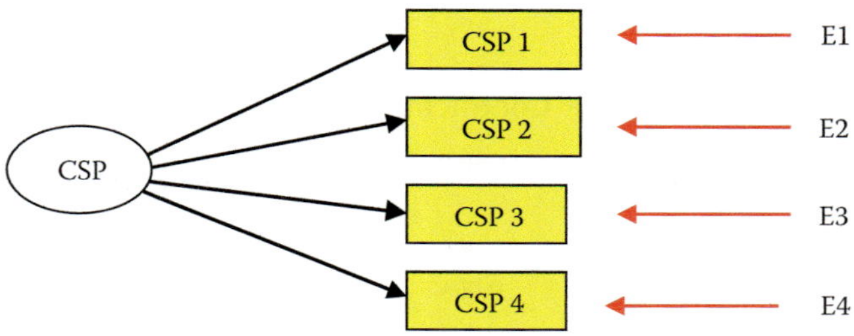

FIGURE 11.3 Measurement model of contractor's safety policy.

In order to establish how well the model fit the sample data and the strength of the hypothesised relationship between the variables, results of the residual covariance matrix (unstandardised and standardised), distribution of standardised residuals, fit statistics and statistical significance at a probability level of 5 percent were examined. In addition, the Cronbach's alpha and the rho coefficient of internal consistency were examined to determine the score reliability. Results of these statistics are presented in the next section for the contractor's safety policy features.

11.4.3.1.3.1 Diagnostic Fit Analysis: Analysis of Residual Covariance Estimate The unstandardised and standardised absolute residual matrix values of the contractor's safety policy features are presented in Tables 11.54 and 11.55. The result reveals that all the absolute residual values and the average off-diagonal absolute residual values were close to zero. The unstandardised average off-diagonal residual was 0.0223, while the standardised average off-diagonal residual was found to be 0.0222. A residual value greater than 2.58 is described as large (Byrne, 2010). The results obtained for the contractor's safety policy features measurement model suggested a fairly acceptable fit to the sample data because the absolute residual were all less than 2.58. In order for a model to be described as well-fitting, the distribution of standardised residuals should be symmetrical and centred around zero (Byrne, 2010).

TABLE 11.54
Residual Covariance Matrix for Contractor's Safety Policy Model (Unstandardised)

	CSP 1	CSP 2	CSP 3	CSP 4
CSP 1	0.001			
CSP 2	0.006	0.000		
CSP 3	0.005	0.005	0.000	
CSP 4	0.020	0.008	0.000	0.000

Note: Average absolute residual = 0.0208. Average off-diagonal absolute residual = 0.0223. % falling between −0.1 and +0.1 = 99.99%.

TABLE 11.55
Residual Covariance Matrix for Contractor's Safety Policy Model (Standardised)

	CSP 1	CSP 2	CSP 3	CSP 4
CSP 1	0.000			
CSP 2	0.006	0.000		
CSP 3	0.004	0.005	0.000	
CSP 4	0.020	0.008	0.000	0.000

Note: Average absolute residual = 0.0206. Average off-diagonal absolute residual = 0.0222. % falling between –0.1 and +0.1 = 99.99%.

Further review of the frequency distribution reveals most residual values (99.99%) fall between –0.1 and 0.1, which is in the acceptable range. Of the remaining residuals, 0.01% fell outside the –0.1 to 0.1 range.

From this information, the results suggested a measurement model that was well-fitting, albeit a minimal discrepancy in fit between the hypothesised model and the sample data. Therefore, since this diagnostic fit analysis indicated a good fit, further tests of goodness of fit were possible to conclusively make a decision on the fit and appropriateness of the measurement model.

11.4.3.1.3.2 Goodness-of-Fit Statistics: Robust Maximum Likelihood (RML) The analysis strategy of goodness of fit for the contractor's safety policy followed a three-statistics strategy of fit indexes as recommended (Hu & Bentler, 1999). The sample data on the contractor's safety policy measurement model yielded the $S - B\chi^2$ of 3249.5 with 1861 degrees of freedom with a probability of $p = 0.0000$. This chi-square value indicated that the departure of the sample data from the postulated measurement model was significant and, hence, indicative of good fit. The chi-square test is very sensitive to sample size and is used more as a descriptive index of fit rather than as a statistical test (Kline, 2005). The normed chi-square value is usually adopted by most researchers. The normed chi-square is the procedure of dividing the chi-square by the degrees of freedom. The normed values of up to 3.0 or even 5.0 are recommended (Kline, 2005).

Values for the NFI range between 0 and 1, whilst recommended values greater than 0.90 indicate a good fit (Bentler & Bonnet, 1980). A cut-off criterion of NFI ≥.95 is also recommended (Hu & Bentler, 1999). This index is sensitive to sample size, underestimating fit for samples less than 200 (Mulaik et al., 1989; Bentler, 1990), and is thus not recommended to be solely relied on (Kline, 2005). The value of the NNFI can indicate poor fit despite other statistics pointing towards good fit if small samples are used (Bentler, 1990; Kline, 2005; Tabachnick & Fidell, 2007). Owing to the non-normed nature, NNFI values can go above 1.0 and can thus be difficult to interpret (Byrne, 2010). The chi-square and degrees of freedom was found to be 1.75. This ratio was lower than the limit of 3.00 or 5.0 advocated by some authors (Kline, 2005).

TABLE 11.56
Robust Fit Indexes for Contractor's Safety Policy Features Construct

Fit Index	Cut-Off Value	Estimate	Comment
$S-B\chi^2$		3249.5	
df	0≥	1861	Good fit
CFI	0.90≥ acceptable; 0.95≥ good fit	0.794	Acceptable
RMSEA 95%	Less than 0.05 with confidence interval (CI) 0.00–0.05 'good fit'	0.051	Good fit
NFI	Greater than 0.90 'good fit'	0.629	Acceptable
NNFI	Greater than 0.80 'good fit'	0.777	Acceptable
RMSEA 95% CI		0.048: 0.054	Acceptable range

The CFI value was found to be 0.794, which was lower than the cut-off limit of 0.95, so the model is described to have an acceptable fit. The NFI value was 0.629, which is within the given range, but the given cut-off value of NFI ≥.95 is shown in Table 11.56. Therefore, the model is acceptable. The NNFI value obtained is 0.777, which is also below the cut-off value of 0.80. These fit indexes for the contractor's safety policy model suggested that the postulated model adequately describes the sample data and could therefore be included in the full latent variable model analysis (Table 11.56).

11.4.3.1.3.3 Statistical Significance of Parameter Estimates Table 11.57 shows the correlation values, standard errors and the test of statistics. All the correlation values were less than 1.00, and all the Z-statistics were greater than 1.96 and show appropriate signs. The estimates were therefore deemed reasonable, as well as statistically significant. The parameter with the highest standardised coefficient was the indicator with variable CSP 2 (safe storage of formwork and false work) and its parameter coefficient was 0.787.

Most of the parameter estimates had high correlation values close to 1.00. The high correlation values suggest a high degree of linear association between the

TABLE 11.57
Factor Loadings and *Z*-Statistics of Contractor's Safety Policy Measurement

Indicator Variable	Unstandardised Coefficient (λ)	Standardised Coefficient (λ)	*Z*-Statistics	R^2	Significant at 5% Level?
CSP 1	0.882	0.471	6.509	0.602	Yes
CSP 2	0.617	0.787	10.710	0.779	Yes
CSP 3	0.778	0.629	10.021	0.381	Yes
CSP 4	0.819	0.573	9.506	0.605	Yes

Note: Robust statistical significance at 5% level.

indicator variables and the unobserved variable (contractor's safety policy). In addition, the R^2 values were also close to the desired value of 1.00 indicating that the factors explained more of the variance in the indicator variables. The results, therefore, suggest that the indicator variables significantly predict the unobserved construct, because all measured variables are significantly associated with the contractor's safety policy features.

11.4.3.1.3.4 Internal Reliability and Validity of Scores The internal consistency and reliability of scores for the contractor's safety policy features construct was determined from the rho and the Cronbach's alpha coefficient. The reliability coefficient should fall between 0 and 1.00, while values close to 1.00 are desired (Kline, 2005). The rho coefficient of internal consistency was found to be 0.964. This value was above the minimum required value of 0.70. Likewise, the Cronbach's alpha was above the minimum acceptable value of 0.70. The Cronbach's alpha was found to be 0.937 (Table 11.58). Both of these values revealed a high level of internal consistency and therefore reliability, suggesting that the indicator variables represent the same latent construct (contractor's safety policy).

Furthermore, construct validity was determined from the magnitude and reasonableness of the parameter coefficients (factor loading). The parameter coefficients represent the magnitude of correlation or covariance between an item and a construct. Higher parameter coefficients show that the indicator variables have a stronger relationship with a construct and thus converge at a common point. Parameter coefficients of greater than 0.5 indicate a close relationship between the construct and an indicator variable. A parameter coefficient of 0.5 is interpreted as 25 percent of the total variance in the indicator variable being explained by the latent variable (factor). Hence, a parameter coefficient should be 0.5 or higher, and ideally 0.7 or greater to explain about 50 percent of the variance in an indicator variable (Hair et al., 1998).

The standardised parameter coefficient presented in Table 11.57 revealed that all coefficients were significantly higher with the lowest being 0.471 for contractor's safety policy features. The magnitude of the parameter estimate was below the 50 percent minimum. This indicates a weak relationship between the indicator variables and the factors of the contractor's safety policy features construct. Therefore,

TABLE 11.58
Reliability and Construct Validity of Contractor's Safety Policy Feature Model

Factor	Indicator Variable	Factor Loading	Cronbach's Alpha	Rho Coefficient
Contractor's safety policy features	CSP 1	0.5632	0.937	0.964
	CSP 2	0.5187		
	CSP 3	0.5898		
	CSP 4	0.4125		

Note: Parameter estimates are based on standardised solutions.

the contractor's safety policy features satisfied both internal reliability and the construct criteria. The rho value was above the minimum value of 0.70, and the magnitude, signs and statistical significance of the parameter estimates were appropriate (Table 11.58).

11.4.3.1.3.5 Summary of Contractor's Safety Policy Feature Measurement Model The CFA revealed that the residual covariance estimates fell within the acceptable range. Likewise, the robust fit indexes met the cut-off index criteria and all the parameter estimates were statistically significant and feasible. Considering these criteria, the measurement model for the contractor's safety policy feature was found to adequately fit the sample data. Therefore, there was no need to improve the measurement model before it could be included in the full latent variable model. Hence, the contractor's safety policy feature construct was adequately measured by the indicator variables and could be used in the analysis of the full latent variable model.

11.4.3.1.4 Measurement Model for Contractor's Organisational Culture Features Construct

The unidimensional model for contractor's organisational culture (COC) features are presented (Table 11.38). From the 289 cases analysed for this construct, eight indicator variables (F5COC 10, F5COC 11, F5COC 6, F5COC 7, F5COC 8, FI2COC 4, F11COC 1 and F11COC 2) made up of three factors were realised and maintained the name COC as one factor and numbered systematically (Table 11.38). All eight indicator variables obtained were used for the CFA. In order for a variable to be included in a CFA, thus enabling the model to be described as well-fitting, the distribution of residuals covariance matrix should be symmetrical and centred around zero (Byrne, 2010). An eight-indicator model provides good measures of residual matrix and evidence of convergent validity.

The CFA results further revealed that the contractor's organisational culture features had eight dependent variables, nine independent variables and sixteen free parameters. The number of fixed non-zero parameters was nine. These are the eight dependent indicator variables for the contractor's organisational culture: consultation on H&S information to workers, update on H&S information to workers, health and safety inspection, company health and safety policy, management commitment in H&S, involve workers in H&S programs, provision of personal protective equipment, and provision of signs and notices on sites. These indicator variables are presented in Table 11.59. The contractor's organisational culture features measurement model shown in Figure 11.4 was analysed before it could be included in the full latent variable model.

In order to establish how well the model fit the sample data and the strength of the hypothesised relationship between the variables, results on residual covariance matrix (unstandardised and standardised), distribution of standardised residuals, fit statistics and statistical significance at a probability level of 5 percent were examined. In addition, the Cronbach's alpha and the rho coefficient of internal consistency were examined to determine the score reliability. Results of these statistics are presented in the next section for the contractor's organisational culture features.

TABLE 11.59
Postulated Contractor's Organisational Culture Features Model

Latent Construct	Indicator Variable	Label
Contractor's organisational culture (COC)	Consultation on H&S information to workers	COC 1
	Update on H&S information to workers	COC 2
	Health and safety inspection	COC 3
	Company health and safety policy	COC 4
	Management commitment in H&S	COC 5
	Involve workers in H&S programmes	COC 6
	Provision of personal protective equipment (PPE)	COC 7
	Provision of signs and notices on sites	COC 8

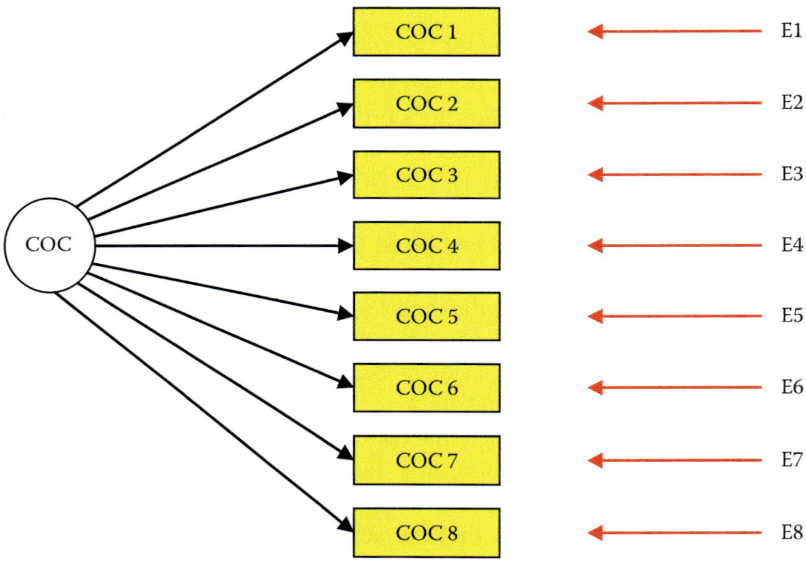

FIGURE 11.4 Measurement model of contractor's organisational culture.

11.4.3.1.4.1 Diagnostic Fit Analysis: Analysis of Residual Covariance Estimate The unstandardised and standardised absolute residual matrix values of the contractor's organisational culture features are presented in Tables 11.60 and 11.61. The result reveals that all the absolute residual values and the average off-diagonal absolute residual values were close to zero. The unstandardised average off-diagonal residual was 0.0223 while the standardised average off diagonal residual was found to be 0.0222. A residual value greater than 2.58 is described as large (Byrne, 2010). The results obtained for the contractor's organisational culture features measurement model suggested a fairly acceptable fit to the sample data because the absolute residuals were all less than 2.58. In order for a model to be described as well-fitting,

TABLE 11.60
Residual Covariance Matrix for Contractor's Organisational Culture Model (Unstandardised)

	COC 1	COC 2	COC 3	COC 4	COC 5	COC 6	COC 7	COC 8
COC 1	0.000							
COC 2	0.049	0.000						
COC 3	–0.009	–0.018	0.002					
COC 4	0.010	–0.018	0.035	0.001				
COC 5	–0.000	–0.006	0.035	0.023	0.000			
COC 6	–0.024	0.005	0.041	0.003	0.027	–0.001		
COC 7	–0.020	–0.009	–0.020	0.010	–0.015	–0.014	0.000	
COC 8	0.004	–0.004	–0.026	0.023	–0.037	–0.024	0.054	0.000

Note: Average absolute residual = 0.0208. Average off-diagonal absolute residual = 0.0223. % falling between –0.1 and +0.1 = 99.99%.

TABLE 11.61
Residual Covariance Matrix for Contractor's Organisational Culture Model (Standardised)

	COC 1	COC 2	COC 3	COC 4	COC 5	COC 6	COC 7	COC 8
COC 1	0.000							
COC 2	0.049	0.000						
COC 3	–0.009	–0.018	0.000					
COC 4	–0.009	–0.018	0.035	0.000				
COC 5	0.001	–0.005	0.034	0.023	0.000			
COC 6	–0.024	0.005	0.037	0.004	0.024	0.000		
COC 7	–0.020	–0.008	–0.020	0.010	–0.014	–0.014	0.000	
COC 8	0.004	–0.003	–0.026	–0.022	–0.036	–0.024	0.054	0.000

Note: Average absolute residual = 0.0206. Average off-diagonal absolute residual = 0.0222. % falling between –0.1 and +0.1 = 99.99%.

the distribution of standardised residuals should be symmetrical and centred around zero (Byrne, 2010).

Further review of the frequency distribution reveals most residual values (99.99%) fall between –0.1 and 0.1, which is in the acceptable range. Of the remaining residuals, 0.01% fell outside the –0.1 to 0.1 range.

From this information, the results suggested a measurement model that was well fitting despite a minimal discrepancy in fit between the hypothesised model and the sample data. Therefore, since this diagnostic fit analysis indicated a good fit, further tests of goodness of fit were possible to conclusively make a decision on the fit and appropriateness of the measurement model.

11.4.3.1.4.2 Goodness-of-Fit Statistics: Robust Maximum Likelihood (RML) The analysis strategy of goodness of fit for the contractor's organisational culture followed a three-statistics strategy of fit indexes as recommended by Hu and Bentler (1999). The sample data on contractor's organisational culture measurement model yielded the $S - B\chi^2$ of 3249.5 with 1861 degrees of freedom with a probability of $p = 0.0000$. This chi-square value indicated that the departure of the sample data from the postulated measurement model was significant and, hence, indicative of good fit. The chi-square test is very sensitive to sample size and is used more as a descriptive index of fit rather than as a statistical test (Kline, 2005). The normed chi-square value is usually adopted by most researchers. The normed chi-square is the procedure of dividing the chi-square by the degrees of freedom. The normed values of up to 3.0 or even 5.0 are recommended (Kline, 2005).

Values for NFI range should be between 0 and 1, with Bentler and Bonnet (1980) recommended values greater than 0.90 indicating a good fit. Hu and Bentler (1999) have given a cut-off criterion of NFI ≥.95. This index is sensitive to sample size, underestimating fit for samples less than 200 (Mulaik et al., 1989; Bentler, 1990), and is thus not recommended to be solely relied on (Kline, 2005). The value of the NNFI can indicate poor fit despite other statistics pointing towards good fit if small samples are used (Bentler, 1990; Kline, 2005; Tabachnick & Fidell, 2007). Owing to the non-normed nature, NNFI values can go above 1.0 and can thus be difficult to interpret (Byrne, 2010). The chi-square and degrees of freedom was found to be 1.75. This ratio was lower than the limit of 3.00 or 5.0 advocated by some authors (Kline, 2005).

The CFI value was found to be 0.794, which was lower than the cut-off limit of 0.95 so the model is described to have an acceptable fit. The NFI value was 0.629, which is within the given range, but the given cut-off value of NFI ≥.95 is shown in Table 11.62. Therefore, the model is acceptable. The NNFI value obtained is

TABLE 11.62
Robust Fit Indexes for Contractor's Organisational Culture Features Construct

Fit Index	Cut-Off Value	Estimate	Comment
$S - B\chi^2$		3249.5	
df	0≥	1861	Good fit
CFI	0.90≥ acceptable; 0.95≥ good fit	0.794	Acceptable
RMSEA 95%	Less than 0.05 with confidence interval (CI) 0.00–0.05 'good fit'	0.051	Good fit
NFI	Greater than 0.90 'good fit'	0.629	Acceptable
NNFI	Greater than 0.80 'good fit'	0.777	Acceptable
RMSEA 95% CI		0.048–0.054	Acceptable range

0.777, which is also below the cut-off value of 0.80. These fit indexes for the contractor's organisational culture model suggested that the postulated model adequately describes the sample data and could therefore be included in the full latent variable model analysis (Table 11.62).

11.4.3.1.4.3 Statistical Significance of Parameter Estimates Table 11.63 shows the correlation values, standard errors and the test of statistics. All the correlation values were less than 1.00, and all the Z-statistics were greater than 1.96 and show appropriate signs. The estimates were therefore deemed reasonable, as well as statistically significant. The parameter with the highest standardised coefficient was the indicator with variable COC 1 (company health and safety policy) and its parameter coefficient was 0.744.

Most of the parameter estimates had high correlation values close to 1.00. The high correlation values suggest a high degree of linear association between the indicator variables and the unobserved variable (contractor's organisational culture). In addition, the R^2 values were also close to the desired value of 1.00 indicating that the factors explained more of the variance in the indicator variables. The results, therefore, suggest that the indicator variables significantly predict the unobserved construct, because all the measured variables are significantly associated with the contractor's organisational culture features.

11.4.3.1.4.4 Internal Reliability and Validity of Scores The internal consistency and reliability of scores for the contractor's organisational culture features construct was determined from the rho and the Cronbach's alpha coefficient. The reliability coefficient should fall between 0 and 1.00, with values close to 1.00 as desired (Kline, 2005). The rho coefficient of internal consistency was found to be 0.964. This value was above the minimum required value of 0.70. Likewise, the Cronbach's alpha was above the minimum acceptable value of 0.70. The Cronbach's alpha was

TABLE 11.63
Factor Loadings and *Z*-Statistics of Contractor's Organisational Culture Measurement

Indicator Variable	Unstandardised Coefficient (λ)	Standardised Coefficient (λ)	*Z*-Statistics	R^2	Significant at 5% Level?
COC 1	0.778	0.629	10.021	0.381	Yes
COC 2	0.819	0.573	9.506	0.605	Yes
COC 3	0.712	0.703	10.500	0.671	Yes
COC 4	0.668	0.744	10.699	0.446	Yes
COC 5	0.748	0.664	10.238	0.560	Yes
COC 6	0.701	0.713	10.580	0.491	Yes
COC 7	0.773	0.634	10.044	0.598	Yes
COC 8	0.756	0.654	10.210	0.572	Yes

Note: Robust statistical significance at 5% level.

TABLE 11.64
Reliability and Construct Validity of Contractor's Organisational Culture Feature Model

Factor	Indicator Variable	Factor Loading	Cronbach's Alpha	Rho Coefficient
Contractor's organisational culture features	COC 1	0.5898	0.937	0.964
	COC 2	0.6730		
	COC 3	0.709		
	COC 4	0.5779		
	COC 5	0.6475		
	COC 6	0.6067		
	COC 7	0.6691		
	COC 8	0.6547		

Note: Parameter estimates are based on standardised solutions.

found to be 0.937 (Table 11.64). Both of these values revealed a high level of internal consistency and therefore reliability, suggesting that the indicator variables represent the same latent construct (contractor's organisational culture).

Furthermore, construct validity was determined from the magnitude and reasonableness of the parameter coefficients (factor loading). The parameter coefficients represent the magnitude of correlation or covariance between an item and a construct. Higher parameter coefficients show that the indicator variables have a stronger relationship with a construct and thus converge at a common point. Parameter coefficients of greater than 0.5 indicate a close relationship between the construct and an indicator variable. A parameter coefficient of 0.5 is interpreted as 25 percent of the total variance in the indicator variable being explained by the latent variable (factor). Hence, a parameter coefficient should be 0.5 or higher, and ideally 0.7 or greater to explain about 50 percent of the variance in an indicator variable (Hair et al., 1998).

The standardised parameter coefficient presented in Table 11.63 revealed that all coefficients were significantly higher with the lowest being 0.573 for contractor's organisational culture features. The magnitude of the parameter estimate was above the 50 percent minimum. This indicates a strong relationship between the indicator variables and the factors of the contractor's organisational culture features construct. Therefore, the contractor's organisational culture features satisfied both internal reliability and the construct criteria. The rho value was above the minimum value of 0.70, and the magnitude, signs and statistical significance of the parameter estimates were appropriate (Table 11.64).

11.4.3.1.4.5 Summary of Contractor's Organisational Culture Feature Measurement Model The CFA revealed that the residual covariance estimates fell within the acceptable range. Likewise, the robust fit indexes met the cut-off index criteria and all the parameter estimates were statistically significant and feasible. Considering these criteria, the measurement model for the contractor's organisational culture feature was found to adequately fit the sample data. Therefore, there was no need to improve the measurement model before it could be included in the full latent variable model. Hence, the contractor's organisational culture feature construct was adequately measured by the indicator variables and could be used in the analysis of the full latent variable model.

11.4.3.1.5 Measurement Model for Adherence to Safety Regulations Features Construct

The unidimensional model for adherence to safety regulations (ASR) features are presented (Table 11.38). From the 289 cases analysed for this construct, three indicator variables (F8RWSC 3, F8RWSC 4 and F8RWSC 5) made up of one factor was realised and renamed ASR and numbered systematically (Table 11.38). All three indicator variables obtained were used for the CFA. In order for a variable to be included in the CFA, thus enabling the model to be described as well-fitting, the distribution of residuals covariance matrix should be symmetrical and centred around zero (Byrne, 2010). The three-indicator model provides good measures of residual matrix and evidence of convergent validity.

The CFA results further revealed that the adherence to safety regulations features had three dependent variables, four independent variables and six free parameters. The number of fixed non-zero parameters was four. These are the three dependent indicator variables for the adherence to safety regulations: adhere to warning signs and notices, follow safety regulations and adhere to company safety policies. These indicator variables are presented in Table 11.65. The adherence to safety regulations features measurement model shown in Figure 11.5 was analysed before it could be included in the full latent variable model.

TABLE 11.65
Postulated Adherence to Safety Regulations Features Model

Latent Construct	Indicator Variable	Label
Adherence to safety regulations (ASR)	Adhere to warning signs and notices	ASR 1
	Follow safety regulations	ASR 2
	Adhere to company safety policies	ASR 3

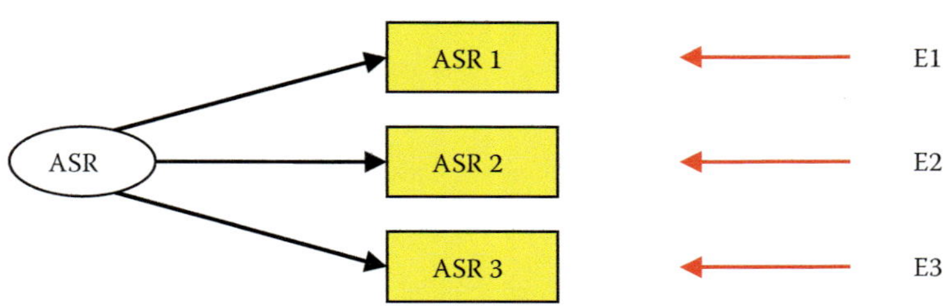

FIGURE 11.5 Measurement model of adherence to safety regulations.

In order to establish how well the model fit the sample data and the strength of the hypothesised relationship between the variables, results on residual covariance matrix (unstandardised and standardised), distribution of standardised residuals, fit statistics and statistical significance at a probability level of 5 percent were examined. In addition, the Cronbach's alpha and the rho coefficient of internal consistency were examined to determine the score reliability. Results of these statistics are presented in the following sections for the adherence to safety regulations features.

11.4.3.1.5.1 Diagnostic Fit Analysis: Analysis of Residual Covariance Estimate The unstandardised and standardised absolute residual matrix values of the adherence to safety regulations features are presented in Tables 11.66 and 11.67. The results reveal that all the absolute residual values and the average off-diagonal absolute residual values were close to zero. The unstandardised average off-diagonal residual was 0.0223, while the standardised average off-diagonal residual was found to be 0.0222. A residual value greater than 2.58 is described as large (Byrne, 2010). The results obtained for the adherence to safety regulations features measurement model suggested a fairly acceptable fit to the sample data because the absolute residuals were all less than 2.58. In order for a model to be described as well-fitting, the distribution

TABLE 11.66
Residual Covariance Matrix for Adherence to Safety Regulations Model (Unstandardised)

	ASR 1	ASR 2	ASR 3
ASR 1	0.000		
ASR 2	0.003	0.005	
ASR 3	–0.003	0.000	–0.002

Note: Average absolute residual = 0.0208. Average off-diagonal absolute residual = 0.0223. % falling between –0.1 and +0.1 = 99.99%.

TABLE 11.67
Residual Covariance Matrix for Adherence to Safety Regulations Model (Standardised)

	ASR 1	ASR 2	ASR 3
ASR 1	0.000		
ASR 2	0.002	0.000	
ASR 3	–0.002	0.000	0.000

Note: Average absolute residual = 0.0206. Average off-diagonal absolute residual = 0.0222. % falling between –0.1 and +0.1 = 99.99%.

of standardised residuals should be symmetrical and centred around zero (Byrne, 2010).

Further review of the frequency distribution reveals most residual values (99.99%) fall between –0.1 and 0.1, which is in the acceptable range. Of the remaining residuals, 0.01% fell outside the –0.1 to 0.1 range.

From this information, the results suggested a measurement model that was well-fitting despite a minimal discrepancy in fit between the hypothesised model and the sample data. Therefore, since this diagnostic fit analysis indicated a good fit, further tests of goodness of fit were possible to conclusively make a decision on the fit and appropriateness of the measurement model.

11.4.3.1.5.2 Goodness-of-Fit Statistics: Robust Maximum Likelihood (RML) The analysis strategy of goodness of fit for the adherence to safety regulations followed a three-statistics strategy of fit indexes as recommended (Hu & Bentler, 1999). The sample data on adherence to safety regulations measurement model yielded the $S - B\chi^2$ of 3249.5 with 1861 degrees of freedom with a probability of $p = 0.0000$. This chi-square value indicated that the departure of the sample data from the postulated measurement model was significant and, hence, indicative of good fit. The chi-square test is very sensitive to sample size and is used more as a descriptive index of fit rather than as a statistical test (Kline, 2005). The normed chi-square value is usually adopted by most researchers. The normed chi-square is the procedure of dividing the chi-square by the degrees of freedom. The normed values of up to 3.0 or even 5.0 are recommended (Kline, 2005).

Values for the NFI range between 0 and 1, with Bentler and Bonnet (1980) recommended values greater than 0.90 indicating a good fit. Hu and Bentler (1999) have given a cut-off criterion of NFI ≥.95. This index is sensitive to sample size, underestimating fit for samples less than 200 (Mulaik et al., 1989; Bentler, 1990), and is thus not recommended to be solely relied on (Kline, 2005). The value of the NNFI can indicate poor fit despite other statistics pointing towards good fit if small samples are used (Bentler, 1990; Kline, 2005; Tabachnick & Fidell, 2007). Owing to the non-normed nature, NNFI values can go above 1.0 and can thus be difficult to interpret (Byrne, 1998). The chi-square and degrees of freedom was found to be 1.75. This ratio was lower than the limit of 3.00 or 5.0 advocated by some authors (Kline, 2005:137).

The CFI value was found to be 0.794, which was lower than the cut-off limit of 0.95, so this is described to have an acceptable fit. The NFI value was 0.629, which is within the given range, but the given cut-off value of NFI ≥.95 is shown in Table 11.68. Therefore, the model is acceptable. The NNFI value obtained is 0.777, which is also below the cut-off value of 0.80. These fit indexes for the adherence to safety regulations model suggested that the postulated model adequately describe the sample data and could, therefore, be included in the full latent variable model analysis (Table 11.68).

11.4.3.1.5.3 Statistical Significance of Parameter Estimates Table 11.69 shows the correlation values, standard errors and the test of statistics. All the correlation values were less than 1.00, and all the Z-statistics were greater than

TABLE 11.68
Robust Fit Indexes for Adherence to Safety Regulations Features Construct

Fit Index	Cut-Off Value	Estimate	Comment
$S - B\chi^2$		3249.5	
df	0≥	1861	Good fit
CFI	0.90≥ acceptable; 0.95≥ good fit	0.794	Acceptable
RMSEA 95%	Less than 0.05 with confidence interval (CI) 0.00–0.05 'good fit'	0.051	Good fit
NFI	Greater than 0.90 'good fit'	0.629	Acceptable
NNFI	Greater than 0.80 'good fit'	0.777	Acceptable
RMSEA 95% CI		0.048–0.054	Acceptable range

TABLE 11.69
Factor Loadings and *Z*-Statistics of Adherence to Safety Regulations Measurement

Indicator Variable	Unstandardised Coefficient (λ)	Standardised Coefficient (λ)	*Z*-Statistics	R^2	Significant at 5% Level?
ASR 1	0.797	0.604	8.279	0.635	Yes
ASR 2	0.786	0.618	8.482	0.618	Yes
ASR 3	0.786	0.618	8.468	0.618	Yes

Note: Robust statistical significance at 5% level.

1.96 and show appropriate signs. The estimates were therefore deemed reasonable, as well as statistically significant. The parameter with the highest standardised coefficient was the indicator with variable ASR 2 and ASR 3 (follow safety regulations and adhere to company safety policies) and its parameter coefficient was 0.618.

Most of the parameter estimates had high correlation values close to 1.00. The high correlation values suggest a high degree of linear association between the indicator variables and the unobserved variable (adherence to safety regulations). In addition, the R^2 values were also close to the desired value of 1.00 indicating that the factors explained more of the variance in the indicator variables. The results, therefore, suggest that the indicator variables significantly predict the unobserved construct, because all the measured variables are significantly associated with the adherence to safety regulations features.

11.4.3.1.5.4 Internal Reliability and Validity of Scores The internal consistency and reliability of scores for the adherence to safety regulations features construct was determined from the rho and the Cronbach's alpha coefficient. According to Kline

TABLE 11.70
Reliability and Construct Validity of Adherence to Safety Regulations Feature Model

Factor	Indicator Variable	Factor Loading	Cronbach's Alpha	Rho Coefficient
Adherence to safety regulations features	ASR 1	0.6565	0.937	0.964
	ASR 2	0.6476		
	ASR 3	0.6476		

Note: Parameter estimates are based on standardised solutions.

(2005:59), the reliability coefficient should fall between 0 and 1.00. Values close to 1.00 are desired. The rho coefficient of internal consistency was found to be 0.964. This value was above the minimum required value of 0.70. Likewise, the Cronbach's alpha was above the minimum acceptable value of 0.70. The Cronbach's alpha was found to be 0.937 (Table 11.70). Both of these values revealed a high level of internal consistency and therefore reliability, suggesting that the indicator variables represent the same latent construct (adherence to safety regulations).

Furthermore, construct validity was determined from the magnitude and reasonableness of the parameter coefficients (factor loading). The parameter coefficients represent the magnitude of correlation or covariance between an item and a construct. Higher parameter coefficients show that the indicator variables have a stronger relationship with a construct and thus converge at a common point. Parameter coefficients of greater than 0.5 indicate a close relationship between the construct and an indicator variable. A parameter coefficient of 0.5 is interpreted as 25 percent of the total variance in the indicator variable being explained by the latent variable (factor). Hence, a parameter coefficient should be 0.5 or higher, and ideally 0.7 or greater, to explain about 50 percent of the variance in an indicator variable (Hair et al., 1998).

The standardised parameter coefficient presented in Table 11.69 revealed that all coefficients were significantly higher with the lowest being 0.604 for adherence to safety regulations features. The magnitude of the parameter estimate was above the 50 percent minimum. This indicates a strong relationship between the indicator variables and the factors of the adherence to safety regulations features construct. Therefore the adherence to safety regulations features satisfied both internal reliability and the construct criteria. The rho value was above the minimum value of 0.70, and the magnitude, signs and statistical significance of the parameter estimates were appropriate (Table 11.70).

11.4.3.1.5.5 Summary of Adherence to Safety Regulations Feature Measurement Model The CFA revealed that the residual covariance estimates fell within the acceptable range. Likewise, the robust fit indexes met the cut-off index criteria and all the parameter estimates were statistically significant and feasible. Considering these criteria, the measurement model for the adherence to safety regulations feature was found to adequately fit the sample data. Therefore, there was no need to improve the measurement model before it could be included in the full latent variable model. Hence the adherence to safety regulations feature construct was adequately measured by the indicator variables and could be used in the analysis of the full latent variable model.

11.4.3.1.6 Measurement Model for Health and Safety Compliance Features Construct

The unidimensional model for health and safety compliance (HSC) features are presented (Table 11.38). From the 289 cases analysed for this construct, four indicator variables (HSC 2, HSC 3, HSC 4 and HSC 5) were realised (Table 11.38). All four indicator variables obtained were used for the CFA (Byrne, 2010). In order for a variable to be included in a CFA, thus enabling the model to be described as well-fitting, the distribution of residuals covariance matrix should be symmetrical and centred around zero (Byrne, 2010). The three-indicator model provides good measures of residual matrix and evidence of convergent validity.

The CFA results further revealed that the health and safety compliance features had four dependent variables, five independent variables and eight free parameters. The number of fixed non-zero parameters was five. These are the four dependent indicator variables for the health and safety compliance: compensations paid on accident victims will be reduced, reduce the cost of training on health and safety (H&S), a limited number of H&S education by government representatives and limited number of H&S monitoring by government representatives. These indicator variables are presented in Table 11.71. The health and safety compliance features measurement model shown in Figure 11.6 was analysed before it could be included in the full latent variable model.

In order to establish how well the model fit the sample data and the strength of the hypothesised relationship between the variables, results on residual covariance matrix (unstandardised and standardised), distribution of standardised residuals, fit statistics and statistical significance at a probability level of 5 percent were examined. In addition, the Cronbach's alpha and the rho coefficient of internal consistency

TABLE 11.71
Postulated Health and Safety Compliance Features Model

Latent Construct	Indicator Variables	Label
Health and safety compliance (HSC)	Compensations paid on accident victims will be reduced	HSC 1
	Reduce the cost of training on health and safety (H&S)	HSC 2
	A limited number of H&S education by government representatives	HSC 3
	Limited number of H&S monitoring by government representatives	HSC 4

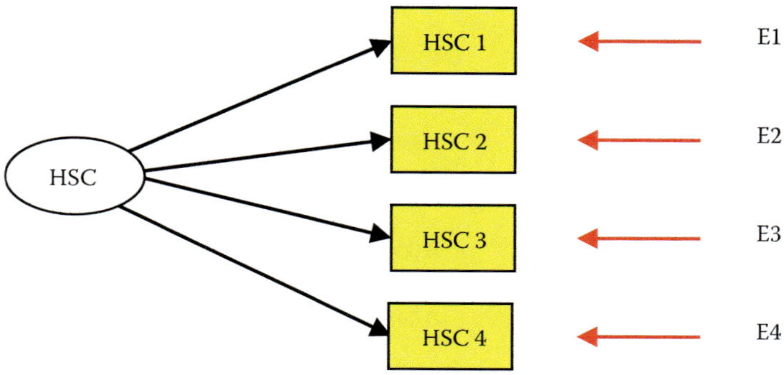

FIGURE 11.6 Measurement model of health and safety compliance.

were examined to determine the score reliability. Results of these statistics are presented in the next section for the health and safety compliance features.

11.4.3.1.6.1 Diagnostic Fit Analysis: Analysis of Residual Covariance Estimate The unstandardised and standardised absolute residual matrix values of the health and safety compliance features are presented in Tables 11.72 and 11.73. The result reveals that all the absolute residual values and the average off-diagonal absolute residual values were close to zero. The unstandardised average off-diagonal residual was 0.0665, while the standardised average off-diagonal residual was found to be 0.0658. A residual value greater than 2.58 is described as large (Byrne, 2010). The results obtained for the health and safety compliance features measurement model suggested a fairly acceptable fit to the sample data because the absolute residuals were all less than 2.58. In order for a model to be described as well-fitting, the distribution of standardised residuals should be symmetrical and centred around zero (Byrne, 2010).

TABLE 11.72
Residual Covariance Matrix for Health and Safety Compliance Model (Unstandardised)

	HSC 1	HSC 2	HSC 3	HSC 4
HSC 1	0.000			
HSC 2	0.184	–0.001		
HSC 3	–0.058	–0.023	0.000	
HSC 4	–0.046	–0.049	0.039	0.000

Note: Average absolute residual = 0.0400. Average off-diagonal absolute residual = 0.0665. % falling between –0.1 and +0.1 = 99.99%.

TABLE 11.73
Residual Covariance Matrix for Health and Safety Compliance Model (Standardised)

	HSC 1	HSC 2	HSC 3	HSC 4
HSC 1	0.000			
HSC 2	0.183	0.000		
HSC 3	–0.058	–0.022	0.000	
HSC 4	–0.046	–0.048	0.039	0.000

Note: Average absolute residual = 0.0395. Average off-diagonal absolute residual = 0.0658. % falling between –0.1 and +0.1 = 99.99%.

Further review of the frequency distribution reveals most residual values (99.99%) fall between –0.1 and 0.1, which is in the acceptable range. Of the remaining residuals, 0.01% fell outside the –0.1 to 0.1 range.

From this information, the results suggested a measurement model that was well-fitting despite a minimal discrepancy in fit between the hypothesised model and the sample data. Therefore, since this diagnostic fit analysis indicated a good fit, further tests of goodness of fit were possible to conclusively make a decision on the fit and appropriateness of the measurement model.

11.4.3.1.6.2 Goodness-of-Fit Statistics: Robust Maximum Likelihood (RML) The analysis strategy of goodness of fit for the health and safety compliance followed a three-statistics strategy of fit indexes as recommended (Hu & Bentler, 1999). The sample data on health and safety compliance measurement model yielded the $S - B\chi^2$ of 780.7 with 449 degrees of freedom with a probability of $p = 0.0000$. This chi-square value indicated that the departure of the sample data from the postulated measurement model was significant and, hence, indicative of good fit. The chi-square test is very sensitive to sample size and is used more as a descriptive index of fit rather than as a statistical test (Kline, 2005). The normed chi-square value is usually adopted by most researchers. The normed chi-square is the procedure of dividing the chi-square by the degrees of freedom. The normed values of up to 3.0 or even 5.0 are recommended (Kline, 2005).

Values for the NFI ranged between 0 and 1, with Bentler and Bonnet (1980) recommended values greater than 0.90 indicating a good fit. Hu and Bentler (1999) have given a cut-off criterion of NFI ≥.95. This index is sensitive to sample size, underestimating fit for samples less than 200 (Mulaik et al., 1989; Bentler, 1990), and is thus not recommended to be solely relied on (Kline, 2005). The value of the NNFI can indicate poor fit despite other statistics pointing towards good fit if small samples are used (Bentler, 1990; Kline, 2005; Tabachnick & Fidell, 2007). Owing to their non-normed nature, NNFI values can go above 1.0 and can thus be difficult to interpret (Byrne, 1998). The chi-square and degrees of freedom were found to be

1.75. This ratio was lower than the limit of 3.00 or 5.0 advocated by some authors (Kline, 2005).

The CFI value was found to be 0.917, which was lower than the cut-off limit of 0.95, so the model is described to have an acceptable fit. The NFI value was 0.793, which is within the given range, but the given cut-off value of NFI ≥.95 is shown in Table 11.74. Therefore, the model is acceptable. The NNFI value obtained is 0.909, which is above the cut-off value of 0.80. These fit indexes for the health and safety compliance model suggested that the postulated model adequately describe the sample data and could, therefore, be included in the full latent variable model analysis (Table 11.74).

11.4.3.1.6.3 Statistical Significance of Parameter Estimates Table 11.75 shows the correlation values, standard errors and the test of statistics. All the correlation values were less than 1.00, and all the *Z*-statistics were greater than 1.96 and show appropriate signs. The estimates were therefore deemed reasonable, as well as statistically significant. The parameter with the highest standardised coefficient was the

TABLE 11.74
Robust Fit Indexes for Health and Safety Compliance Features Construct

Fit Index	Cut-Off Value	Estimate	Comment
$S-B\chi^2$		680.7	
df	0≥	449	Good fit
CFI	0.90≥ acceptable; 0.95≥ good fit	0.917	Good fit
RMSEA 95%	Less than 0.05 with confidence interval (CI) 0.00–0.05 'good fit'	0.042	Good fit
NFI	Greater than 0.90 'good fit'	0.793	Acceptable
NNFI	Greater than 0.80 'good fit'	0.909	Good fit
RMSEA 95% CI		0.036–0.049	Acceptable range

TABLE 11.75
Factor Loadings and *Z*-Statistics of Health and Safety Compliance Measurement

Indicator Variable	Unstandardised Coefficient (λ)	Standardised Coefficient (λ)	*Z*-Statistics	R^2	Significant at 5% Level?
HCS 1	0.576	0.817	10.740	0.332	Yes
HSC 2	0.713	0.702	9.298	0.508	Yes
HSC 3	0.812	0.583	6.913	0.660	Yes
HSC 4	0.772	0.636	8.065	0.595	Yes

Note: Robust statistical significance at 5% level.

indicator with variable HSC 3 (limited number of H&S monitoring by government representatives) and its parameter coefficient was 0.812.

Most of the parameter estimates had high correlation values close to 1.00. The high correlation values suggest a high degree of linear association between the indicator variables and the unobserved variable (health and safety compliance). In addition, the R^2 values of three (HSC 2, HSC 3 and HSC 4) indicator variables were close to the desired value of 1.00 indicating that the factors explained more of the variance in the indicator variables. The results, therefore, suggest that the indicator variables significantly predict the unobserved construct, because all the measured variables are significantly associated with the health and safety compliance features.

11.4.3.1.6.4 Internal Reliability and Validity of Scores The internal consistency and reliability of scores for the health and safety compliance features construct was determined from the rho and the Cronbach's alpha coefficient. According to Kline (2005), the reliability coefficient should fall between 0 and 1.00. Values close to 1.00 are desired. The rho coefficient of internal consistency was found to be 0.964. This value was above the minimum required value of 0.70. Likewise, the Cronbach's alpha was above the minimum acceptable value of 0.70. The Cronbach's alpha was found to be 0.937 (Table 11.76). Both of these values revealed a high level of internal consistency and therefore reliability, suggesting that the indicator variables represent the same latent construct (health and safety compliance).

Furthermore, construct validity was determined from the magnitude and reasonableness of the parameter coefficients (factor loading). The parameter coefficients represent the magnitude of correlation or covariance between an item and a construct. Higher parameter coefficients show that the indicator variables have a stronger relationship with a construct and thus converge at a common point. Parameter coefficients of greater than 0.5 indicate a close relationship between the construct and an indicator variable. A parameter coefficient of 0.5 is interpreted as 25 percent of the total variance in the indicator variable being explained by the latent variable (factor). Hence, a parameter coefficient should be 0.5 or higher, and ideally 0.7 or greater, to explain about 50 percent of the variance in an indicator variable (Hair et al., 1998).

TABLE 11.76
Reliability and Construct Validity of Health and Safety Compliance Feature Model

Factor	Indicator Variable	Factor Loading	Cronbach's Alpha	Rho Coefficient
Health and safety compliance	HSC 1	0.5763	0.808	0.806
	HSC 2	0.7127		
	HSC 3	0.8122		
	HSC 4	0.7716		

Note: Parameter estimates are based on standardised solutions.

The standardised parameter coefficient presented in Table 11.75 revealed that all coefficients were significantly higher with the lowest being 0.583 for health and safety compliance features. The magnitude of the parameter estimate was above the 50 percent minimum. This indicates a strong relationship between the indicator variables and the factors of the health and safety compliance features construct. Therefore, the health and safety compliance features satisfied both internal reliability and the construct criteria. The rho value was above the minimum value of 0.70, and the magnitude, signs and statistical significance of the parameter estimates were appropriate (Table 11.76).

11.4.3.1.6.5 Summary of Health and Safety Compliance Measurement Model The CFA revealed that the residual covariance estimates fell within the acceptable range. Likewise, the robust fit indexes met the cut-off index criteria and all the parameter estimates were statistically significant and feasible. Considering these criteria, the measurement model for the health and safety compliance feature was found to adequately fit the sample data. Therefore, there was no need to improve the measurement model before it could be included in the full latent variable model. Hence the health and safety compliance feature construct was adequately measured by the indicator variables and could be used in the analysis of the full latent variable model.

11.4.3.2 Structural Model: Testing of the Hypothesised Structural Equation Model

Each of the five latent factors to be included in the full latent model was working very well based on the test indexes and the statistical significance of the parameter estimates. Then the full structural model was tested, which included all five factors (with tested indicator variables) and the health and safety compliance outcome (manifest) variables for the study. Once again, the CFA measurement model for latent constructs were tested in order to confirm whether the indicators that have been used to measure one or more latent factors hold. Thus, loadings of the indicators on the specific factors were examined to see how well each factor has been specified in the context of the others.

Covariances between the latent factors are added to the model for any relationship that will be examined when the structural model is tested. Also, covariances between the latent factors and outcome variables are also added to rule out the possibility that any of them may serve as an indicator of any of the proposed factors. As already indicated (analysis of the measurement models), the measurement models indicated that the models (latent variables CFAs) worked well and it was therefore feasible to test the full latent variable model. The question of whether measurement models should be checked before analysing the full SEM is simply a strategy a researcher adopts (Hayduk & Glaser, 2000). Similarly, the question of how many factors a construct should have is also debatable (Bollen, 1989; Hayduk & Glaser, 2000). However, assessing the measurement models first has an advantage. The first merit of analysing the latent variable measurement models separately before analysing the full SEM model is that the research is assured of a proper working measurement model before analysing the full SEM latent model.

Hence, the researcher avoids the frustration of re-specifying the full model if a solution cannot be obtained. If a CFA model cannot be satisfactorily fitted, moving to the structural model will provide no additional guidance or benefit. However, these observations as presented in the current study were a pure confirmatory analysis and therefore recommendations were based on whether the postulated priori model fit the sample data. Hence, not all the initially derived indicator variables from the literature that were on the questionnaires were tested in the CFA, as the preliminary residual covariance matrix (factor loadings) for some indicator variables of some latent constructs were more than the recommended value. A residual covariance matrix value greater than 2.58 is described as large (Byrne, 2010). In order for a model to be described as well fitting, the distribution of the residuals should be symmetrical and centred around zero (Byrne, 2010).

11.4.3.2.1 Hypothesised Relation for the Structural Model

The hypothesised model (Model 2.0) was tested, in which safe act and working condition, government support, contractor's safety policy, contractor's organisational culture, and adherence to safety regulations were expected to define health and safety compliance. The hypothesised model was fitted to the data for the entire sample and, as is the norm, covariances for all the exogenous factors and variables were specified. The five-factor model was fitted to the data with the robust maximum likelihood (RML) method of EQS and the model converged. As with all of the analyses presented in this study, the testing of this model was based on the RML estimation and robust statistics were used to ascertain the fit of the model. The robust solution adjusts for non-normality in the data. As is the norm in SEM analyses (Kline, 2005), one variable loading per latent factor was set equal to 1.0 in order to set the metric for that factor (Figure 11.7).

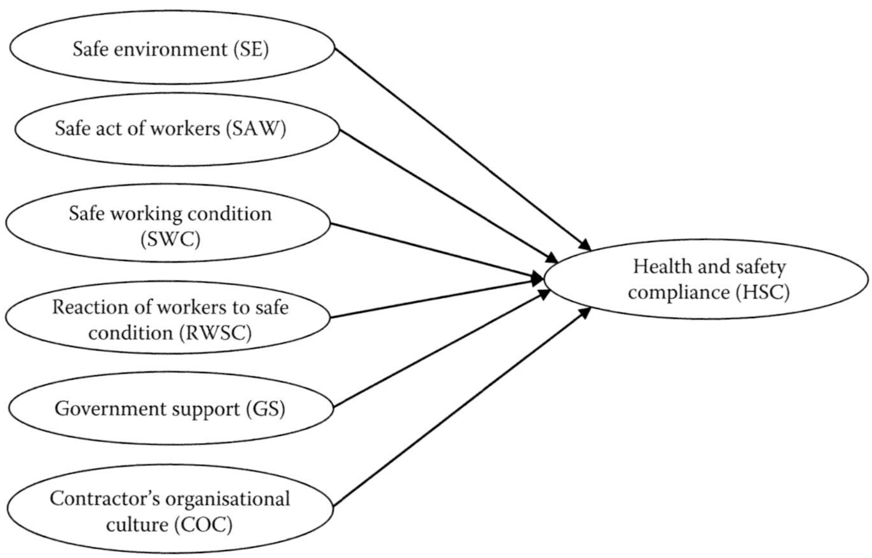

FIGURE 11.7 Hypothesised model of health and safety compliance (Model 1.0).

11.4.3.2.2 Fit Statistics on the Structural Model

A confirmatory factor analysis of the full latent model was conducted. The full structural model hypothesised that safe act and working condition features, government support features, contractors' safety policy features, contractors' organisational culture features and adherence to safety regulations features define health and safety compliance in small and medium-sized enterprise (SME) contractors. The structural equation model is presented in Figure 11.8 (Model 2.0).

The Model 2.0 is founded on the general hypothesis for the study, which is based on the fact that overall health and safety compliance is directly related to the influence of the exogenous variables in predicting or determining overall health and safety compliance of small to medium-sized construction companies. The theory and basis of the model was presented in Chapter 10. The number of cases that were analysed for the full latent variable Model 2.0 was 558. Out of the total sample size, all the 588 cases had positive weights, and 28 had missing variables, which were corrected with the maximum likelihood method of missing data correction. The model had 64 dependent variables and 78 independent variables. It also had 219 free parameters and 78 numbers of fixed nonzero parameters. The covariance matrix of the model was analysed using the robust maximum likelihood estimation method. Raw data was used for the analysis and an estimation approach available in EQS (robust maximum likelihood) as already discussed, which adjusts the model fit chi-square test statistics and standard errors of individual parameter estimates were used.

11.4.3.2.3 Analysis of Residual Covariance Estimate

Investigation of the average absolute residual values of the structural model revealed that all the absolute residual values and the average off-diagonal absolute residual values were close to zero. The unstandardised average off-diagonal residual was 0.0665, while the standardised average off-diagonal residual was found to 0.0658. These residual values were considered small, as they were all less than 2.58 (Byrne, 2010). In addition, 99.99 percent of the residuals fell within the acceptable range of –0.1 and 0.1. The significance of this distribution is that for a structural model to be described as well-fitting, the distribution of residuals should be symmetrical and centred around zero (Byrne, 2010), which the analysed data has displayed. From this information, the results suggest that the hypothesised structural model had a good fit to the sample data; overall, the model was quite well-fitting. Therefore, since this initial assessment of the structural model residuals indicated a good fit, further tests of goodness of fit were justified.

11.4.3.2.4 Structural Model Goodness-of-Fit Statistics: Robust Maximum Likelihood

The test of the hypothesis that small to medium-sized construction companies health and safety compliance is a six-factor structure, as depicted in Figure 11.7 (Model 1.0) via the sample data on the model, yielded a robust likelihood ratio test ($S - B\chi^2$) of 680.7 with 449 degrees of freedom. The associated p-value was less than 0.0001 ($p =$ 0.0000) with a sample of 588 cases. The chi-square index suggested that the difference between the hypothesised model and the sample data matrix was significant

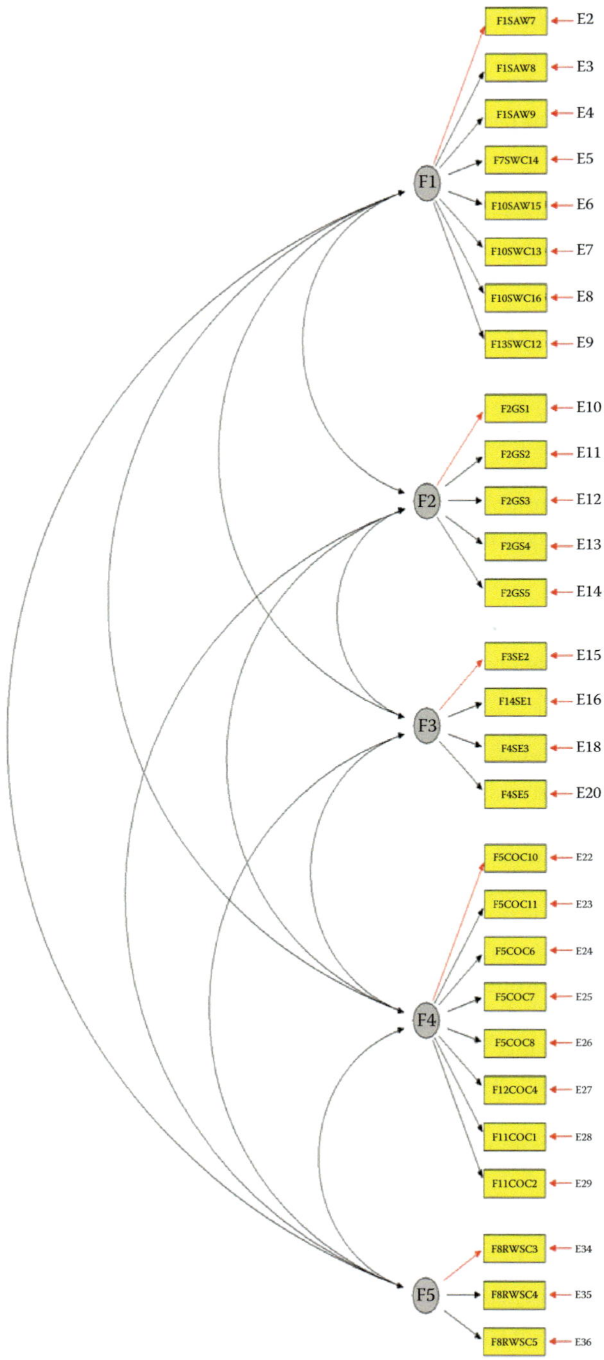

FIGURE 11.8 Model 2.0. An integrated health and safety compliance model. Model parameters (from left to right): ASR (3 indicator variables), COC (8 indicator variables), CSP (4 indicator variables), GS (5 indicator variables) and SAWC (8 indicator variables).

but not entirely adequate. Interpreted literally, this test statistic indicates that given the present sample data, the hypothesis bearing on small to medium-sized construction companies health and safety compliance relates as summarised in the model, represents an unlikely event (i.e. occurring less than one time in a thousand under the hypothesis) and should be rejected. However, the chi-square test (likelihood ratio test) of fit is very sensitive and therefore could not be relied upon to determine model fit. The analysis of covariance structure (SEM) is grounded in large sample size theory. As such, large sample sizes are critical to obtaining precise parameter estimates, as well as to the tenability of asymptotic distribution approximations (MacCallum et al., 1996; Byrne, 2010). Therefore, a normed chi-square value is usually adopted by most researchers (Bentler, 1999; MacCallum et al., 1996; Kline, 2005; Byrne, 2006). Normed chi-square is the procedure of dividing the $S - B\chi^2$ by the degrees of freedom. The normed values of up to 3.0 or even 5.0 are recommended (Kline, 2005). From the above chi-square and degrees of freedom values the ratio was found to be 8.45:1. This ratio was within the limit of 5.0 advocated for by some authors (Kline, 2005:137; Byrne, 2010) and therefore indicative of a reasonable fit of the model. However, other fit indexes indicated a good fit of the model to the latent variables. The robust CFI was found to be 0.917. The CFI was greater than 0.90, which is the upper limit value for model acceptance. Moreover, a two-strategic approach is considered satisfactory to accept or reject a model (Hu & Bentler, 1999). Hence, RMSEA and SRMR statistics were further used to decide on the acceptability of the model.

The robust RMSEA with a 95 percent confidence interval (lower bound value = 0.036 and the upper bound value = 0.049) was found to be 0.042. The RMSEA index was just below the upper limit of 0.05 for the model to be described as good. However, the value of 0.042 indicated that the model has good fit. An RMSEA of between 0.08 and 0.10 provides a poor fit and below 0.08 shows a good fit (MacCallum et al., 1996). Also, an RMSEA with a 95 percent confidence interval (CI) in a well-fitting model lower limit should be close to 0.00, while the upper limit should be less than 0.08 (Hooper, Coughlan & Mullen, 2008). Hence, the structural model RMSEA with a 95 percent CI met the above criteria and the model could be considered a good fit. In addition, the absolute fit index, NFI, was found to be 0.793, while the NNFI was found to be 0.909. The NFI and the NNFI absolute fit index indicated an adequate fit of the full structural model to the sample data. Therefore, the goodness-of-fit statistics indexes (CFI, NFI, NNFI, RMSEA, RMSEA at 95% and $S - B\chi^2$) met the condition for model acceptance (Table 11.77). However, the Lagrange multiplier (LM) test conducted on the full latent model sample data did not reveal any significant indicators of model mis-specification of the hypothesised parameters. In EQS, a model can be said to be mis-specified if there are any mis-fitting parameters using the LM test (Byrne, 2010). The criterion is to identify any significant drop in the χ^2 values of parameters. Also, in univariate and multivariate analysis, the probability that a parameter estimate is equal to zero should be less than 0.05 in order to be rejected. This is also an indication of mis-specification according to Byrne (2010). Hence, inspection of the LM test output revealed that there were no significant mis-fitting variables that would have warranted model re-specification.

TABLE 11.77
Robust Fit Indexes for Structural Model 2.0

Fit Index	Cut-Off Value	Estimate	Comment
$S-B\chi^2$		680.7	
df	0≥	449	Good fit
CFI	0.90≥ acceptable; 0.95≥ good fit	0.917	Good fit
RMSEA 95%	Less than 0.05 with confidence interval (CI) 0.00–0.05 'good fit'	0.042	Good fit
NFI	Greater than 0.90 'good fit'	0.793	Acceptable
NNFI	Greater than 0.80 'good fit'	0.909	Good fit
RMSEA 95% CI		0.036–0.049	Acceptable range

11.4.3.2.5 Internal Reliability and Construct Validity of the Structural Model

The rho coefficient and the Cronbach's alpha coefficient were examined in order to establish score reliability for the structural model. The reliability coefficient should fall between 0 and 1.00, while values close to 1.00 are desired (Kline, 2005). The rho coefficient of internal consistency was found to be 0.964. This value was above the minimum required value of 0.70. Similarly, the Cronbach's alpha was above the minimum acceptable value of 0.70 at 0.937. Both of these values indicated a high degree of internal consistency and homogeneity (Table 11.75). These findings informed that the degree to which responses are consistent across all indicator variables was statistically significant, indicating that the measures of the latent variables' total scores are the best possible unit of analysis for the exogenous variables, which thus predict the endogenous variable (health and safety compliance). The construct validity for the structural model was determined by examining the magnitude of the parameter coefficients. High parameter coefficients of greater than 0.5 indicate a close relation between the factor and an indicator variable. However, a parameter coefficient has to be between 0.5 and 0.7 or greater to explain about 50 percent of the variance in an indicator variable (Hair et al., 1998). Inspection of the standardised parameter coefficient presented in Table 11.78 shows that they were significantly high with a maximum of 0.7090. The parameter estimate of 0.7090 meant that the health and safety compliance accounted for about 70.9 percent of the variance COC 3 and was therefore indicative of a good fit between the indicator variable and the factor, likewise in the other factors.

11.4.3.3 Structural Model Hypothesis Testing

Besides assessing the goodness of fit of the structural model, the feasibility of a model can be judged by a further inspection of the obtained solution. However, this involves inspection of the statistical significance of the parameter estimates, standard errors and the test statistics (Raykov, Tomer & Nesselroade, 1991). Therefore, the rejection of the hypothesis depends on how reasonable parameter estimates were

TABLE 11.78
Reliability and Construct Validity of the Latent Variables

Latent (Exogenous) Factor	No. of Indicator Variables	Indicator Variable	Parameter Coefficient	Rho Coefficient	Cronbach's Alpha
Safe act and working condition	8	SAWC 1	0.6189	0.980	0.973
		SAWC 2	0.6013		
		SAWC 3	0.5799		
		SAWC 4	0.5370		
		SAWC 5	0.5917		
		SAWC 6	0.6437		
		SAWC 7	0.694		
		SAWC 8	0.692		
Government support	5	GS 1	0.5335		
		GS 2	0.6390		
		GS 3	0.6440		
		GS 4	0.6047		
		GS 5	0.5964		
Contractor's safety policy	4	CSP 1	0.5632		
		CSP 2	0.5187		
		CSP 3	0.5898		
		CSP 4	0.4125		
Contractor's organisational culture	8	COC 1	0.5898		
		COC 2	0.6730		
		COC 3	0.7090		
		COC 4	0.5779		
		COC 5	0.6475		
		COC 6	0.6067		
		COC 7	0.6691		
		COC 8	0.6547		
Adherence to safety regulations features	3	ASR 1	0.6565		
		ASP 2	0.6476		
		ASP 3	0.6476		
Health and safety compliance	4	HSC 1	0.5763	0.806	0.808
		HSC 2	0.7127		
		HSC 3	0.8122		
		HSC 4	0.7716		

in terms of their magnitude, signs and statistical significance. In addition, if the output showed estimates that had correlation values greater than 1.00, had negative variances and the correlation or covariances were not definite positive, then they were said to be displaying unreasonable estimates (Byrne, 2010). Likewise, the test statistics had to be greater than 1.96 based on the probability level of 5 percent before the hypothesis can be rejected (Byrne, 2010). The test statistic reported was the parameter estimate divided by its standard error and therefore it functions as a Z-statistic to test that the estimate is statistically different from zero. Hence, the test was used to evaluate the hypothesis.

11.4.3.3.1 Testing the Influence of the Exogenous Variables on Overall Health and Safety Compliance

It was a general hypothesis that small and medium-sized construction companies' health and safety compliance is related to the influence of the exogenous variables in predicting the overall health and safety compliance in developing countries using Ghana as a case study. Results from the SEM analysis yielded support for the hypothesis. The hypothesised relationships between all exogenous factors and the endogenous factor were found to be significant and they all had definite positive directions. Inspection of the correlation values, standard errors and the test statistics in Table 11.77 revealed that all standardised coefficient correlation values were not greater than 1.00. All test statistics (Z-values) were greater than 1.96 ($p < 0.05$) and the signs were appropriate. They all have positive values (refer to Tables 11.79 and 11.80), suggesting that all latent variables measured the overall health and safety compliance of small and medium-sized construction companies in Ghana. The estimates were therefore reasonable as well as statistically significant. Therefore, the general hypothesis that small and medium-sized construction companies' health and safety compliance is directly related to the influence of the exogenous variables in predicting overall health and safety compliance in Ghana small and medium-sized construction companies could not be rejected. The relationship between health and safety compliance and the indicators was found to be the most significant. The parameter with the highest standardised coefficient for this factor was the indicator variable COC 3. The parameter coefficient was found to be 0.7090. The indicator variable COC 3, which asked whether health and safety inspection should be included in the contractor's organisational culture to enhance health and safety compliance, was found to be more associated with overall health and safety compliance than any other indicator variable.

However, in order to determine whether each exogenous variable considerably predicted the endogenous construct, an inspection of the interfactor correlation (R^2) values were examined, thus establishing the exogenous variables' direct influence on the dependent variable (presented in the subsequent sections). However, the overall results therefore suggest that the exogenous variables considerably predict the endogenous variable (health and safety compliance). Further assessment of the outcome variables of overall health and safety compliance revealed that all standardised factor loadings values were generally large and statistically significant (values ranged from 0.384 to 0.817). However, the interfactor correlation (R^2) values were all statistically significant (values ranged from 0.0.332 to 0.853), as shown in Table 11.80. The variance accounted for in each measure by the endogenous variable revealed that the scores were significant at the 5 percent level. The score results suggested that the interfactor relationship between the manifest variables is strong and has significant level of correlations.

TABLE 11.79
Model 2.0 Factor Loadings and *Z*-Statistics

Indicator Variable	Unstandardised Coefficient (λ)	Standardised Coefficient (λ)	*Z*-Value	Significant at 5% Level?
SAWC 1	0.766	0.642	9.990	Yes
SAWC 2	0.745	0.668	10.183	Yes
SAWC 3	0.718	0.696	10.363	Yes
SAWC 4	0.665	0.747	10.603	Yes
SAWC 5	0.733	0.681	10.240	Yes
SAWC 6	0.797	0.604	9.540	Yes
SAWC 7	0.720	0.694	10.290	Yes
SAWC 8	0.722	0.692	10.259	Yes
GS 1	0.923	0.384	7.733	Yes
GS 2	0.867	0.499	9.565	Yes
GS 3	0.855	0.519	9.767	Yes
GS 4	0.843	0.539	7.922	Yes
GS 5	0.776	0.631	9.422	Yes
CSP 1	0.882	0.471	6.509	Yes
CSP 2	0.617	0.787	10.710	Yes
CSP 3	0.778	0.629	10.021	Yes
CSP 4	0.819	0.573	9.506	Yes
COC 1	0.778	0.629	10.021	Yes
COC 2	0.819	0.573	9.506	Yes
COC 3	0.712	0.703	10.500	Yes
COC 4	0.668	0.744	10.699	Yes
COC 5	0.748	0.664	10.238	Yes
COC 6	0.701	0.713	10.580	Yes
COC 7	0.773	0.634	10.044	Yes
COC 8	0.756	0.654	10.210	Yes
ASR 1	0.797	0.604	8.279	Yes
ASR 2	0.786	0.618	8.482	Yes
ASR 3	0.786	0.618	8.468	Yes
HSC 1	0.576	0.817	10.740	Yes
HSC 2	0.713	0.702	9.298	Yes
HSC 3	0.812	0.583	6.913	Yes
HSC 4	0.772	0.636	8.065	Yes

Notes: Robust Statistical Significance at 5% level. The SEM analysis norm (Kline, 2005) is when one variable loading per latent factor is set equal to 1.0 in order to set the metric for that factor.

11.4.3.3.1.1 Testing the Direct Influence of Safe Act and Working Condition Features on Overall Health and Safety Compliance Results from the confirmatory factor analysis of the full structural model, presented in Tables 11.79 and 11.80, yield support for the general hypothesis. The relationship between the factors and the endogenous variable (dependent variable) was found to be statistically significant at

TABLE 11.80
Model 2.0 Factor Loadings, *Z*-Statistics, Variance Accounted for and Reliability and Construct Validity

Indicator Variable	Standardised Coefficient (λ)	*Z*-Value	R^2	Cronbach's Alpha	Rho Coefficient
SAWC 1	0.642	9.990	0.587	0.973	0.980
SAWC 2	0.668	10.183	0.554		
SAWC 3	0.696	10.363	0.516		
SAWC 4	0.747	10.603	0.442		
SAWC 5	0.681	10.240	0.537		
SAWC 6	0.604	9.540	0.537		
SAWC 7	0.694	10.290	0.635		
SAWC 8	0.692	10.259	0.519		
GS 1	0.384	7.733	0.839		
GS 2	0.499	9.565	0.853		
GS 3	0.519	9.767	0.751		
GS 4	0.539	7.922	0.731		
GS 5	0.631	9.422	0.710		
CSP 1	0.471	6.509	0.602		
CSP 2	0.787	10.710	0.779		
CSP 3	0.629	10.021	0.381		
CSP 4	0.573	9.506	0.605		
COC 1	0.629	10.021			
COC 2	0.573	9.506	0.605		
COC 3	0.703	10.500	0.671		
COC 4	0.744	10.699	0.446		
COC 5	0.664	10.238	0.560		
COC 6	0.713	10.580	0.491		
COC 7	0.634	10.044	0.598		
COC 8	0.654	10.210	0.572		
ASR 1	0.604	8.279	0.635		
ASR 2	0.618	8.482	0.618		
ASR 3	0.618	8.468	0.618		
HSC 1	0.817	10.740	0.332	0.808	0.806
HSC 2	0.702	9.298	0.508		
HSC 3	0.583	6.913	0.660		
HSC 4	0.636	8.065	0.595		

Note: Robust statistical significance at 5% level.

the 5 percent probability level. On the other hand, all standardised parameter estimates showed high correlation values close to 1.00 suggesting a high degree of linear association between the indicator variables and the endogenous construct. Inspection of the R^2 values for the safe act and working condition variables revealed that the values were above 0.50 (i.e. SAWC 1 ($R^2 = 0.587$), SAWC 2 ($R^2 = 0.554$), SAWC 3 ($R^2 = 0.516$), SAWC 5 ($R^2 = 0.537$), SAWC 6 ($R^2 = 0.537$), SAWC 7 ($R^2 = 0.635$) and SAWC 8 ($R^2 = 0.519$)) and close to the desired value of 1.00. The exception is the indicator variable SAW 4 ($R^2 = 0.442$), which is the weakest among the variables. The result of R^2 for SAW 4 suggests that this indicator variable did not considerably predict the endogenous factor construct. Despite the non-coherent level of the interfactor correlation within the indicator variables, the direct influence of safe act and working condition factor on overall compliance is statistically significant as shown in Table 11.80.

11.4.3.3.1.2 Testing the Direct Influence of Government Support Features on Overall Health and Safety Compliance Inspection of the R^2 values for the government support indicators revealed that all five indicator variables that were used to measure the latent factor had values close to the desired value of 1.00. All the five variables – GS 1($R^2 = 0.839$), GS 2 ($R^2 = 0.853$), GS 3 ($R^2 = 0.751$), GS 4 ($R^2 = 0.731$) and GS 5 ($R^2 = 0.710$) – were very strong in predicting the endogenous variable (Table 11.80). This suggests that the interfactor relationship of these variables and other indicators in determining the overall health and safety compliance is major. Furthermore, assessment of the variance accounted for in each measure by the endogenous variable revealed that all scores were significant at the 5 percent level. The reported parameter coefficient explained more than 25 percent of the variance in the latent variable, which was indicative of an adequate fit between the latent variables and the endogenous construct. Thus, the score results suggested that the influence of this latent factor on the endogenous variable was direct and significant.

11.4.3.3.1.3 Testing the Direct Influence of Contractor's Safety Policy Features on Overall Health and Safety Compliance Inspection of the score values for this factor revealed that all standardised factor loadings were generally large and statistically significant (values ranged from 0.471 to 0.787) with the exception of one indicator variable: CSP 1 = 0.471. However, the interfactor correlation (R^2) values were also large and statistically significant (values ranged from 0.602 to 0.779) with the exception of CSP 3 = 0.381, as shown in Table 11.80. Also, the variances accounted for in each measure by the endogenous variable revealed that the scores were significant at the 5 percent level. The values were above the minimum required value of 25 percent. Hence, the score results suggested that the influence of the contractor's safety policy put in place by the contractor's on the endogenous variable was direct and statistically significant.

11.4.3.3.1.4 Testing the Direct Influence of Contractor's Organisational Culture Features on Overall Health and Safety Assessment of the standardised factor loadings revealed that all values were generally moderate and statistically significant (values ranged from 0.5779 to 0.744). Also, the interfactor correlation (R^2) values were also moderate and statistically significant (values ranged from 0.560 to 0.671) with the exception of three, as their interfactor correlation (R^2) had lower values (ranged from 0.381 and 0.491), as shown in Table 11.80. The total variances accounted for in each indicator variable by the endogenous variable revealed that the scores were significant at the 5 percent level. The score results suggested that the influence of contractor's organisational culture in determining overall SME contractors' health and safety compliance was direct and statistically significant.

11.4.3.3.1.5 Testing the Direct Influence of Adherence to Safety Policy Features on Overall Health and Safety Compliance The inspection of the standardised factor loadings revealed that all values were generally large and statistically significant (Table 11.80). The R^2 values were large and statistically significant (values ranged from 0.618 to 0.635). This suggests that the interfactor relationship between the variables is significant. The variance accounted for in each measure by the endogenous variable revealed that the scores were significance at the 5 percent level. The score results suggested that the direct influence of adherence to safety regulations in determining SME contractors' overall health and safety is statistically significant.

11.4.3.4 Summary of Structural Model

Results from the EQS output revealed that the robust fit indexes, CFI, NFI, NNFI and the RMSEA values met the cut-off index criteria and the parameter estimates were found to be statistically significant and reasonable. The postulated model, which hypothesised that the overall H&S compliance of small and medium-sized construction companies is directly related to the influence of the exogenous variables in predicting or determining whether the overall H&S compliance fit the sample data was adequate. In view of the fact that the analysis was both an exploratory and confirmatory of a prior model, there was no need to further improve the structural model. Investigation of alternative models, such as the reduction of latent variables, could be a matter for further studies, as the current study was both an exploratory and confirmatory analysis of the priori. However, the Lagrange multiplier test did not unveil significant indication of model mis-specification demanding a re-specification. For most models, model enhancement is purely a process that attempts to fine-tune small features of the sample and does not essentially add value to an already fitted model, such as the current model. Likewise, when an initial model fits well, it is probably unwise to modify it to achieve even better fit because modifications may simply be fitting idiosyncratic characteristics of the sample (MacCallum et al., 1996). Hence, the presented model (Model 2.0) was therefore accepted with its level of fit. The lines of covariances (Figure 11.9) indicate that the integrated holistic influence of the latent variables determines overall H&S compliance because they were all statistically significant.

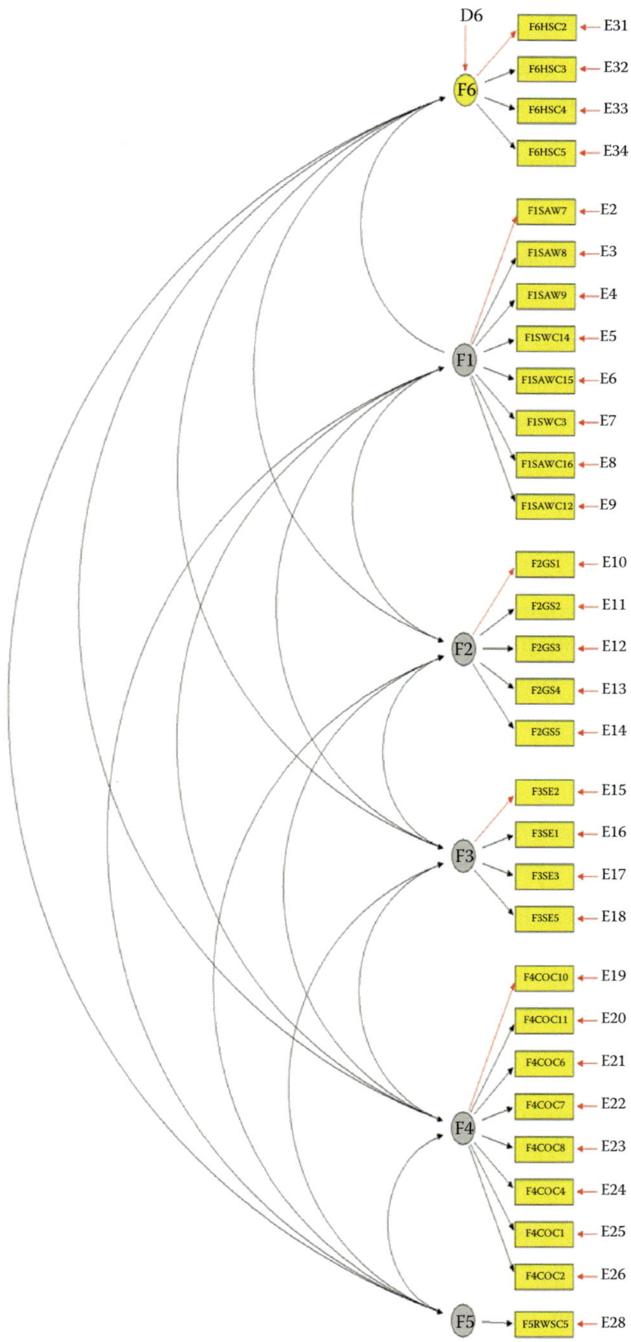

FIGURE 11.9 Model 2.0. An integrated health and safety compliance model. Covariance relationship (from left to right): ASR (3 indicator variables), COC (8 indicator variables), CSR (4 indicator variables), GS (5 indicator variables), SAWC (8 indicator variables) and HSC (4 indicator variables).

11.5 SUMMARY

The postulation for the overall model was that the overall H&S practice is directly related to the influence of the exogenous (latent) variables in predicting or determining overall H&S compliance. The SEM results of the measurement model were presented in this chapter. These results were obtained from an analysis of SEM to determine whether the indicator variables (questionnaire items) actually measured the constructs that they were supposed to measure. In addition, results were presented in order to establish whether the statistically significant number of factors for the latent models were feasible. Likewise, the measurement model reliability and construct validity were also reported. The analysis of the structural model (full latent model-SEM) was conducted, which validates the hypothesised integrated holistic health and safety compliance model. The influence of the latent variables on the endogenous variable was also reported. It was concluded that there is, therefore, no need to further improve the fit of the structural model. Further findings from the SEM results revealed that the exogenous variables' influences determine the H&S compliance of SME contractors in the Ghanaian construction industry. Further, it was found that four exogenous variables have a significant direct influence on the endogenous variables, while one had a weak (indirect) influence in determining health and safety compliance among SME contractors. It can be concluded that the five-factor model schematically portrayed in Figure 11.7 (Model 1.0) represents an adequate description of H&S compliance among small to medium-sized construction companies in Ghana.

REFERENCES

Bentler, P. and Bonnett, D. (1980). Significance tests and goodness of fit tests in the analysis of covariance structures. *Psychological Bulletin*, 88(3): 588–606.

Bentler, P.M. (1998). Causal modeling via structural equation systems. In Nesselroade, J.R. and Cattell, R.B. (Eds.), *Handbook of multivariate experimental psychology: Perspectives on individual differences*. New York: Plenum Press, 317–335.

Bentler, P.M. and Chou, C.P. (1987). Practical issues in structural modeling. *Sociological Methods and Research*, 16(1): 78–117.

Boomsma, A. (2000). Reporting analyses of covariance structures. *Structural Equation Modeling*, 7(3): 461–483.

Byrne, M.B. (2010). *Structural equation modeling with AMOS: Basic concepts, applications and programming* (2nd edn.). New York: Routledge Taylor & Francis Group.

Field, A. (2005). *Discovering statistics using SPSS (and sex, drugs and rock 'n' roll)* (2nd edn.). London: Sage.

Hair, J.F., Anderson, R.E., Tatham, R.L. and Black, W.C. (1998). *Multivariate data analysis* (5th edn.). Englewood Cliffs, NJ: Prentice Hall.

Hair, J.F., Black, W.C., Babin, J.B., Anderson, R.E. and Tatham, R.L. (2006). *Multivariate data analysis* (6th edn.). Upper Saddle River, NJ: Pearson/Prentice Hall.

Hooper, D., Coughlan, J. and Mullen, M. (2008). Structural equation modelling: Guidelines for determining model fit. *Electronic Journal of Business Research Methods*, 6(1): 53–60.

Hoyle, R. (1995). *SEM concepts, issues and applications*. London: Sage Publications.

Hsu, I.-Y., Su, T.-S., Kao, C.-S., Shu, Y.-L., Lin, P.-R. and Tseng, J.-M. (2012). Analysis of business safety performance by structural equation model. *Safety Sciences*, 50(1): 1–11.

Hu, L. and Bentler, P.M. (1999). Cut-off criteria for fit indices in covariance structure analysis: Conventional criteria versus new alternatives. *Structural Equation Modeling*, 6(1): 1–55.

Kline, R.B. (2005). *Principle and practice of structural equation modeling* (2nd edn.). New York: Guilford Press.

MacCallum, R.C., Browne, M.W. and Sugawara, H.M. (1996). Power analysis and determination of sample size for covariance structure modelling. *Psychological Methods*, 1: 130–149.

McDonald, R.P. and Ho, M.-.R. (2002). Principles and practice in reporting statistical equation analyses. *Psychological Methods*, 7(1): 64–82.

Mulaik, S.A., James, L.R., Van Alstine, J., Bennet, N., Lind, S. and Stilwell, C.D. (1989). Evaluation of goodness-of-fit indices for structural equation models. *Psychological Bulletin*, 105(3): 430–445.

Nunnally, J.C. and Bernstein, I.H. (1994). *Psychometric theory.* New York: McGraw-Hill.

Raykov, T., Tomer, A. and Nesselroade, J.R. (1991). Reporting structural equation modelling results in psychology and aging: Some proposed guidelines. *Psychology and Aging*, 6: 499–503.

Schreiber, B.J., Stage, K.F., King, J., Nora, A. and Barlow, A.E. (2006). Reporting structural equation modelling and confirmatory factor analysis results: A review. *The Journal of Educational Research*, 99(6): 323–337.

Tabachnick, B.G. and Fidell, L.S. (2007). *Using multivariate statistics* (5th edn). New York: Allyn & Bacon.

Tong, D.Y. (2007). An empirical study of E-recruitment technology adoption in Malaysia: Assessment of modified technology acceptance model. Multimedia University, Malaysia.

Section X

Discussion of Findings

12 Discussion of Results

12.1 INTRODUCTION

The quantitative research findings, with reference to the descriptive and inferential statistics, have been discussed in detail and were presented in the previous chapter. Also, the research hypotheses were tested based on the structural equation modelling (SEM) result analysis, validating the assumption that health and safety (H&S) compliance of small to medium-sized construction companies is a five-factor model, as schematically portrayed in Figure 11.8 (Model 2.0) (see Chapter 11). The findings from the SEM analysis showed that the factors of safe act and working condition, government support, contractor's safety policy, contractor's organisational culture, and adherence to safety regulations were found to have a significant influence in predicting the health and safety compliance of small to medium-sized construction companies. However, the adherence to state policy feature had a weak (indirect) influence in predicting health and safety compliance. Notwithstanding, the covariation with the other exogenous construct to determine health and safety compliance was found to be statistically significant.

12.2 QUESTIONNAIRE SURVEY RESULTS

The structural model results of this book revealed that the general hypothesis, which states that safe act and working condition, government support, contractor's safety policy, contractor's organisational culture, and adherence to safety regulations jointly predict health and safety compliance among small to medium-sized construction companies in Ghana, could not be rejected. In view of the hypothesis, the discussion section will be structured in order to respond to the research questions. Compliance with occupational health and safety (OHS) regulations is the effort made by management to determine whether it correlates with OHS performance. Compliance with OHS regulations brings about benefits not limited to avoiding direct and indirect costs (Smallwood, Haupt & Shakantu, 2008; Okeola, 2009; Idoro, 2011), but also contributes to organisations' competitive advantages (Windapo & Oladipo, 2012). However, adequate OHS training and education enhance the OHS performance (e.g. compliance with OHS regulation). Also, compliance with H&S will reduce accidents, as revealed from the findings (Chapter 11, Table 11.14). Findings from this aspect revealed that 89 percent of the respondents indicated that compliance with H&S will contribute to a reduction in accident rates. Therefore, it is clear the participants in the survey have good knowledge about their work environment and the result of the study could thus be used to make an inference.

12.2.1 Safe Environmental Features' Influence on Health and Safety Compliance of Small to Medium-Sized Construction Companies

The research question was put forward to determine the extent to which H&S compliance amongst small to medium-sized construction companies is influenced by the safe environmental features among small to medium-sized enterprise (SME) contractors. First, a descriptive assessment of the safe environmental features revealed that 92.4 percent of the respondents indicated that a safe and healthy work environment will contribute to H&S compliance. Another 92 percent of the respondents indicated that the safe storage of equipment will contribute to H&S compliance. Also, 91.6 percent of the respondents indicated that the provision of a warning system will contribute to H&S compliance. A further assessment of the safe environmental features revealed that 90.3 percent of the respondents indicated safe transportation of equipment, 90.2 percent of the respondents indicated safe transportation of materials and 90.1 percent of the respondents indicated safe storage of materials as contributors to H&S compliance. In the other two safe environmental features, 86 percent of the respondents indicated safe transportation of formworks and false work, and 82.8 percent of the respondents indicated safe storage of formworks and false work as contributors to H&S compliance (see Chapter 11, Table 11.15). Furthermore, results from the structural model revealed that the relationship between the safe environmental features and the endogenous variable (health and safety compliance) was found to be statistically significant at a 5 percent probability level.

On the other hand, all standardized parameter estimates showed high correlation values, suggesting a high degree of linear association between the indicator variables and the endogenous construct. Also, the interfactor values for this variable were considerable, suggesting that more than 50 percent of the latent variable predicted the endogenous factor construct. The summarised result for this variable revealed that the latent factor has a direct influence in determining overall H&S compliance. The results suggest that most variables included in the model have a significant effect on H&S compliance. Further findings suggest that safe environmental features are a significant determinant of H&S compliance. In addition, this result shows that a safe work environment can determine how issues of compliance with OHS regulations are taken care of by construction firms (Windapo & Oladapo, 2012). Therefore, the lower the level of safe environmental features, the less likely the small to medium-sized construction companies will comply with H&S regulations. Also, it should be noted that small to medium-sized construction companies differ significantly from large firms, and these factors affect their compliance level.

The implication of these findings is that overall H&S compliance is a product of the direct influence of safe environmental features. Therefore, the H&S compliance of small to medium-sized construction companies in the Ghanaian construction industry can be improved through safe environmental features, such as safe and healthy work environment, safe storage of equipment, the provision of warning systems, safe transportation of equipment, safe transportation of materials, safe storage of materials, safe transportation of formworks and safe storage of formworks. Research among the small to medium-sized construction companies in Ghana has shown that despite the acknowledged significance of the safe environmental features to the contractors, the Public Works Department (PWD), the Ministry of Water Resources and Works and Housing (MWRWH), and other stakeholders responsible for the provision of H&S regulations in the construction industry have not been able to give solutions to the OHS challenges in Ghana (Frempong & Essegbey, 2006; Kheni, Dainty & Gibb, 2007, 2008; Ofori & Toor, 2012; Kheni & Braimah, 2014; Annan, Addai & Tulashie, 2015). The findings emanating from the assessment of safe environmental features were therefore significant, and when attention is given to the issues of OHS challenges regarding the safe environmental features, the much-desired H&S compliance of small to medium-sized construction companies will be achieved. Furthermore, the findings make it possible for policy makers to address factors of safe environmental features. Most important, the number of unsafe environmental features observed within the SME contractors' firms and the construction industry in general will be addressed to ensure that all workers in the construction industry comply with H&S.

12.2.2 Safe Act of Workers Influence on Health and Safety Compliance of Small to Medium-Sized Construction Companies

The findings for this variable as measured in the study suggested that safe act of workers features have a direct influence on the prediction of H&S compliance of small to medium-sized construction companies. Findings from the interfactor relationship revealed that safe environmental features had a significant association with the latent variables in predicting the endogenous variable (Chapter 11, Table 11.16). From the assessment of the variance accounted for in each measure by the endogenous variable, it was revealed that all scores were statistically significant at the 5 percent level. The reported parameter coefficient explained more than the baseline level of the variance in the latent variable, which was indicative of an adequate prediction of the endogenous construct. Hence, these results suggested that the influence of this latent factor on the endogenous variable was direct and statistically significant. The sixteen indicator variables used in measuring the safe act of workers features construct were highly effective, as shown in Table 11.16. For instance, ensure the use of personal protective equipment (PPE) was the most effective item with 94.4 percent followed by work in good physical conditions with 94 percent. The lowest effective items were ensure proper positioning of tasks with 86.7 percent, avoid annoyance and horseplay at the workplace with 86.5 percent and do not service equipment that is in operation with 77.8 percent.

The findings suggest that the respondents were in agreement with the safe act of workers features, but they disagreed with some safe act of workers features such as annoyance and horseplay at the workplace, and do not service equipment that is in operation. These findings support the views of Mansingh and Haupt (2008), who opined that accidents do not only happen as a result of operatives' unsafe actions and unsafe site conditions. However, from the perspective of the domino theory by Heinrich (1930) accidents may also be due to a lack of management control and organisational failures. Further causes of accidents, as asserted by Petersen (2000) and Chua and Goh (2004), are management system failure and human errors. This theory holds that the cause of accidents lies in the organisational and management processes (Behm, 2008; Bellamy, Geyer & Wilkinson, 2008). Therefore, an effort aimed at addressing H&S should be directed more at addressing organisational and project management factors. These should include management at the industry, project, and company or organisation level because accidents are prevalent in the construction industry. Mansingh and Haupt indicated a different view on the causes of accident and emphasised that they were due to workers' unsafe acts. Moreover, Mansingh and Haupt supported the argument as viewed from the perspective of the domino theory by Heinrich.

During the questionnaire survey, the researchers observed that most of the small to medium-sized construction companies had problems with the safe acts of workers on sites. Most of the firms visited had different complaints from respondents. The most highly rated variable was ensuring the use of personal protective equipment (PPE). A lack of adequate PPE and most often not put into proper use led to high levels of non-compliance because workers did not bother to report minor accidents on sites to the authorities. This is clearly evident from the results of this study, although the respondents' found individual aspects of the safe act of workers and the overall level of non-compliance to be at a very high rate. Previous research informs that a safe work environment with safe acts of workers are important indicators that determine the health and safety compliance of small to medium-sized construction companies (Hosseinian & Torghabeh, 2012; Abdul Hamid, Yusuf & Singh, 2003). The reason for the cause of the accident is people and the way management handles the prevention of accident, as indicated in the theory. The majority of the accidents that take place are due to human error, and the accidents can only be prevented if management provides a conducive environment for the employees in which to work. Findings on this exogenous construct revealed that the workers' compliance as the measure of covariance and interfactor association with other indicators was average.

The implications of these findings are that overall health and safety compliance is a product of the direct influence of the safe act of workers features and that the health and safety compliance of SME contractors in the Ghanaian construction industry is paramount. The findings originating from the safe act of workers features assessment were therefore significant because when attention is given to issues of non-compliance regarding safe act of workers features, the health and safety compliance of small to medium-sized construction companies' contractors in the major cities and regions in Ghana will be achieved. Besides, the findings make it possible for policy makers to address factors of safe act of workers in order to comply with health and safety regulations in their firms.

12.2.3 Safe Working Condition Features Influence on Health and Safety Compliance of Small to Medium-Sized Construction Companies

The SEM analysis for this variable indicator (safe working condition) revealed that only one indicator was closely associated with the dependent variable. The other variables were weak in predicting the endogenous variable. However, a further assessment of the variance accounted for in each measure by the endogenous variable revealed that the scores were significant, as the values were above the minimum required value. The statistical assessment suggests that the direct influence of this factor on the endogenous variable was weak (indirect). However, the total variance accounted for revealed that it has a good indirect association with the other latent variables in the prediction of overall health and safety compliance. Similarly, descriptive assessment of the safe working condition features revealed that the following were present in the firm: provide safety regulations of equipment (92.0%); safe movement around workplace (92.0%); workers should be given adequate ventilation (91.9%); provision of facilities that are clean, safe and accessible to all workers (91.2%); provision of break periods for workers to access the facilities (90.8%); and good inspection programme (90.2%). Further findings from the descriptive statistics show the presence of the following safe working conditions within the firms: availability of facilities within a reasonable distance from the work area (89.8%), provision of sufficient lighting system for enclosed areas (89.5%), provision of safe means of facilities all the time (89.4%), good company safety policies (89.2%), and provision of adequate facilities (toilet, drinking water, washing and canteen) (89.1%).

These findings agree with the work of Idubor and Oisamoje (2013), who posit that adequate OHS training and education enhance the OHS performance at work (e.g. compliance with OHS regulations). Most of the respondents were in agreement that safe working conditions among small to medium-sized construction companies will enhance health and safety compliance. Othman (2012) further asserted that technical failure and inadequate training coupled with harsh work environment and unsafe methods of working inter alia are among the causes of non-compliance with OHS regulations (Adenuga, Soyingbe & Ajayi, 2007; Windapo & Oladapo, 2012; Idubor & Oisamoje, 2013). Safe working condition features have been observed as a vital determinant of health and safety compliance. Safe working conditions have been found by most of the respondents to be the least significant factor in determining the level of health and safety compliance in small to medium-sized construction companies. Overall, the factors of safe working conditions were found to have minimum impact on the overall health and safety compliance of small to medium-sized construction companies.

Hence, Ofori and Toor (2012) posit that because the Ghanaian construction industry is dominated by small to medium-sized construction companies, this has manifested in their non-compliance with health and safety. The Department of Occupational Safety and Health (2008) report also indicated that the compliance of small to medium-sized construction companies have an effect on their behaviour and compliance levels within individual employees in the construction industry. The findings on the hypothesis indicate that safe working conditions features do not significantly influence health and safety compliance of small to medium-sized construction companies. The findings offer little information that could be used by the Ministry of Water Resources and Works and Housing (MWRWH) to influence health and safety compliance in small to medium-sized construction companies. A checklist of items defining the factors of safe working condition features could not enable stakeholders to meet the basic required criteria to influence health and safety compliance through the ongoing Social Security and National Insurance Trust (SNNIT) housing projects in the country, for instance.

12.2.4 Reaction of Workers to Safe Condition on Health and Safety Compliance

The reaction of workers to safe condition enables employers to have an idea on how workers relate to their work environment. Findings suggest that when reaction of workers to safe condition is incorporated into construction projects, the outcomes are more likely to suit surrounding circumstances, and this will eventually lead to health and safety compliance. Descriptive assessment of the reaction of workers to safe condition features revealed the following: adhere to warning signs and notices (92.6%), follow safety regulations (92.6%), adhere to company safety policies (91.2%) and attend safety training programme (90.2%).

Further findings from the descriptive statistics show the presence of the following reaction of workers to safe condition: adhere to guidance on recommended illumination level for various tasks (89.5%), attend safety education programme (87.4%) and adhere to regular use of provided change room (87%). Most times, the promotion of reaction of workers to safe condition is difficult to realise. Small to medium-sized construction companies are predominantly non-unionised, a factor that is relevant to H&S compliance. Hence, there is no input of safety representatives or union pressure for improvements in safety in small to medium-sized construction companies (Health and Safety Executive, 2005). These factors negate the promotion of required compliance programmes among small to medium-sized construction companies.

12.2.5 Contractor's Organisational Culture on Health and Safety Compliance in Small to Medium-Sized Construction Companies

The importance of including this variable in the model is to determine the influence of contractor's organisational culture on health and safety compliance among small to medium-sized construction companies in the Ghanaian construction industry. This is because contractor's organisational culture accountability can easily be measured by the extent to which it involves health and safety compliance in decision-making and having control over resources that affect work. It is therefore necessary for planners and policy makers to include accountability in their development process.

The SEM results for this exogenous variable revealed that the standardised factor values and interfactor correlations for the contractor's organisational culture latent factor were large and statistically significant (Chapter 11, Table 11.69). Inspection of the total variances accounted for in each measure by the endogenous variable revealed that the scores were also significant. The relationship between contractor's organisational culture and health and safety compliance was found to be the most significant amongst all tested latent variables. The parameter with the highest standardised coefficient for this factor was the indicator variable COC 2. The parameter coefficient was found to be 0.6968. The indicator variable COC 2, which asked the update on H&S information to workers, was found to be more associated with overall H&S compliance than the remaining four indicator variables. The finding was that the contractor's organisational culture has a direct positive influence on health and safety compliance. The finding was consistent with the view of other researchers that culture creates a homogeneous set of assumptions and decision premises in which compliance occurs without surveillance (Grote, 2007). The Institution of Occupational Safety and Health (IOSH, 2004) contends that it is insufficient, for example, to provide safe equipment, systems and procedures if the culture is not conducive to a healthy and safe working environment.

The most significant findings from the SEM results highlighted the fact that only one indicator variable (COC 1), which had an unacceptably high unstandardised and standardised residual covariance matrix greater that 2.58, was dropped from the model. The Cronbach's alpha was found to be 0.872 (Table 11.69), which was above the minimum acceptable value of 0.70. Likewise, previous research revealed that contractor's organisational culture has a significant effect on overall health and safety compliance (Fong & Kwok, 2009). This is because compliance normally occurs based on a comparison of that which is expected with that which is received. It should be noted that compliance is not the only direct outcome, but prior exposure to what is to be received has also been found to directly affect non-compliance.

12.2.6 Government Support Influence on Health and Safety Compliance

The finding was that government support had a direct positive influence on the health and safety compliance in small to medium-sized construction companies. Government support (GS) was defined by five indicator variables: formulate H&S policy for construction, implementation of H&S policy by government representatives, monitoring of H&S policy implementation by government representatives, provision of H&S policy update by government representatives and provide H&S training by government representatives. The findings suggested that government support has a direct positive influence on the overall health and safety compliance of small to medium-sized construction companies. The finding was consistent with that of Ofori (2012), who found that government support had a direct positive link with the government policy of as undertaken in small to medium-sized construction companies. There is considerable investment in the construction industry because of its contribution to the national socio-economic development and the physical infrastructure of the country. In other words, the findings revealed that government support among small to medium-sized construction companies in Ghana will influence and increase the health and safety compliance of workers' overall compliance. This is apparent from the standardised factor loadings and interfactor correlations with other indicator variables, which were statistically significant (Chapter 11, Table 11.46). Also, the variances accounted for in each measure by the endogenous variable revealed that the scores were statistically significant and the values were above the minimum required value of 25 percent to be ascribed as an influence on health and safety compliance.

The finding was significant in that it provides the Ministry of Water Resources, Works and Housing, and other stakeholders with knowledge of the fact that the government support of small to medium-sized construction companies influences their compliance with health and safety. The overall results suggested that the influence of government support in determining health and safety compliance is direct and statistically more significant than any other factor. Similarly, the findings are significant because government support could constitute a feasible checklist of health and safety compliance that could be provided to small to medium-sized construction companies to ensure their workers comply with the stipulated health and safety regulations.

12.3 DELPHI SURVEY AND QUESTIONNAIRE RESULTS

Findings from the Delphi study were the exploration of the factors considered to be the principal determinants of health and safety compliance in small to medium-sized construction companies. These include safe environment, safe act of workers, safe working condition, reaction of workers to safe condition, government support and contractor's organisational culture. The results revealed that small to medium-sized construction companies are likely to comply with H&S owing to the following factors: reduction in accident rate (89.0%), available skills to identify hazards or risk (86.8%), available knowledge to interpret H&S rules (86.8%), available H&S experts (85.3%), interest in compliance with environmental health regulations (85.3%), good track of H&S regulations (83.7%) and enough capacity (81.5%). Other findings from the Delphi survey on the factors that influence small to medium-sized construction companies H&S compliance are as follows: good track records of H&S regulations (80.7%), available personnel to monitor changing legal requirements (80.4%) and reduction in accident compensation (80.1%) (see Chapter 11, Table 11.14).

However, the Delphi study was validated by a field questionnaire survey. The results suggested that the identified factors from the Delphi study have direct and indirect influence in determining health and safety compliance. In the questionnaire survey, the hypothesis that the exogenous factors had a direct and positive influence on health and safety compliance could not be rejected. The exogenous variables of safe environment, safe act of workers, safe working conditions, reaction of workers to safe condition, government support and contractor's organisational culture were found to have a statistically significant influence in predicting health and safety compliance among small to medium-sized construction companies in the Ghanaian construction industry. The findings from both the Delphi and the questionnaire survey, therefore, suggested that the exogenous variables influenced the determination of the endogenous variable (small to medium-sized construction companies H&S compliance). The merit of using structural equation modelling to validate the Delphi findings was that it was possible to specifically ascertain which of the exogenous factors had a significant influence on H&S compliance of small to medium-sized construction companies. Therefore, instead of making a general statement that the exogenous variables had an influence on determining the H&S compliance, it was possible to precisely state that the factors – safe environment, safe act of workers, safe working conditions, government support and contractor's organisational culture – had a direct (stronger) statistically significant influence on the SME contractors. Reaction of workers to safe working condition features had an indirect (weak) influence in determining H&S compliance with SME contractors. Safe act of workers and contractor's organisational culture were found to exert a more profound influence on H&S compliance amongst small to medium-sized construction companies. It appears to enhance health and safety compliance significantly, based on the SEM results. Government support and safe working condition were also strongly associated with increased compliance.

12.4 SUMMARY

In conclusion, the findings from the questionnaire survey generally supported the predictions that were made by the experts from the Delphi study. The validated predictions were those of safe act of workers, safe working condition, government support, contractor's safety policy, contractor's organisational culture and adherence to safety regulations. In addition, the existing literature lends support to the findings of the current study. The supported findings were that safe act of workers, safe working condition, government support, contractor's safety policy, contractor's organisational culture and adherence to safety regulations are fundamental to H&S compliance by small to medium-sized construction companies' contractors in Ghana.

REFERENCES

Abdul Hamid, A.R., Yusuf, W.Z.W. and Singh, B. (2003). Hazards at construction sites. Proceedings of the 5th Asia-Pacific Structural Engineering and Construction Conference (APSEC 2003), 26–28 August, Johor Bahru, Malaysia.

Adenuga, O.A., Soyingbe, A.A. and Ajayi, M.A. (2007). A study on selected safety measures on construction companies in Lagos, Nigeria. RICS (Cobra).

Annan, J.S., Addai, E.K. and Tulashie, S.K. (2015). A call for action to improve occupational health and safety in Ghana and a critical look at the existing legal requirement and legislation. *Safety and Health at Work*, 6(2): 146–150. Available from: http://www.e-shaw.net [Accessed 20 May 2015].

Behm, M. (2008). Linking construction fatalities to the design for construction safety concept. *Safety Science*, 43(8): 589–611.

Bellamy, L.J., Geyer, T.A.W. and Wilkinson, J. (2008). Development of a functional model that integrates human factors, safety management systems and wider organisational issues. *Safety Science*, 46: 461–492.

Chua, D.K.H. and Goh, Y.M. (2004). Incident causation model for improving feedback of safety knowledge. *Journal of Construction Engineering and Management*, 130(4): 542–551.

Department of Occupational Safety and Health (DOSH). (2008). Annual report: Labour and human resources statistics. Available from: http://www.mohr.gov.my/ [Accessed February 2015].

Fong, P. and Kwok, C. (2009). Organisational culture and knowledge management success at project and organisational levels in construction firms. *Journal of Construction Engineering Management*, 135(12): 1348–1356.

Frempong, G. and Essegbey, G. (2006). Towards an African e-Index, SME e-ACCESS and USAGE across 14 African countries, 25–27. Available from: http://www.researchictafrica.net [Accessed 30 September 2014].

Grote, G. (2007). Understanding and assessing safety culture through the lens of organizational management of uncertainty. *Safety Science*, 45: 637–652.

Health and Safety Executive (HSE). (2005) Making an impact on SME compliance behaviour: An evaluation of the effect of interventions upon compliance with health and safety legislation in small and medium-sized enterprises.

Heinrich, H.W. (1930). *Industrial accident prevention*. New York: McGraw-Hill.

Hosseinian, S.S. and Torghabeh, Z.J. (2012). Major theories of construction accident causation models: A literature review. *International Journal of Advances in Engineering & Technology*, 4(2): 53–66.

Idoro, G.I. (2011). Comparing occupational health and safety (OHS) management efforts and performance of Nigerian construction contractors. *Journal of Construction in Developing Countries*, 16(2): 151–173.

Idubor, E.E. and Oisamoje, M.D. (2013). An exploration of health and safety management issues in Nigeria's effort to industrialise. *European Scientific Journal*, 9(12): 154–169.

Institution of Occupational Safety and Health IOSH. (2004). Promoting a positive culture: A guide to health and safety culture. Leicestershire, UK: IOSH.

Kheni, N.A. and Braimah, C. (2014). Institutional and regulatory frameworks for health and safety administration: Study of the construction industry of Ghana. *International Refereed Journal of Engineering and Science (IRJES)*, (3)2: 24–34. Available from: http://www.irjes.org [Accessed 04 July 2015].

Kheni, N.A., Dainty, A.R.J. and Gibb, A.G.F. (2007). Influence of political and socio-cultural environments on health and safety management within SMEs: A Ghana case study. In: Boyd, D. (Ed.), *Proceedings of the 23rd Annual Association of Researchers in Construction Management (ARCOM) Conference*, 3–5 September, Belfast, UK, 159–168.

Kheni, N.A., Dainty, A.R.J. and Gibb, A.G.F. (2008). Health and safety management in developing countries: A study of construction SMEs in Ghana. *Construction Management and Economics*, 26(11): 1159–1169. Available from https://dspace.lboro.ac.uk [Accessed 10 April 2015].

Mansingh, K.S. and Haupt, T.C. (2008). Construction accident causation: An exploratory analysis. In: Hinze, J., Boehner, S. and Lew, J. (Eds.), *Evolution of and directions in construction safety and health*. CIB W099, Rotterdam, 465–482.

Ofori, G. (2012). Developing the construction industry in Ghana: The case for a central agency. National University of Singapore.

Ofori, G. and Toor, S.R. (2012). Leadership development for construction SMEs. Proceedings POC 2012 Conference, Working Paper Proceedings, Engineering Project Organizations Conference, 10–12 July, Rheden, The Netherlands.

Okeola, O.G. (2009). Occupational health and safety (OHS) assessments in the construction industry. 1st Annual Civil Engineering Conference, 26–28 August, University of Ilorin, Nigeria.

Othman, A.A.E. (2012). A study of the causes and effect of contractors' noncompliance with the health and safety regulations in the South African construction industry. *Architectural Engineering and Design Management*, 8: 180–191.

Petersen, D. (2000). The behavioural approach to safety management. *Professional Safety*, 37–39.

Smallwood, J., Haupt, T. and Shakantu, W. (2008). Construction health and safety in South Africa: Status and recommendation. CIDB report.

Windapo, A. and Oladapo, A. (2012). Determinant of construction firms' compliance with health and safety regulations in South Africa. In: Smith, S.D. (Ed.), *Proceedings of 28th Annual Association of Researchers in Construction Management (ARCOM) Conference*, 3–5 September, Edinburgh, UK, 433–444.

Section XI

Conclusions and Recommendations

13 Conclusions and Recommendations

13.1 INTRODUCTION

The general overall objective of the current study was to develop a health and safety (H&S) compliance model for small to medium-sized construction companies in developing countries. Ghana was used as a case study to identify the determinant attributes which collectively predict small to medium-sized construction companies' H&S compliance. The study adopted a mixed-methods approach to achieve the general objectives of the study. An extensive literature review, a Delphi survey and a field questionnaire survey were carried out. The field questionnaire survey was conducted in order to validate findings from the Delphi study with regard to the factors which predict health and safety compliance. The final results were analysed using structural equation modelling (SEM). Conclusions regarding the study are presented relative to the objectives of the study.

13.2 CONCLUSIONS

13.2.1 Research Objective Conclusion 1

The first objective of the study was to establish the factors that determine health and safety compliance in the construction industry, based on a literature review. In order to achieve this objective, a review of literature was conducted. Findings are that health and safety compliance of small to medium-sized construction companies is a product of multi-faceted construct. Findings further reveal that health and safety in small to medium-sized construction companies is about using appropriate means to ensure workers are both safe and healthy (Finneran & Gibb, 2013). Health and safety compliance means conforming to established guidelines (policy, legislation and standard) (Petersen, 2000; Wiegmann, Zhang, Von Thaden, Sharma & Mitchell, 2002; Chua & Goh, 2004; Health and Safety Executive [HSE], 2005, 2009; Lingard & Rowlinson, 2005; Dingsdag, Biggs, Sheahan & Cipolla, 2006; Adenuga, Soyingbe & Ajayi, 2007; Grote, 2007; Behm, 2008; Bellamy, Geyer & Wilkinson, 2008; Hughes & Ferrett, 2008; Idubor & Oisamoje, 2013; Yu, 2013).

It was also found that health and safety research deals with the construction industry health and safety compliance and aims to inform policy and planning intervention. The literature also informs that health and safety compliance is recognized as an important component of a policy statement which specifies aims and objectives of the health and safety behaviour in construction companies. The policy statement must be dated and signed by the most senior person in the organisation (Lingard & Rowlinson, 2005; Hughes & Ferrett, 2008; HSE, 2009; Yu, 2013). Findings from the literature were that more research and effort are required to address the problem of health and safety compliance in small to medium-sized companies in the construction industry.

13.2.2 Research Objective Conclusion 2

The second objective of the research was to establish the current theories and literature that have been advanced on health and safety compliance and to identify the gaps that needed consideration. A review of literature was carried out to achieve this objective. The findings revealed that health and safety compliance research has not been studied with an all-inclusive construct in the development of the previous models and theories. The identified gaps from the extensive literature review were government support and contractor's organisational culture. The identified gaps formed the new constructs in the current study's conceptual framework (Model 1.0). These gaps were considered essential because contractors have various safety practices that cannot be said to comply with the health and safety regulations. The current study offers a synthesised classification of the constructs, which should be collectively considered to predict health and safety compliance. From the synthesised literature, this current study argues that health and safety compliance is a five-factor construct.

13.2.3 Research Objective Conclusion 3

The third research objective of the study was to determine the main and sub-attributes that bring about health and safety compliance and to examine whether the attribute that determines compliance in other cultural contexts is the same as in Ghana. A Delphi study was conducted in order to achieve these objectives. Findings were that a number of factors that were considered to be important in determining health and safety compliance were identified and amplified by the Delphi study. The factors considered to be paramount determinants of health and safety compliance were safe environment, safe act of workers, safe work condition, reaction of workers to safe condition, government support and contractor's organisational culture. Further, health and safety compliance is ensured if there is a consideration of these factors in the development of SME contractors in Ghana. These factors were collectively considered for the development of the all-inclusive (integrated holistic) health and safety compliance model.

13.2.4 Research Objective Conclusion 4

The fourth research objective was to evaluate the critical factors and issues that affect the small to medium-sized construction companies' non-compliance with health and safety in Ghana. A Delphi study was conducted in order to achieve these objectives. Findings were that a number of factors that were considered to be important in evaluating the critical factors and issues that affect the small to medium-sized construction companies' non-compliance with health and safety were identified and amplified by the Delphi study. Findings from the Delphi survey further reveal that there was a lack of knowledge to identify hazards or risks and interest in compliance with the regulations. Others are a lack of enforcement from the legislative bodies responsible for the implementation of the Occupational Health and Safety Act.

13.2.5 Research Objective Conclusion 5

The fifth research objective of the study was to develop an integrated health and safety compliance model for small and medium-seized enterprise (SME) contractors based on both literature and the Delphi study. A synthesis of the reviewed literature together with the findings from the Delphi study was used to achieve this objective. The conceptual model theorised that health and safety compliance is a five-factor construct. These factors were safe environmental features, safe act of workers, safe working condition, reaction of workers to safe condition, government support and contractor's organisational culture, which jointly predict health and safety compliance among small to medium-sized construction companies. The hypothesis of the factor model was validated by conducting a questionnaire survey and analysing it using the SEM software EQS, version 6.2. The theorised five-factor model was taken through both exploratory factor analysis (EFA) and confirmatory factor analysis (CFA).

Findings from the SEM analysis, which models small to medium-sized construction companies' health and safety compliance as a five-factor model, showed that the factors of safe act of workers, safe working condition, government support, contractor's safety policy, contractor's organisational culture and adherence to safety regulations were found to have a significant influence in determining the small to medium-sized construction companies' health and safety compliance. However, the adherence to safety regulations had a weak (indirect) influence in predicting health and safety compliance. However, its covariance with the other exogenous construct to determine health and safety compliance was found to be statistically significant. These findings validated the conceptually integrated holistic model developed from literature and the Delphi study.

13.3 CONTRIBUTION AND VALUE OF THE BOOK

The value and contribution of the current study is described at three levels. These are the theoretical, methodological and practical levels of the research findings. However, it is pertinent to note that the outstanding contribution of the book is the revelation of the influence of the identified exogenous variables in predicting small to medium-sized construction companies' health and safety compliance.

13.3.1 Theoretical Contribution

The results of the SEM analysis indicated that an integrated health and safety compliance model is a six-factor model. This was achieved through the following processes:

1. The compilation of historical documentation on health and safety practices, policy, legislation and standards in developed and developing countries.
2. The use of a mixed-methods approach made up of a Delphi study and a questionnaire survey. The Delphi survey was used to arrive at a theoretical

model. Furthermore, a questionnaire survey was employed to ascertain the variables in the theoretical model. The SEM software EQS, version 6.2, was used to obtain the required variables on health and safety compliance model.

3. Finally, an integrated health and safety compliance model for SME contractors was developed.

Findings from the literature review did not reveal evidence of a similar study to the current one and this, therefore, suggested that this type of study has not yet been conducted in health and safety studies in Ghana and in order developing countries. Moreover, there was no evidence that suggested that a mixed method using Delphi and SEM had been used among small to medium-sized companies' H&S compliance in the Ghanaian construction industry. Therefore, this study offers a base for other studies to use for other follow-up studies. Likewise, the current study modelled health and safety compliance as a five-factor construct. Apart from the study contributing to theoretical knowledge, it also contributes to methodological advance in terms of the approach used in conducting the study as indicated in the next section.

13.3.2 Methodological Contribution

Most studies have used univariate statistical methods such as analysis of variance (ANOVA), multivariate analysis of variance (MANOVA) or regression modelling to model health and safety compliance. However, the current study used SEM, which is more robust and superior to the methods mentioned to determine causality of factors in a model and their direction of influence (Kline, 2005; Aigbavboa, 2013). SEM is most commonly thought of as a hybrid between some form of ANOVA or regression and some form of factor analysis. In general, it can be remarked that SEM allows one to perform some type of multilevel regression/ANOVA on factors. With SEM analysis, it was possible to identify the factors of health and safety compliance, which had significant effect and hence influence SME contractors' compliance in the construction industry as opposed to a general blanket statement that there are numerous constructs, which influence health and safety compliance. The questionnaire survey instrument had high internal reliability values and, therefore, could be used in similar studies to validate the current study or for similar purposes. Findings from the Delphi study and the conceptual model developed from both the literature review and the Delphi study was validated by conducting a questionnaire survey. Data from the questionnaire survey were analysed using EQS. As a result of this mixed method, a parsimonious model was developed. Aside from this contribution and value to the body of knowledge in terms of the methodological approach, a contribution to practice and the general construction industry was also achieved.

13.3.3 Practical Contribution and Value

The Delphi results have indicated that SME contractors in Ghana have not realised the significance of their involvement in health and safety compliance. However, Delphi results have indicated that small to medium-sized construction companies were likely to implement health and safety compliance elements. Further, SEM results indicated that SME contractors had an influence on health and safety compliance in the construction industry. Moreover, the knowledge of the influence of the six-factor construct could help the contractors to plan, organize, coordinate and control all aspects relating to the health and safety issues. The Ministry of Water Resources, Works and Housing (MWRWH) could use this knowledge to help with decisions on how best they can allocate finances towards the development of policies on health and safety. The practical significance of the study is further elaborated as follows.

13.3.3.1 Significance to the Ministry of Water Resources, Works and Housing

The study results have also demonstrated the level of compliance among small to medium-sized construction companies in the construction industry. The output of the study will help the MWRWH in making decisions about the criteria to be given priority in appropriate health and safety issues. The findings will help the MWRWH to plan programmes for SME contractors, as well as plan effective health and safety policies. This will ultimately enable the MWRWH to know where to commit resources, so that there is health and safety compliance among small to medium-sized construction companies. The government should monitor small to medium-sized construction companies' health and safety practices in order to comply with the laid down regulations. However, the first building block of the successful occupational health and safety (OHS) model is policy. Therefore, the policy should be appropriate to the nature and scale of the organisations' OHS risks. The integrated health and safety compliance model should be used as a guide to ensure that all SME contractors abide by it. The study offers an opportunity for further research to improve the model developed in this study and probably refine indicator variables to suit specific environments. Therefore, the recommendations and policy implications for practice of all these areas in which the current study may add value and contribute are presented next.

13.4 RECOMMENDATIONS

Recommendations are made from the methodological, theoretical and practical points of view.

13.4.1 Methodological

It is recommended that a similar study should be conducted in other developing countries and in established (large) construction companies to improve general application of H&S practices among all contractors in Ghana and specifically in other developing countries. Also, further research should be conducted on the indicator variables to establish any improvement in model fit, as the current study was purely an exploratory and confirmatory factor analysis to select the variables. There is the possibility that health and safety compliance could be defined by more indicator variables. Recognition should be given, however, to the fact that there is no such thing as a perfect model. Moreover, there should be a move to try to improve on the current model rather than invent a new model. The recommended method could commence with a Delphi study followed by a questionnaire survey or vice versa in order to improve its generalisability. Most of the studies in social science and most especially health and safety studies used standard statistical procedures such as ANOVA or MANOVA, and multiple regressions, which do not offer an appropriate and a straightforward way to test a hypothesis at a higher level of abstraction. Therefore, for similar studies, such as the current one, SEM with EQS is recommended to be used as the analysis technique for better results and abstraction.

13.4.2 Theoretical

It was observed from the literature that there were different definitions and understanding of how H&S compliance is formed. This has led in the past to a limited view and narrow conceptualization of H&S compliance mostly in small to medium-sized construction companies. Besides, there has not been consensus on how H&S compliance among small to medium-sized construction companies in Ghana should be measured. However, in the current study, literature was reviewed and synthesised on the determinants of H&S compliance. In conjunction with the experts' knowledge obtained through the Delphi study, a six-factor H&S compliance model was arrived at for small to medium-sized construction companies. These factors were identified as safe environment, safe act of workers, safe working condition, reaction of workers to safe condition, government support and contractor's organisational culture. Two of the factors were renamed and another two were combined to form one factor after the six factors had been taken through CFA. The final five factors for the model are safe act and working condition, government support, contractor's safety policy, contractor's organisational culture and adherence to safety regulations. It is therefore recommended that the developed model and theory of H&S compliance, with particular emphasis on operationalization, should form the basis for further refinement of the concept and thereby make it useful to small to medium-sized construction companies in Ghana and other developing countries.

13.4.3 Practical and Policy Implications

As a result of the identified contributions that the current study makes, as revealed by the findings, the following policy implications and practical recommendations have been identified:

- The policy implication suggests that health and safety compliance can be enhanced through the improvement of the safety practices in and among small to medium-sized construction companies.
- The MWRWH and other stakeholders responsible for health and safety in Ghana and other developing countries should adopt effective management strategies to minimise accidents in the construction industry.
- Future small to medium-sized construction companies' health and safety compliance in the Ghanaian construction industry should contain the five factors.

- It is recommended that small to medium-sized construction companies should know which indicator variables constitute their own health and safety compliance.
- Planning, organising, monitoring, measurement and control of health and safety compliance would be feasible if the MWRWH and stakeholders are aware of indicator variables that define H&S compliance.

13.5 LIMITATIONS

Interesting and valuable findings have emerged from this study. However, the following limitations regarding the current study should be considered. First, the research was only conducted in the major cities in Ghana, namely Accra, Tema, Kumasi and Takoradi. This is because the major cities have most of the small to medium-sized construction companies in Ghana. Given enough resources, it would be preferable to conduct a similar research study among large construction companies in Ghana. Also, the consideration of other developing countries could be included. Second, the SEM software with EQS, version 6.2, methodology used in data analysis may be construed as a limitation.

The results presented herein are based on the analysis of a causal model with raw data. Hence, the results are intended to support the a priori causal model. Third, the use of additional items or constructs might improve the inherent reliability and validity of the measures used. Fourth, several nested models, especially the measurement models, could have been evaluated to check the suitability of other alterative models. The current study was purely exploratory and confirmatory in nature. Fifth, although the internal reliability tests indicated high internal consistency and therefore a well-constructed research tool, some constructs revealed high correlational values. This may be due to the fact that only one questionnaire was used to collect information among small to medium-sized construction companies. A review of the research tool would have benefited findings in this study. A final limitation is related to the sample in addition to the aforementioned limitations the study has shown that some of the SEM measures may have been influenced by the sample size of the study. All empirical studies are limited by the nature of the sample studied. The exploration of the dependent variable (health and safety compliance) has shown that it has a very complex organisation (multi-faceted) and claims for further interpretations.

13.6 SUMMARY

An integrated health and safety compliance model for SME contractors was developed based on the existing health and safety issues, practices and grounded theories. It was postulated that overall, small to medium-sized construction companies' compliance is directly related to the influence of the exogenous (latent) variables in predicting or determining overall health and safety compliance. The postulated model was analysed with the use of the structural equation modelling software EQS, version 6.2. The fit statistics for the measurement and structural models had an adequate fit to the sample data. The finalized empirical model revealed that the exogenous variables (safe act and working condition, government support, contractor's safety policy, contractor's organisational culture and adherence to safety regulations) had a statistically significant influence in determining health and safety compliance. Specifically, the exogenous variables, such as adherence to safety regulations features, had a weak (indirect) influence on determining health and safety compliance among SME contractors in Ghana. From the findings, it is therefore concluded that the five-factor model schematically portrayed in Figures 11.7 and 11.8 (Model 2.0) (see Chapter 11) represents an adequate description of health and safety compliance among SME contractors in Ghana.

The results of this study have theoretical, methodological and policy (practical) value because respondents for the Delphi study were drawn from academics and construction professionals. The respondents for the questionnaire survey were administered among employees of small to medium-sized construction companies. Furthermore, the respondents had a good working knowledge of the studied environment. In addition, the questionnaire survey and the results modelled using structural equation modeling was a validating study of a conceptual model developed from synthesised theories established from literature and more important from the Delphi study. Hence, it is considered that the presented model for health and safety compliance interpretation maintains its validity. The result of the study provided information that can inform governmental, corporate, institutional and community policy makers, as they plan for and implement health and safety compliance among SME contractors. Second, the study provides indicators that will be a baseline for implementing health and safety compliance. Consequently, small to medium-sized construction companies and other stakeholders will be able to contribute to the ways of solution to minimise non-compliance of health and safety among these grade of contractors. Stakeholders and institutions who are involved in the planning process should be able to influence the factors that will bring about health and safety compliance. Also, the conceptual model of health and safety compliance among the level of contractors which has been formulated in this study will provide a reference to researchers who will study health and safety in the near future. The current study lends support to other studies that have utilised alternative methods to establish the factors which influence health and safety compliance among SME contractors.

These studies thus conclude that small to medium-sized construction companies' health and safety compliance is multi-faceted as also claimed in the current study. The current study utilises a more robust modelling method of SEM. By adopting the methodology, the current study was able to model the influence of the selected multifaceted variables and the constructs, which were statistically significant. The practical implication is that the SME contractors' health and safety compliance model be enhanced by improving on the factors of safe act and working condition, government support, contractor's safety policy, contractor's organisational culture and adherence to safety regulations to enable their application by SME contractors. Moreover, the MWRWH and other stakeholders responsible for government projects in Ghana can adopt effective management strategies to improve on the health and safety practices of SME contractors. Future government projects should adapt the five-factor model to minimise accidents in the construction industry.

REFERENCES

Adenuga, O.A., Soyingbe, A.A. and Ajayi, M.A. (2007). A study on selected safety measures on construction companies in Lagos, Nigeria. RICS (Cobra).

Aigbavboa, C.O. (2013). An integrated beneficiary centred housing satisfactory model for publicly funded housing schemes in South Africa. Unpublished DPhil in engineering management, University of Johannesburg, South Africa. Available from: www.ujdigispace.uj.ac.za [Accessed 15 June 2014].

Behm, M. (2008). Linking construction fatalities to the design for construction safety concept. *Safety Science*, 43(8): 589–611.

Bellamy, L.J., Geyer, T.A.W. and Wilkinson, J. (2008). Development of a functional model that integrates human factors, safety management systems and wider organisational issues. *Safety Science*, 46: 461–492.

Chua, D.K.H. and Goh, Y.M. (2004). Incident causation model for improving feedback of safety knowledge. *Journal of Construction Engineering and Management*, 130(4): 542–551.

Dingsdag, D.P., Biggs, H.C., Sheahan, V.L. and Cipolla, C.J. (2006). A construction safety competency framework: Improving OHS performance by creating and maintaining a safety culture. Brisbane: Cooperative Research Centre for Construction Innovation.

Finneran, A. and Gibb, A. (2013). CIB W099 – Safety and health in construction: Research roadmap report for consultation. CIB publication 376. Rotterdam, The Netherlands: CIB General Secretariat.

Grote, G. (2007). Understanding and assessing safety culture through the lens of organizational management of uncertainty. *Safety Science*, 45: 637–652.

Health and Safety Executive (HSE). (2005) Making an impact on SME compliance behaviour: An evaluation of the effect of interventions upon compliance with health and safety legislation in small and medium-sized enterprises.

Health and Safety Executives (HSE). (2009). Safety signs and signals: The Health and Safety (Safety Signs and Signals) Guidance on Regulations. Available from: http//:www.hsebooks.co.uk [Assessed 20 September 2014].

Hughes, P. and Ferrett, E. (2008). *Introduction to health and safety in construction*. Oxford: Elsevier Butterworth-Heinemann.

Idubor, E.E. and Oisamoje, M.D. (2013). An exploration of health and safety management issues in Nigeria's effort to industrialise. *European Scientific Journal*, 9(12): 154–169.

Kline, R.B. (2005). *Principle and practice of structural equation modeling* (2nd edn.) New York: Guilford Press.

Lingard, H. and Rowlinson, S. (2005). *Occupational health and safety in construction project management*. Abingdon: Spon Press.

Petersen, D. (2000). The barriers to safety excellence. *Occupational Hazards*, 37–42 (Dec. 1).

Wiegmann, D.A., Zhang, H., Von Thaden, T.L., Sharma, G. and Mitchell, A.A. (2002). A synthesis of safety culture and safety climate research. University of Illinois Aviation Research Lab Technical Report ARL-02-03/FAA-02-2.

Yu, W.-H. (2013). It's who you work with: Effects of workplace shares of nonstandard employees and women in Japan. *Social Forces*, 92(1): 25–57.

Index

Page numbers followed by f and t indicate figures and tables, respectively.

G

H

L

M

N

O

T

For Product Safety Concerns and Information please contact our EU representative GPSR@taylorandfrancis.com Taylor & Francis Verlag GmbH, Kaufingerstraße 24, 80331 München, Germany